BEYOND SILENT SPRING

BEYOND SILENT SPRING

Integrated pest management
and chemical safety

HELMUT F. van EMDEN
Professor of Horticulture
The University of Reading
Berkshire
UK

and

DAVID B. PEAKALL
Senior Research Fellow
Monitoring and Assessment Research Centre
King's College
The University of London
UK

UNEP

icipe

CHAPMAN & HALL

London · Glasgow · Weinheim · New York · Tokyo · Melbourne · Madras

Published by Chapman & Hall, 2–6 Boundary Row, London SE1 8HN

Chapman & Hall, 2–6 Boundary Row, London SE1 8HN, UK

Blackie Academic & Professional, Wester Cleddens Road, Bishopbriggs, Glasgow G64 2NZ, UK

Chapman & Hall GmbH, Pappelallee 3, 69469 Weinheim, Germany

Chapman & Hall USA, 115 Fifth Avenue, New York, NY 10003, USA

Chapman & Hall Japan, ITP-Japan, Kyowa Building, 3F, 2–2–1 Hirakawacho, Chiyoda-ku, Tokyo 102, Japan

Chapman & Hall Australia, 102 Dodds Street, South Melbourne, Victoria 3205, Australia

Chapman & Hall India, R. Seshadri, 32 Second Main Road, CIT East, Madras 600 035, India

First edition 1996

© 1996 United Nations Environment Programme

Typeset in 10½/12 Photina by WestKey Ltd, Falmouth, Cornwall
Printed in Great Britain by St Edmundsbury Press, Bury St Edmunds, Suffolk

ISBN 0 412 72800 1 (HB)   0 412 72810 9 (PB)

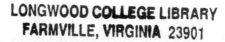Printed on permanent acid-free text paper, manufactured in accordance with ANSI/NISO Z39.48–1992 and ANSI/NISO Z39.48–1984 (Permanence of Paper).

Contents

Foreword

More than 32 years ago, Rachel Carson's *Silent Spring* appeared upon the scene as a landmark of literary achievement which contributed greatly to the foundation of the modern environmental movement.

Rachel Carson had designed *Silent Spring* to shock the public into action against the misuse of chemical pesticides. More than anything else, the book also served as an ecological primer, demonstrating the interrelationship of all things and the dependence of each on a healthy environment for survival.

Today, *Silent Spring* is generally credited with providing impetus to the whole range of anti-pollution laws that came into force in the 1970s. It is also perceived as having played a crucial role in the eventual banning of DDT as well as in the restricted use or total phasing out of the most notorious hard pesticides identified in the book.

The vigorous growth of the chemical industry geared to the production of newer and ever more powerful pesticides can be traced to the introduction of the organochlorine insecticide DDT in the 1940s. These pesticides were meant not only to control insects but also animal pests, disease and weeds. Initially their development was based on the belief that they would provide a definitive solution to pest and vector problems.

On the contrary, they created many problems, which have been documented profusely in the scientific literature, including development of pest resistance to pesticides; existence of toxic residues in food, feeds, soils and animals; lethal effects on nontarget organisms (plants, livestock, wildlife, humans); and damage to the environment.

Today, we can add vastly to Rachel Carson's list. Humankind is exposed to thousands of other chemical substances in ever increasing quantity and variety. Of the 11 million substances known, some 60 000–70 000 are in regular use. Yet toxicological data are only available for a fraction of the more than 3000-odd chemicals which account for 90% by mass of the total used. The data on the environmental and ecotoxicological properties of such substances are even more scanty.

The final chapter of *Silent Spring* is entitled 'The other road'. It gives the results of Carson's research into alternative, more selective ways of controlling pests with biological methods. 'All have this in common,' she wrote, 'they are biological solutions, based on the understanding of the living organisms they seek to control, and of the whole fabric of life to which these organisms belong.'

An important positive consequence of this statement is the concept of integrated pest management (IPM). This approach involves a judicious combination of nonpesticidal tactics with minimal pesticide use. While chemical pesticides aim at the elimination of the maximum proportion of pests, IPM stresses maintaining pests at low levels that cause no economic damage but provide hosts to sustain natural control agents.

With the scientific and technological advances made in recent years, it has become necessary to develop new paradigms for IPM. Contemporary trends point towards plant health management and biointensive IPM, where the problem-solving equation includes not only the pest but also the agro/sylon-ecosystem relevant to particular pest problems. The concept of plant health now increasingly carries the message of prevention not cure. Similarly, there is a need felt to examine the impact of increasing use of other chemicals and of discovering ways and means for their environmentally sound management.

It is in the above context that the United Nations Environment Programme (UNEP) and the International Centre of Insect Physiology and Ecology (ICIPE) thought it appropriate to revisit *Silent Spring* to evaluate the progress made in the past 30 years in the field of chemical safety and IPM in particular.

The two international organizations have taken up the joint challenge to produce this fitting sequel to *Silent Spring*. This volume, with a particular focus on the tropical world, has two main objectives:

1. to enhance and update public awareness of the potential adverse impact of pesticides and other chemicals on the environment, animals and public health; and
2. to promote and foster acceptance by governments, international organizations, donor agencies, farmers and consumers of environmentally safe practices for the management of pests and chemicals, but above all of the need to invest in the development of truly sustainable alternatives to these chemicals.

To achieve these objectives, UNEP and ICIPE jointly invited various experts (Table 1) to prepare reviews on different topics in order to assemble:

1. a survey of IPM-related activities and their impact during the past two decades in tropical regions of Asia, Africa and Latin America;
2. an appraisal of the scientific and technological advances in areas that

Table 1 Resource papers presented at the Nairobi workshop on IPM and *Beyond Silent Spring*

A critique of the pesticide age; its philosophy, its approaches and its death-knell
  T. R. Odhiambo (ICIPE) and A. A. Abdelrahman (Agricultural Research Corporation, Wad Medani, Sudan)

Concept of integrated pest management
  R. C. Saxena and S. Mihok (ICIPE)

Changing role of pesticides in pest management
  G. A. Schaefers (Cornell University, USA)

Pesticide application technology in pest management
  G. A. Matthews (Imperial College, London, UK)

Population biology in integrated pest management
  J. C. van Lenteren (Wageningen Agricultural University, The Netherlands) and W. A. Overholt (ICIPE)

Social and economic aspects of integrated pest management
  G. Goodell (The Johns Hopkins University, Washington, DC, USA), J. Ssennyonga and G. T. Lako (ICIPE) and S. Tadla (Addis Ababa University, Ethiopia)

Practice of integrated pest management in the tropics and subtropics, 1970–90
  Africa – O. Zethner (DANAGRO, Glostrup, Denmark)
  South and South-East Asia – A. K. Raheja (Indian Council of Agricultural Research, New Delhi, India)
  South America – C. Campanhola (EMBRAPA, Jaquaraiuna, SP, Brazil)

Growing role of biotechnology in integrated pest management
  F. Gould (University of Raleigh, USA) and E. O. Osir (ICIPE)

Issues of biodiversity in pest management
  H. F. van Emden (University of Reading, UK) and Z. T. Dabrowski (ICIPE)

Environmentally sound management of chemicals and toxic wastes
  D. Peakall (King's College, London, UK)

---

could set new trends for IPM that would be environmentally friendly and sustainable; and

3. activities in the field of chemical safety in general and pesticide-related problems, for suggesting approaches to environmentally sound management of toxic chemicals and wastes.

The papers of these experts were presented and reviewed in a workshop convened in Nairobi in September 1992 (Table 2). Based on the reviews presented at the Nairobi Workshop and on the suggestions made there, the framework for the volume was prepared by Professors K.N. Saxena (ICIPE) and H.N.B. Gopalan (UNEP). UNEP and ICIPE entrusted the task of compiling the book based on the framework to Professor Helmut van Emden and Dr David Peakall. Professor Pimentel graciously peer reviewed the entire book. UNEP and ICIPE are grateful to Professor van Emden, Dr

Table 2 List of resource persons, consultants and reviewers at the Nairobi workshop, September 6–8, 1992

A. A. Abdelrahman (Agricultural Research Corporation, Wad Medani, Sudan)
K. Ampong-Nyarko (ICIPE)
C. Campanhola (EMBRAPA, Jaquaraiuna, SP, Brazil)
Z. T. Dabrowski (ICIPE)
H. N. B. Gopalan (UNEP, Nairobi, Kenya)
J. W. Huismans (UNEP, Geneva, Switzerland)
J. H. Koeman (Wageningen Agricultural University, The Netherlands)
G. A. Matthews (Imperial College, London, UK)
S. Mihok (ICIPE)
A. Ng'eny-Mengech (ICIPE)
T. R. Odhiambo (ICIPE)
E. O. Osir (ICIPE)
W. A. Overholt (ICIPE)
D. Peakall (King's College, London, UK)
A. K. Raheja (Indian Council of Agricultural Research, New Delhi, India)
A. Renzoni (University di Siena, Italy)
K. N. Saxena (ICIPE)
R. C. Saxena (ICIPE)
G. A. Schaefers (Cornell University, USA)
S. Tedla (Addis Ababa University, Ethiopia)
B. Waiyaki (UNEP, Nairobi, Kenya)

Peakall, Professor Pimentel, the invited reviewers (Table 1) and the participants at the Nairobi Workshop (Table 2) for making their joint revisit to *Silent Spring* most rewarding and exciting.

We hope that this volume will lead to an enhanced appreciation of the problems posed by excessive use of synthetic chemicals, and that the solutions offered can, through practical implementation, mitigate their adverse effects.

Hans R. Herren
Director-General,
International Centre of Insect
    Physiology and Ecology,
Nairobi,
Kenya

Ms. Elizabeth Dowdeswell
Executive Director,
United Nations Environment
Programme,
Nairobi,
Kenya

Preface

We are grateful to Elizabeth Dowdeswell of UNEP and to Dr Hans Herren of ICIPE for writing the foreword to this book and explaining the background to *Beyond Silent Spring*. We would like to join them in thanking all those who have contributed to the gestation of this volume, and especially those who gave of their expertise to provide the review papers originally designed to form the basis of the book.

The Nairobi Workshop in September 1992, however, changed our remit. Instead of editing the various review papers, we were asked to write the book on a new framework agreed at that meeting, although much of the material provided in the reviews proved indispensable to that task. We can only apologize to colleagues who feel their contributions have not been utilized as they would have wished. We have therefore avoided attributing information to individual reviews, and the names of those providing material are instead listed as part of the Foreword.

We would also join the writers of the Foreword in expressing our personal thanks to Professor David Pimentel for reviewing our draft manuscript and making very useful suggestions, especially in relation to omissions in our coverage of the subject. His suitability as a peer reviewer was evident in the valuable comments he was able to make across all the many subjects covered in the book. However, we do owe it to Professor Pimentel to make it clear that the opinions expressed in the final version of the book are not necessarily his, but our own. He clearly feels that the future of pest management involves considerably less use of pesticide than we envisage, and would not share our view that much successful pest management will continue to require the use of some pesticide. He would also disagree with another major point we make, that biological control (with the accent on control) is not a natural phenomenon, though this point of debate is perhaps more semantic than fundamental in nature.

ICIPE and UNEP felt it was important to limit the authorship of the book to two persons in order to maintain continuity and ensure smooth development of the theme. It will soon become obvious to the reader that we have somewhat distinct and contrasting styles of writing. Neither of us felt it was possible to adapt to the style of the other, neither did we feel either style should be edited to the alternative form. So the book has been written by the two of us, in harmony but not in unison.

We are grateful to Professor Sir Colin Spedding for the East Asian proverb 'He who takes the middle of the road is likely to be crushed by two rickshaws.' Neither environmentalists nor the chemical industry are likely to be satisfied with the middle road we have taken in this book. Those who hold the middle position are not usually as vociferous as those who put forward more extreme arguments; hence the phrase 'the silent majority'. However, the middle road is but an alternative and equally valid one; one can espouse it just as fervently as others hold their opinions.

We are the first to admit, and clearly state in Chapter 3, that the 'integrated' of integrated pest management is an integration of the disciplines of zoology (especially entomology), plant pathology and weed science. So we must also admit that we have, to a large extent, been blinkered, and used entomology for the main examples both of the principles and practice of pest management. We have been restricted by our lack of competence in plant pathology and weeds and by an emphasis on entomology in most of the review material we had to work with. To a large extent, therefore, 'pest' in this book relates more to the English than the American usage of the word. The alternative would have been to involve more authors in drafting the book. We take refuge in the perhaps arrogant notion that pest management is still dominated in its development by pests in the English sense; nevertheless we hope workers emphasizing other disciplines will not find our basic concepts and principles irrelevant.

H.v E.
D.P.
November 1994

Table of acronyms

ABCP	Africawide Biological Control Programme
AChE	acetylcholinesterase
AVRDC	Asian Vegetable Research and Development Centre
BBC	British Broadcasting Corporation
BHC	benzene hexachloride
BPH	brown planthopper
Bt	*Bacillus thuringiensis*
BYDV	barley yellow dwarf virus
CBD	coffee berry disease
CDA	controlled droplet application
CFCs	chlorofluorocarbons
CGIAR	Consultative Group on International Agricultural Research
CICIU	Center for Introduction and Production of Beneficial Insects
CIKARD	Center for Indigenous Knowledge for Agriculture and Rural Development
CILSS	Comité permanent Interétats de Lutte contre la Secheresse dans le Sahel
CITES	Convention on International Trade in Endangered Species of Wild Fauna and Flora
CNPSo	Centro Nacional de Pesquisa de Soja
CNPT	National Wheat Research Center (Brazil)
CO_2	carbon dioxide
2,4-D	2,4-dichlorophenoxyacetic acid
DBCP	dibromochloropropane
DBM	diamond-back moth
DDE	dichlorodiphenyldichloroethylene
DDT	dichlorodiphenyltrichloroethane
DIMBOA	2,4-hydroxy-7-methoxy-1,4-benzoxazin-3-one
DNA	deoxyribonucleic acid

EDB	ethylene dibromide
EEC	European Economic Community
EMBRAPA	Empresa Brasiliera de Pesquisa Agropecuária
FAO	Food and Agriculture Organization
GATT	General Agreement on Tariffs and Trade
GEMS	Global Environmental Monitoring System
GESAMP	Group of Experts on the Scientific Aspects of Marine Pollution
GNP	gross national product
GRID	Global Resource Information Database
HCH	hexachlorocyclohexane
HEALS	Human Exposure Assessment Locations
IBPGR	International Board for Plant Genetic Resources
ICA	Columbia Agricultural Institute
ICIPE	International Centre of Insect Physiology and Ecology
ICRISAT	International Crops Research Institute for the Semi-Arid Tropics
IGRs	Insect growth regulators
IIBC	International Institute of Biological Control
IITA	International Institute of Tropical Agriculture
ILCA	International Livestock Centre for Africa
ILEIA	Information Centre for Low External Input Agriculture
INFOTERRA	International Referral System for Sources of Environmental Information
INIA	Chilean National Institute for Agricultural Research
IOBC	International Organization for Biological Control
IPM	integrated pest management
IPMR	Integrated Pest Management for Rice Network
IRPTC	International Register of Potentially Toxic Chemicals
IRRI	International Rice Research Institute
JH	juvenile hormone
LCA	life-cycle assessment
LC_{50}	lethal concentration for 50% of population
LD_{50}	lethal dose for 50% of population
LRTAP	long-range transport of air pollutants
MAFF	Ministry of Agriculture, Fisheries and Food (UK)
MPD	minimum pre-market data
MRR	marginal rate of return
NARS	National Agricultural Research Services
NATO	North Atlantic Treaty Organization
NGOs	non-governmental organizations
NO	nitric oxide
NO_2	nitrogen dioxide
N_2O	nitrous oxide

NO_x	oxides of nitrogen
NTE	neurotoxic esterase
ODA	Overseas Development Administration (UK)
OECD	Organization for Economic Co-operation and Development
OPs	organophosphates
PCBs	polychlorinated biphenyls
PCDDs	polychlorinated dibenzodioxins
PCDFs	polychlorinated dibenzofurans
PEGS	Pesticide Exposure Group Sufferers
PVC	polyvinyl chloride
SIRM	sterile insect release method
SO_x	oxides of sulfur
SO_2	sulfur dioxide
SSSIs	sites of special scientific interest
2,4,5-T	2,4,5-trichlorophenoxyacetic
TCDD	tetrachlorodibenzodioxin
T-DNA	transferred DNA
TEF	toxic equivalent factor
TPM	total population management
ULV	ultra-low volume
UNCED	United Nations Conference on Environment and Development
UNDP	United Nations Development Programme
UNEP	United Nations Environment Programme
UNSD	United Nations Statistical Office
USAID	United States Agency for International Development
USEPA	United States Environmental Protection Agency
UV	ultraviolet
UVB	biologically active UV
WARDA	West African Rice Development Association
WHO	World Health Organization

CHAPTER 1

Introduction

Previous books reviewing *Silent Spring* have been *Since Silent Spring* published in 1970 and *Silent Spring Revisited* published in 1987. Since *Silent Spring* was published in 1962 environmental concerns have become much broader, and this book reflects this change by examining the integrated approach, often at an international level, to the use and control of chemicals.

This is, as far as we know, the third book to use 'Silent Spring' in the title since Rachel Carson's original *Silent Spring* burst on the scene in 1962. Our predecessors have been *Since Silent Spring* by Frank Graham Jr (1970) and *Silent Spring Revisited* published by the American Chemical Society in 1987.

Frank Graham's book is written very much along the lines of *Silent Spring* and gives the impression of being the work of a devoted disciple. The style is shown by phrases such as 'many Americans live perpetually in a sea of pesticides' and 'the spectre of cancer has hung about pesticide use ever since *Silent Spring* was published.' The first six chapters are taken up with descriptions of the controversy over the publication of *Silent Spring*. Then there are several chapters on the problems caused by the organochlorine pesticides, such as 'Miss Carson's "nightmares" unfold'. This is an account of the poisoning of fish in the Mississippi by endrin. Another such chapter is 'The human toll' which covers such instances as endrin in flour in the Middle East and parathion-contaminated sugar in Mexico. Graham's book dated very rapidly as it was written at the very end of the era of organochlorine usage (at least in the Western world). Bans on most of these materials came in the early 1970s.

Silent Spring Revisited, edited by Marco, Hollingworth and Durham and published by the American Chemical Society in 1987, is a very different style of book. It consists of a series of papers by distinguished scientists and is, considering the publisher, remarkably sympathetic to *Silent Spring*. The changes in regulation and outlook since 1962 are described; Jack Moore of the United States Environmental Protection Agency (USEPA) describes the changes in regulations in the US since *Silent Spring* and discusses the slow progress of alternatives to pesticides. David Pimentel answers the question 'Is Silent Spring behind us?' with a qualified 'Yes'. He notes that the most serious problems which were caused by organochlorine pesticides have decreased during the past two decades. However, he notes that some pesticide problems have increased, and cites reduced populations of natural enemies, the increase of resistance of pest species to insecticides, fishery losses and poor performance of integrated pest management as examples. He concludes that crop losses continue to increase despite the increased usage of pesticides.

The chapters are individual units with little sign of general editing to make a unified assessment. This is clearly shown in opinions expressed in *Silent Spring Revisited* about *Silent Spring* itself. Shirley Briggs, the driving force behind the Rachel Carson Trust, says many reviews included pat phrases, such as 'Of course she exaggerated and made mistakes, but in general she was on the right track.' These face-saving phrases have taken on a kind of immortality; they turn up repeatedly from people who admit, when questioned, that they do not know of any inaccuracies or mistakes, but 'so many people said there were some.' She continues 'We at the

Rachel Carson Council have yet to be shown a valid example.' Two chapters later Chris Wilkinson writes 'Despite many scientific inaccuracies and broad unsubstantiated conclusions, *Silent Spring* had an enormous impact on the way that pesticides were viewed.' Kohn of the Zoecon Corporation makes the following assessment, 'Not all Carson's predictions have come true. Birds do sing and we are living longer than ever before. The legacy of Carson may be found in the legislation, some good and some not so good and the public belief, well justified, in the need to defend the environment against further deterioration.' He does mention specific inaccuracies, 'With the advent of man the situation began to change, for man alone of all forms of life can create cancer-producing substances' (p. 219); 'Natural carcinogenic agents are few in number' (p. 220); and 'We are now aware of an alarming increase in malignant disease' (p. 221). These were, apparently, not passed on by the editors to Shirley Briggs for comment.

The last chapter includes the editors' (one each from industry, academia and government) final comment:

> Was Rachel Carson right? In many respects, yes. In her time, the environment was relentlessly assaulted by a society hoping for total control. Nature was not as self-cleansing as we believed. Many of Carson's predictions about environmental toxicity, human health effects, water contamination, and waste site problems have proved correct. Was Rachel Carson wrong? In fewer respects, yes. Nature, not just humans, generates its share of carcinogens and other poisons. Nature and humans both use chemicals for their own advantage. Human life span is still increasing. Society has responded, and, of course, birds still sing. Biological controls alone have been able to replace chemicals only in a few species cases.

Silent Spring was based on those studies that had been made and were available at that time. This may sound like an obvious statement, but it is surprising how many persons think that environmental studies started with *Silent Spring*. Certainly *Silent Spring* had a major influence on the environmental movement. Thirty years after its initial publication it is still available. It was on sale at a bookshop at Heathrow Airport recently and the total sales are around the 2 million mark, surely much greater than any environmental book published since then. The 25th Anniversary Edition sold 130 000 copies. One of the most interesting stories on the circulation of *Silent Spring* is that it was immediately translated into Russian and 200 numbered copies were run off for distribution to high officials for secret use only. Although *Silent Spring* was the first book written to expose the overuse of chemicals, it is surprising that it made such an impact and was raised to the position of a cult. The impacts of Einstein's theory of relativity in the 1930s and of Steven Hawkings' *Brief History of Time* in the late 1980s on a public that did not understand either theory are other examples.

In the present book, which has been entitled *Beyond Silent Spring*, we look at a broader picture, emphasizing the integrated approach. We are looking both at pesticides and at other industrial chemicals. In both cases we are interested in ways of reducing their impact on man and the environment.

Public awareness of environmental issues has changed markedly since 1962 and this has been reflected in international activities. The two most important events on the international scene have been the United Nations Conference on the Human Environment in Stockholm in 1972 and the United Nations Conference on Environment and Development (UNCED) held in Rio de Janeiro in 1992.

The basic Declaration of the Stockholm Conference was that 'the capacity of the earth to produce vital renewable resources must be maintained and, wherever practicable, restored and improved'. The importance of the Stockholm Conference was that it provided directives for national and international action. It brought into being the United Nations Environment Programme (UNEP). This included such important programs as the Global Environmental Monitoring System (GEMS), the International Register of Potentially Toxic Chemicals (IRPTC) and the International Referral System for Sources of Environmental Information (INFOTERRA). During the same period, concern for marine pollution led to the London Convention on Dumping. Conventions on wetland conservation, world heritage sites and the control of trade in endangered species (CITES) were also signed in the early 1970s.

The UNCED conference reaffirmed the Stockholm Declaration and sought to build on it by 'establishing a new and equitable global partnership through the creation of new levels of cooperation among states, key sectors of societies and people' to work towards 'international agreements which respect the interests of all and protect the integrity of the global environmental and developmental system'. The basic theme was sustainable development, expressed in Principle 3 of the Rio Declaration as 'the right to development must be fulfilled so as to equitably meet developmental and environmental needs of present and future generations'. A detailed document, Agenda 21, was produced. Those sections of Agenda 21 relevant to chemicals are considered in more detail in Chapter 10.

There has been a dramatic change in the role of pesticides in the 30 years since *Silent Spring*, though not the same in different parts of the world. In developed countries, pesticide use continued to rise steeply during the 1960s and 1970s. For example in Japan, there was a sevenfold increase between 1960 and 1970. That this was a 'pesticide treadmill' phenomenon is evident from the fact that no yield increase was associated with the rise. During a similar period, insecticide expenditure in the USA doubled, yet again losses from pest attack did not decrease. Resistance to pesticides has been the principal driving force behind the introduction of

IPM methods in the developed countries, with cotton and glasshouse and orchard crops among the first areas where resistance to insecticides has forced the abandonment of routine reliance on pesticides. The first case of resistance to pesticides was detected in 1914; by 1990 nearly 500 cases of resistance in insects and mites could be listed (Roush and Tabashnik, 1990). To these cases can be added ever increasing numbers of plant diseases, weeds, nematodes and rodents that are also showing pesticide resistance.

The development of resistance affects integrated pest management (IPM) by reducing the number of pesticides that remain effective, and by applying pressure for application at higher dosages and frequencies (with potentially devastating implications for biological control). There is also a decreased incentive for industry to market more highly specific compounds. This, coupled with the escalating costs of pesticide development, suggests there will be a fall in the number of materials available for use in new IPM programs. Metcalf (1980) claims that resistance to insecticides is the greatest single problem facing applied entomology. Since the judicious use of pesticides is a critical part of IPM, it is vital that the current pesticides are not allowed to be lost through resistance. This danger is perhaps the most cogent reason of all for using IPM methods to reduce the quantity and frequency of current insecticide applications.

More recently, the banning of several compounds has removed the possibility of routine chemical control of some pests. Increasingly, public pressure is influencing both pesticide legislation and grower attitudes in favor of IPM. Long delays are now occurring in pesticide registration and re-registration, to the point where many growers are severely limited in the pesticides they may legally apply. As soon as choice of materials becomes very limited, growers themselves see the danger of pesticide resistance as a motive for switching to IPM (p. 62). The result has been an increasing implementation of IPM and a decrease in pesticide use in developed countries. Even so, examples of practical IPM remain few and it is clearly still in its early days. However, one has the feeling that, at long last, IPM is taking off. Pesticides are, however, still very much a part of many IPM programmes, but increasingly used as a stiletto instead of as a scythe. As pointed out earlier, there are excellent pesticides if used judiciously. IPM will develop more rapidly and successfully if they are accorded a proper role among the IPM strategies available for integration than if attempts are made to 'go it alone' with more biological methods.

At the time of *Silent Spring*, most pesticides were highly persistent (e.g. the organochlorines) and were applied either according to the calendar or when the pest was observed. The first major change was the introduction of 'economic thresholds', when treatment was only applied when warranted by established relationships between pest densities and crop yield. For a while, much of the practice of what we now call IPM was little more than modified spray programs involving the more intelligent use of pesticide. There is still probably an over-reliance on the use of pesticides in IPM programs. This is because the third phase, in which the maximum use is made of more biological tactics, is still only in late infancy. Pesticides can often cope with pest problems as a sole control measure, but at an environmental cost and with the danger of the development of pesticide-resistant pest strains. Other control measures (e.g. host-plant resistance, biological control etc.) often may not provide an acceptable level of control. The remaining gap then has to be filled. Thus pesticides are still, and are likely to remain, a component in the majority of IPM programs where the economic threshold for pest damage is low. The key to their use in IPM is that the chemical input must be designed so as not to prevent the operation of the more biological control components. How this may be achieved will be discussed in Chapter 5. It then ceases to be necessary for the chemical to provide the total required control on its own. Thus less toxic and more environmentally friendly chemicals are likely to be effective and used less frequently and even at reduced dosages. However, some countries (we think mistakenly) legislate against reducing the manufacturer's recommended application rate by making it an offence to deviate from that recommendation. The accent will often be on selective compounds and on ingenious solutions for applying broad-spectrum compounds selectively (p. 77–81). The shift away from emphasis on chem-icals in IPM does not detract from the logic and appropriateness of their use in a multiple, all-suitable-techniques strategy for sustainable pest control.

Thus IPM, including some pesticide use, is the route by which it has been possible to reverse the pesticide treadmill in developed countries. The same strategy is also appropriate to preventing the pressures for greater food production in developing countries leading inexorably to the same pesticide problems that have been experienced elsewhere. Here, as de-scribed in Chapters 3 and 4, international agencies are being at least partially successful at introducing and encouraging the use of insecticides in an IPM context. Here pressures for yield increases and the lack of regulation make routine pesticide applications the easy option.

The IPM philosophy also makes it possible to envisage very worthwhile increases in yield by introducing minimal insecticide use for that third of the world agricultural hectarage that currently receives no pesticides. Whereas full protection with pesticides would be prohibitively expensive there, the increases in yield with IPM could justify a limited pesticide input.

Another major change since *Silent Spring* was published is the nature of chemical pollution problems. In 1962, problems tended to be acute and local. Now more subtle but widespread effects on populations, communi-ties and ecosystems caused by acid rain, global warming and the ozone hole are our chief concerns. This makes it more than ever important not only to view the life cycle of the chemical from production to ultimate destruction, but to take account of the interaction of the chemical and its breakdown products with other processes. Various names have been proposed, 'integrated life-cycle management', 'integrated substance chain management', etc. The newsletter of the Society of Environmental Toxicology and Chemistry, has a large section headed 'LCA-News' without any definition of LCA. Presumably, if one does not know what LCA stands for, then one will not be interested in reading that section of the newsletter. Still the term 'life-cycle assessment' does seem to be most widely used. Therefore, although it would be convenient to use the word 'integrated' in the title to show the relationship to IPM, we have decided to use LCA.

Ideally one should take a 'cradle to grave' approach for each chemical. Regrettably we still do not have an adequate information base to make this feasible. The Organization for Economic Co-operation and Development (OECD) has estimated that, of the 10 000 chemicals for which the overall production is above 1000 tons yr^{-1}, we have ecotoxicological information on only 10%. The first question to be asked is 'How many of the remaining 90% are likely to pose an environmental threat?' One approach is a desk exercise examining the manufacture, usages and disposal of each chemical to decide which are likely to pose a threat and, if so, at which stage. As the horror of Bhopal showed, the initial chemicals from which plastics are made can be highly toxic while the finished products in use would seem to pose little threat to the environment. However, at the time of disposal there is the possibility of material being slowly leached out or toxic compounds created by combustion.

Another possible approach is to look, in the most likely places, to see if specific chemicals are in the environment. Their presence should trigger detailed investigations, whereas their absence would increase our confidence that the *status quo* was acceptable. Dioxins were discovered in the environment by deliberate looking, and polychlorinated biphenyls (PCBs) fortuitously while examining for other organochlorines. These families of compounds are considered in more detail later (pp. 30, 248–9).

International cooperation can play an important role. UNEP maintains a register (IRPTC) of the basic information on the toxicity of environmentally important chemicals. The OECD has developed a registry to list hazard assessments that are under way by national organizations. These can be exchanged and thus duplication avoided. The OECD itself is undertaking a study of 147 compounds with an annual production of over 1000 tons yr^{-1} on which there are no good toxicological data. Freedom of information would greatly increase the amount of data that are available. The rights of the companies producing the data need to be protected, but some way should be devised to make this information available to regulatory authorities. Determined efforts have to be made to reduce our ignorance on many widely used chemicals.

The experience of the industrial world makes it clear that prevention of pollution is cheaper than clean-up. This is true of local problems such as effluents from mining, of regional problems such as acid rain, and also of more global problems such as the chemicals that cause ozone layer depletion. While prevention is cost-effective it may well be difficult to find the money. Take, for example, the position of a country producing its electricity using soft coal with power plants that do not have the means of removing acidic oxides. The cost of altering existing plants is enormous, while switching to other fuels may require foreign currency that is not available. It is important that pollution standards are realistic. Recent water quality regulations issued by the European Economic Community (EEC) have been criticized as too stringent. This is not the forum to debate any specific issue. However, even in the most affluent country it does not make sense to spend large sums of money to clean water beyond what is necessary. It can be a difficult balance, and there is the obvious tendency to make regulations very tough so that there is no possibility of being wrong from the viewpoint of safety to human or environmental health. But the wasting of resources can also have a negative impact on human and environmental health, since funds are then no longer available for more worthwhile programs. Realistic pollution standards that can be met world-wide should be the objective.

While we do not disagree with the principle that the 'polluter pays', we would point out that it has serious limitations regarding determining blame and extracting payment. Additionally, although the polluter may pay, the polluter then passes on the cost. A fuller discussion of the 'polluter pays' principle will be found in Chapter 8.

This international aspect of pollution control is going to become increasingly important with the ratification of the Uruguay Round of GATT (General Agreement on Tariffs and Trade). The proposals of this round of negotiations are designed to strengthen GATT's basic commitment to free trade. 'Free trade' is an attractive slogan, but 'deregulated international commerce' maybe a better description (Daly, 1993). From an environmental point of view it is vital that the environmental costs of manufacture are included in the pricing of the goods. In the jargon of economists, costs should be 'internalized' (borne by the producer); if they are 'externalized' (borne by someone else), then not only is the competition basically unfair, but there is the potential for major environmental problems. A specific example is that if factory wastes are allowed to be dumped untreated then the cost is externalized; it is borne by the persons and wildlife around the factory. If the waste is cleaned up before dumping by the manufacturer then the costs are internalized. Even if it is agreed that costs should be internalized and that countries that do not comply are excluded under the GATT agreement, there is still a problem. While standards are set nationally, the firms that produce under the most permissive standards are at the greatest advantage. Countries with higher standards are at a disadvantage and thus free or unrestricted trade has the potential to support lower standards. Internationally agreed standards, such as those put forward by the World Health Organization (WHO), are the best way to mitigate this problem. The question of international cooperation is discussed more fully in Chapter 10.

In this book, drawing on the resources of many international agencies, we try to give a holistic view of where we are 30 years after *Silent Spring*. The emphasis is on the integrated or life-cycle approach.

References

Daly, H.E. (1993) The perils of free trade. *Scientific American*, **November**, 24–9.

Graham, F., Jr (1970) *Since Silent Spring*, Houghton Miflin, Boston.

Marco, G.J., Hollingworth, R.M. and Durham, W. (1987) *Silent Spring Revisited*, American Chemical Society, Washington, DC.

Metcalf, R.L. (1980) Changing role of insecticides in crop protection. *Ann. Rev. Entomol.*, **25**, 136–42.

Roush, R.T. and Tabashnik, B.E. (eds) (1990) *Pesticide Resistance in Arthropods*, Chapman & Hall, New York.

CHAPTER 2

The world of chemicals

A vast number of different chemicals have been synthesized, although the number in production of over 10 000 tons yr^{-1} is only 1000–2000. Chemicals may be considered both in specific classes, the most important of which from environmental considerations are the heavy metals and the petroleum hydrocarbons, and by use pattern. In the latter category, pesticides, by-products of energy production and those highly stable synthetic compounds, the polychlorinated biphenyls and the chlorofluorocarbons, which have escaped into the environment, are the most important. All these categories of environmentally important chemicals are considered in this chapter.

Introduction

We live in a world of chemicals. The chemical industries supply us with the materials that are the main basis of our comfortable way of life. In considering the problems of environmentally sound management of chemicals and wastes one has to have some idea of the scope of the problem. This chapter has no equivalent in *Silent Spring*, which was concerned exclusively with pesticides. In recent times it has become clear that it is not practical to treat pesticides and nonpesticides in isolation. The discovery by Jensen that polychlorinated biphenyls (PCBs) occurred widely in the environment meant that the actions of these chemicals had to be considered in relation to those of dichlorodiphenyltrichloroethane (DDT). PCBs, although not used as a pesticide, are similar in structure and in many ways act like DDT. Acid rain is a by-product of energy production and use, but it, like pesticides, has effects on forestry and agriculture (Chapter 9).

Natural and unnatural chemicals

There is a tendency to relate 'unnatural' with 'harmful', whereas there is no such relationship. Some natural chemicals can be very harmful and many unnatural chemicals do not pose a threat.

Harmful chemicals have always been present in the environment. The radioactive gas, radon, has been formed by natural processes and has been leaking from the ground since the beginning of time. The heavy metals, such as lead and mercury, also occur naturally. Most of the adverse effects of these metals are due to man's use. These uses go back to ancient times. Lead poisoning in workers extracting this metal was described by Hippocrates in the second century BC, but it greatly increased with the industrial revolution. Petroleum products are also naturally occurring compounds, but until the development of the internal combustion engine most of them remained locked below the surface of the earth. Now huge amounts of them are extracted, some are spilled in transportation and the rest are burned to alter the composition of the atmosphere. The 'green revolution' in farming in the 1940s produced an explosion of chemical use. This has involved the use of many chemicals. Some, such as pesticides, can be natural but are mostly synthetic. Natural compounds, such as phosphate and nitrate as fertilizers, are used in large amounts.

Numbers of chemicals and information available

The total number of chemical compounds that have been synthesized is enormous, perhaps as many as 10 million. The European Inventory of Existing Commercial Substances lists 110 000 chemical compounds that are produced commercially, and 1000–2000 are added to this list annually. Most of these are produced in small amounts and many are used only as intermediates in chemical processes. For new industrial chemicals, many countries have a requirement for a basic package of data to be produced; often this is the minimum pre-market data (MPD) set proposed by OECD.

In 1984 the US National Research Council reported that sufficient information available for a complete hazard assessment to be made was available for only 2% of chemicals produced commercially. Even for a partial hazard assessment this was only 14%. These are low figures, although it should be said that their data requirements for hazard assessment are rather high. In 1990 the OECD produced a list of 1338 chemicals that are produced in high volumes (defined as more than $10\,000$ tons yr^{-1} in any member country) (Brydon *et al.*,1990). These high-volume chemicals account for 90–95% of the total global chemical production. An examination by the OECD, completed in 1990, revealed that there was sufficient information for a detailed hazard assessment with only 434 of the 948 organic chemicals on the list. For 147 organic chemicals, no safety data were available. The OECD is now undertaking a detailed review on these 147 chemicals. The initial hope is to find the data necessary for hazard assessment. Where they are not available, it will be necessary to generate these essential data.

Many lists of toxic chemicals that assign a priority to the degree of hazard have been compiled. The EEC divides priority chemicals into a 'black list' and a 'grey list'. These lists are given in Table 2.1. It will be seen that some categories, for example 'persistent synthetic substances', are quite vague. The US Environmental Protection Agency (USEPA) Priority List gives all the compounds of concern by their chemical names. It includes many organohalogen compounds, but does not list any organophosphorus compounds. The USEPA List contains 13 metals, compared with 22 in the EEC list.

The situation with the information available on pesticides is better. For this group of compounds there are always some toxicity data, since they are developed to kill something, and some information on exposure, since they are added to the environment during their normal usage. While the exposure side of the question is not straightforward as pesticides are transported, accumulated, metabolized and degraded, nevertheless it is normally better known than for nonpesticidal chemicals.

Hazard assessment

How many chemicals in commercial production in the world are actually hazards? Hazard is generally considered to have two components, toxicity and exposure. Put in the simplest terms, if there is no exposure there is

Table 2.1 'Black' and 'Gray' lists of the European Economic Community

Black list
1. Organohalogen compounds and substances that may form such compounds in the aquatic environment
2. Organophosphorus compounds
3. Organotin compounds
4. Substances, the carcinogenic activity of which is exhibited in or by the aquatic environment
5. Mercury and its compounds
6. Cadmium and its compounds
7. Persistent mineral oils and hydrocarbons of petroleum
8. Persistent synthetic substances

Gray list
1. The following metalloids/metals and their compounds:
   1. Zinc 2. Copper 3. Nickel 4. Chromium 5. Lead 6. Selenium
   7. Arsenic 8. Antimony 9. Molybdenum 10. Titanium 11. Tin
   12. Barium 13. Beryllium 14. Boron 15. Uranium 16. Vanadium
   17. Cobalt 18. Thallium 19. Tellurium 20. Silver
2. Biocides and their derivatives not appearing above
3. Substances that have a deleterious effect on the taste and/or smell of products for human consumption
4. Toxic or persistent organic compounds of silicon
5. Inorganic compounds of phosphorus and elemental phosphorus
6. Nonpersistent mineral oils and hydrocarbons of petroleum origin
7. Cyanides, fluorides
8. Certain substances which may have an adverse effect on the oxygen balance, particularly ammonia and nitrites

no hazard and if there is no toxicity there is no hazard. To be accurate, there is no such thing as zero toxicity. As the sixteenth century Swiss physician Bombastus von Hohenheim put it '*sola dosis facet veninum*' (the dose is the poison). Furthermore if you give an analytical chemist a large enough budget, he will devise an instrument that will find incredibly small amounts of almost any compound.

Another approach to the hazard assessment of chemicals is to look carefully at the exposure side of the equation. The Great Lakes of North America have some 36 million people and a vast array of different industries around their shores. Extensive analytical work on fish from the Great Lakes has identified nearly 500 different man-made chemicals in their flesh. While this list is probably not complete, and certainly additional compounds could be expected to be found in other contaminated areas of the world, it does suggest that chemicals present in the environment in detectable amounts represent only a small proportion of the chemicals in use.

A hazard ranking of these compounds has been made by the US Fish and Wildlife Service (Passino and Smith, 1987). They divided them into 19 chemical classes, testing the toxicity of representative compounds and combining this toxicity with the amounts present in the fish and source ranking. The source ranking was to separate the degree that the chemicals were estimated to have come from man-made sources compared with natural ones. This gave the highest contribution to the hazard ranking, which was therefore greatest for purely anthropogenic compounds such as the organochlorines. By contrast, the polycyclic aromatic hydrocarbons that occur naturally, but also have anthropogenic inputs, were ranked lower.

The top five classes of chemicals on their hazard ranking list are:

1. DDT and PCBs;
2. phthalate esters;
3. toxaphene;
4. polyaromatic hydrocarbons; and
5. chlorinated polycyclics.

The important role of organochlorines, with the stable carbon–chlorine bond, in environmental problems is clearly shown by this listing. Numbers 1, 3 and 5 are all organochlorines, and all except the PCBs are used as pesticides. The least investigated of these five classes of compounds is the phthalates. Their high hazard ranking comes from their moderate toxicity, moderate occurrence levels and a high source ranking. The appearance of DDT in the first category may come as a surprise considering its moderate toxicity and the fact that it has been banned in North America for many years. It is, in fact, likely that now the greater contribution comes from the PCBs. One of the most exciting developments – the result of a combination of the fields of molecular biology and wildlife toxicology – has been to express the toxic effects of these compounds in terms of dioxin equivalents. This concept is discussed in Chapter 8.

Classes of chemicals

Chemicals can be classified either by their chemical composition or by the uses to which they are put. Both approaches have their merits and here a combination of the two is used. Chemicals of very diverse structure, many synthetic, some natural, have been used as pesticides, and it is convenient to consider them together. Other pollutants, such as the heavy metals and the petroleum hydrocarbons, are more conveniently considered as groups of chemicals.

From the viewpoint of the environmental problems, the three usage patterns of greatest concern are pesticides, by-products of energy production and highly stable chemicals used to resist high temperatures and pressures. 'Pesticide' is a broad term, literally meaning 'pest-killer'. Pesticides can be subdivided into insecticides, fungicides, herbicides, rodenticides and other '-icides'. Pesticide use has given rise to serious concerns for humans and wildlife; as pointed out above they are poisonous by definition, at least to some forms of life, and are deliberately added to the environment.

The use of pesticides by regions, in the mid-1970s and early 1980s (the latest data that are available) is given in Table 2.2 (UNEP, 1991). There was, overall, a small decrease in the usage of insecticides between the two dates. This decrease resulted from significant reductions in the Americas which counterbalanced the increases in other parts of the world. Some decreases may have been due to using more toxic chemicals in smaller amounts and to better delivery systems, such as ultra-low-volume techniques. These reduce the amount of the pesticide used, although not necessarily the area sprayed. The largest increase was in herbicides, with increases in all areas except North America; growth was particularly marked in the former Soviet Union. This region also showed the largest proliferation in the use of fungicides: the future pattern of usage in that area will be watched with interest. However, the pesticide usage pattern given by Pryde (1991) for the Soviet Union over the period 1976–88 is rather different. The insecticide usage is quite similar, but Pryde's figures

Table 2.2 Consumption of pesticides by region. Comparison of mean value in tons active ingredient per year for 1975–77 with 1982–84 (from UNEP (1991/92) *Environmental Data Report*, Blackwell, Oxford, with permission)

	All pesticides		Insecticides		Fungicides		Herbicides	
	75–77	82–84	75–77	82–84	75–77	82–84	75–77	82–84
Africa	68 181	66 608 − 2.3%	25 570	30 362 + 18.7%	34 758	34 362 0.0%	n/a	n/a
Asia	284 476	315 910 + 11.0%	126 429	140 229 + 10.9%	123 087	128 306 + 4.2%	31 530	39 663 + 25.8%
Europe	506 830	585 405 + 8.6%	53 154	57 806 + 7.1%	258 795	277 178 + 27.3%	191 562	243 878 + 15.4%
North America	529 194	484 052 − 8.5%	114 986	113 519 − 21.7%	143 753	145 408 + 1.1%	238 616	224 764 − 0.6%
Oceania	62 289	66 993 + 7.6%	3 487	4 667 + 33.8%	12 988	13 963 + 7.5%	45 814	48 217 + 5.2%
South America	108 324	99 350 − 8.3%	48 325	31 945 − 33.9%	29 525	29 159 − 1.2%	27 271	38 199 + 37.8%
USSR	348 767	535 400 + 53.5%	65 500	69 333 + 5.9%	186 333	230 067 + 23.5%	86 933	236 000 + 143.5%
World	1 908 059	2 153 718 + 12.9%	467 451	448 008 − 4.2%	789 239	858 713 + 8.8%	632 526	832 336 + 31.6%

do not show the massive increase in fungicides. His figures for 1988 are 68 000 tons of insecticides, 242 500 tons of fungicides and 155 600 tons of herbicides.

Another point that can be made from Table 2.2 is that the use of pesticides in the developing world, especially in Africa, is small compared to that in the developed world. The differences in the classes of chemicals between North America and Europe are interesting; the amounts of insecticides used in North America are much higher and the amounts of fungicides much lower compared with Europe.

Silent Spring is, essentially, a book about organochlorine pesticides. An examination of the index shows many entries under all the well-known organochlorines – aldrin, BHC (benzene hexachloride), chlordane, DDT, dieldrin, endrin, heptachlor and lindane – whereas there are only brief entries under carbamates and organophosphates (OPs). Only two OPs, malathion and parathion, are mentioned specifically. Data on the trends of pesticide use by chemical class are hard to obtain. Most organochlorines are now banned in the developed world, but even elsewhere the usage of this class of pesticides has decreased. In Egypt the use of DDT stopped in 1971, lindane in 1978 and endrin in 1981. By 1990 the pesticides used most heavily were organophosphates and pyrethroids (El-Sabae, 1989). Nevertheless, organochlorines continue to be used extensively in some parts of the world. DDT and BHC still comprise over 50% of the pesticides used in India, and their use is still increasing. In the late 1980s, 10 000 tons of DDT and 47 000 tons of BHC were used annually (Ramesh *et al.*, 1993).

Even though *Silent Spring* was largely about the organochlorine pesticides, there is a section on the mode of action of the organophosphorus pesticides. Later in the book, some concerns are expressed regarding the long-term effects of these materials.

Organophosphates

Although the organophosphates and carbamates are often called the 'second generation' of pesticides, actually they were developed over the same period as the organochlorines. Organophosphates were first developed by the military as nerve gases and only subsequently as insecticides. The mechanism of action of the OPs and carbamates is well established. Impulses of the nervous system pass from nerve to nerve across minute gaps between them, known as 'synapses', with the aid of the chemical acetylcholine. This compound is formed, performs its function of transmitting nervous impulses and then is destroyed by an enzyme called acetylcholinesterase (AChE). The action of OPs and carbamates is to inhibit this enzyme so that acetylcholine is no longer destroyed. Thus the nerve fires continuously; this can lead to tremors, convulsions and death.

Compared to the organochlorine pesticides, the organophosphates are often said to be more toxic, less persistent and their effects more transient. While these points are true, in the main, they are an oversimplification. The range of toxicity of OPs is considerable. The acute oral LD_{50} (the lethal dose for 50% of the population) of parathion to mallard is 2 mg kg^{-1} whereas that of malathion is 1500 mg kg^{-1}. Endrin, one of the most toxic of the organochlorines, has an LD_{50} of 5.5 mg kg^{-1} whereas for DDT the value is greater than 2000 mg kg^{-1} (Hudson, Tucker and Haegele, 1984). Thus the range of toxicity of both groups of compounds is great and there is a large degree of overlap.

Although organophosphates can under some circumstances (on the surface of leaves, in the soil) persist for some time, they are for the most part metabolized and eliminated rapidly. This usually makes their detection by chemical analysis difficult, and it is normal to assay the degree of inhibition of AChE of an animal to test for exposure to the compounds. The initial recovery of AChE in mammals and birds following exposure is rapid to 50–60% of the normal level, followed by slower further recovery. In birds the time to recover to normal levels of AChE is about a month (Fleming and Grue, 1981). There are concerns about possible chronic effects, both in humans and in wildlife. These are considered later.

A number of esterases other than acetylcholinesterase are inhibited by organophosphorus compounds and carbamates. For example, butyrylcholinesterase has sometimes been studied in parallel with acetylcholinesterase. However, the precise physiological role of butyrylcholinesterase is unknown, although it is often regarded as a marker enzyme for the glial cells and other non-neuronal elements. The neurotoxic esterase is also inhibited, and the interaction of OPs with this enzyme has been extensively studied by Johnson and co-workers (reviewed in Johnson, 1990). While the covalent binding of some organophosphorus esters to the neurotoxic esterase (NTE) has been shown to lead to irreversible polyneuropathy, the physiological function of the esterase is unknown. Johnson himself, after 20 years' research, merely concludes that 'it seems likely that the whole NTE protein which is tightly bound to neuronal membranes does serve a physiological function.'

A large number of instances of mortality of birds caused by OPs are known. In North America three-quarters of the incidents reported involve diazinon, fenthion, parathion or phosphamidon, whereas in Europe carbophenothion and chlorofenvinphos are the principal pesticides causing problems (Grue *et al.*, 1983). Among the carbamates, carbofuran has caused the most problems. Some instances of bird mortality involve considerable numbers. For example 10 000 American robins on berry fields in Florida and 500 greylag geese in Scotland.

In these cases the mortality was confirmed by direct counts, but potentially more serious are much larger losses that have been extrapolated from losses in small areas to the total area sprayed. Pearce, Peakall and Erskine (1976) used the reduction of singing males to estimate that 2.9 million songbirds were killed following spraying of the forests of New Brunswick with phosphamidon. Based on the number of carcasses found at specific sites and multiplying this by the total area treated, the USEPA estimated that the use of carbofuran on corn in the United States killed 1.3–1.6 million birds per year. Pimentel *et al.* (1993) have estimated that 67 million birds are killed annually in the United States. There is obviously considerable uncertainty, but the figures do suggest that pesticide usage could be a significant mortality factor.

Are there long-term effects on organisms from exposure to OPs? In *Silent Spring* Miss Carson gives some cases in which persons exposed to this class of chemical suffered from prolonged ill-effects. She concludes that 'all these consequences of organic phosphate poisoning, if survived, may be a prelude to worse. In view of the severe damage they inflict upon the nervous system, it was perhaps inevitable that these insecticides would eventually be linked with mental disease.' It can be argued that the few cases cited in *Silent Spring* 30 years ago are anecdotal. An article in the *New Scientist* (Bartle, 1991) suggests that many persons in Britain are suffering from serious chronic effects because of exposure to organo-phosphates. The writer interviewed 46 persons and detailed their failure to get help or recognition of the problem from British bureaucracy. An organization, PEGS (Pesticide Exposure Group of Sufferers), has been set up in the UK, but its function is to provide support for sufferers and it is not involved in scientific studies. PEGS states that it has case histories on several hundred persons and that these are to be made into a database. However, no firm scientific data are available at this time.

Concerns have been raised about sheep-dips, but again scientific data are hard to obtain. In the UK, the pesticide officer of the National Union of Farmers was unable to supply any definite information on the adverse effects of OPs on their members. Recently (March 1994) the UK Ministry of Agriculture, Fisheries and Food (MAFF) announced the formation of a Medical and Scientific Panel to look into reports of long-term ill health effects claimed to arise from exposure to organophosphorous sheep-dips. At the moment the emphasis of MAFF has been on the use of the correct protective clothing. However, from April 1995 operators will also be required to have a certificate of competence. In short, the evidence that OPs can cause long-term ill health is as anecdotal now as it was when *Silent Spring* was written.

Although the organophosphates had been available long before the problems with the organochlorine pesticides were recognized, their higher cost and lower safety (from the human health viewpoint) had restricted their use. Sometimes environmental problems caused switches from organochlorines to organophosphates before formal bans were introduced.

For example, in the forest spraying programs in New Brunswick to control spruce budworm, DDT was the insecticide chosen from the start of the program in 1952, and remained the only one used until 1957. After this initial period, it was used together with carbaryl and malathion until 1968. The pattern of insecticide use in the United States to control the gypsy moth was similar (Peakall and Bart, 1983). In recent years fenitrothion has been the main insecticide used, although the pathogen *Bacillus thuringiensis* has been used operationally since 1984. Between 1988 and 1992 it had been used on up to a third of the area sprayed. Initially the main problem with DDT was the loss of the important salmon fisheries. The problems with raptorial birds were not recognized until the late 1960s, although they formed part of the evidence that was to lead to the banning of DDT in the United States in 1972.

Another reason for the shift from organochlorine pesticides was the onset of resistance. For example, by 1960 two major cotton pests (bollworm and tobacco budworm) had become difficult to control with DDT: by 1965 they could no longer be controlled by DDT or a range of other pesticides (Bottrell and Adkisson, 1977). The problems of resistance and the strategies used to combat this problem are considered in Chapter 4.

Pyrethroids

Although natural pyrethroids have been in use from as far back as 1828, synthetic pyrethroids have only been in wide-scale use since the early 1980s (Hirano, 1989). The pyrethroids are highly toxic to insects. Dosages of the formulated material are typically less than 200 g ha^{-1} and with the most active compounds such as deltamethrin may be as low as 10–25 g ha^{-1}. They are quickly metabolized and have low toxicity to warm-blooded animals. The negative side of pyrethroids is their high toxicity to beneficial insects, fish and aquatic arthropods. In a review of the impact of insecticides on the ecology of ducks in the Canadian prairies, Sheehan *et al.* (1987) showed that synthetic pyrethroids can substantially damage aquatic ecosystems at rates recommended for control of cereal crop pests. A substantial decrease in invertebrate species diversity and abundance with little recovery during the breeding season was noted. These workers calculate that the loss of aquatic biomass is so great that it is impossible for ducks to obtain enough food to raise normal broods. In contrast, fish mortality has, despite the low LC_{50}s, rarely been observed following application of pyrethroids. It is considered that this is due to adsorption of the pyrethroids to other materials (Demoute, 1989). Regulators face a difficult choice; pyrethroids are among the safest pesticides from the viewpoint of vertebrate toxicity, but can be highly hazardous to the breeding habitat of waterfowl and beneficial insects including bees.

Herbicides

Herbicides are widely used in large amounts and their usage is increasing at a greater rate than that of other categories of pesticides (Table 2.2). The phenoxyacetic acids such as 2,4-D (2,4-dichlorophenoxyacetic acid) and 2,4,5-T (2,4,5-trichlorophenoxyacetic acid), the ureas (e.g. diuron) and the triazines (e.g. simazine) typically have low toxicity to warm-blooded animals unless contaminated (see below). Some other herbicides such as the bipyridinium group (e.g. paraquat) can be extremely toxic. Another widely used herbicide, glyphosate, is an organophosphorus compound.

The greatest concern from a wildlife point of view about the usage of herbicides is the alteration of habitats. In *Silent Spring* there is a strong attack on the use of herbicides and some examples of successful alternatives are given. However, the progress towards alternative controls has not been great in the period since *Silent Spring*. The use of herbicides has increased greatly and they are currently the most widely used type of pesticide in many parts of the world. The most basic question is how far we are prepared to manipulate habitats for the benefit of wildlife rather than agriculture, and the extent to which it is possible to accommodate both needs. This fundamental question is outside the remit of this volume. How to alter the habitat once the basic question has been answered is a secondary consideration. However, in much of the world there is no possibility of returning to labor-intensive practices.

The concern about habitat changes extends to entire ecosystems. Detailed, long-term studies on the grey partridge in the British Isles (Potts, 1986) have shown that the decreased availability of food caused by the use of herbicides is the main cause of the marked decline of this species. World-wide, Potts estimates that the population of this species has decreased from 20 million before 1940 to some 4 million by 1984.

Decreases have been noted for a wide variety of birds on farmland in the UK (Gibbons, Reid and Chapman, 1993). Their *Atlas*, compiled over the period 1988–91, compares the bird fauna of the UK with that in the first *Atlas* project compiled from 1968 to 1972. Significant declines have been found for several species for which open farmland is an important habitat. Declines were as great as 75% for the corn bunting and 50% for the skylark. The causes of these declines are complex. However, alteration of habitat (from herbicide use and changes in other farming practices) is likely to have been more important than direct mortality from chemicals. Current 'set-aside' programs give the opportunity for some interesting studies of this problem.

The expanding use of herbicides (Table 2.2) needs to be monitored carefully. A recent survey of herbicide contamination of Mediterranean estuarine waters, sponsored by the Food and Agriculture Organization

(FAO), has shown the presence of several herbicides in estuaries on the European side (Readman *et al.*, 1993). These authors consider that the concentrations of herbicides detected would be unlikely to have acute effects on most organisms, including man, but did consider that there could be changes in phytoplankton communities.

In *Silent Spring* it is stated that it is a matter of controversy whether or not herbicides such as 2,4,5-T are actually toxic to humans. The evidence that they are is, not surprisingly, anecdotal. The strongest evidence that human health problems can be related to the usage of herbicides comes from the massive use of herbicides in the Vietnam War, although even here the issue was not clear-cut. However, it is likely that the problems were caused by dioxin impurities (Chapter 8). On physiological grounds, herbicides (other than glyphosate) are far less likely to cause long-term problems to humans than would organophosphorus compounds which are known to act by attacking the nervous system.

Fungicides

Fungicides were not considered in *Silent Spring*, which is surprising since the early fungicides caused considerable environmental concern. One of the earliest fungicides was Bordeaux mixture which contained copper sulfate and quicklime. Runoff in rivers has caused local problems, and orchards treated over a long period with copper sulfate were almost devoid of life (Mellanby, 1967).

Several organomercury compounds have been used as fungicides. Studies in Sweden have shown mortality of wildlife associated with the use of these compounds. These studies and other problems associated with mercury are considered in more detail in Chapter 8. The modern synthetic fungicides have much lower toxicity to higher organisms.

Chemicals associated with energy production

Energy use is fundamental both to industry and for domestic comfort and yet it cannot be produced without environmental cost. An important consideration is the changing pattern of energy use. The changes between 1965 and 1990 for the least and most developed countries are given in Table 2.3.

Several interesting points emerge from the table. The percentage increase has been much greater in the developing world and it is this figure that expresses the improvement to the standard of living. An increase of two and half times suggests a substantial improvement. The ratio of energy used in the developed and developing countries has also changed considerably, with the ratio being only half what it was 25

Table 2.3   Changes in energy use in the least and most developed countries

	1965	1990	% increase	Absolute increase
Least developed countries	125	330	264	205
Highly developed countries	3640	4870	134	1130
Ratio	×29	×14		

Units: kg oil equivalents person^{-1}.

years ago. In the industrialized world the rate of increase of the use of energy is slower. Yet when one looks at the figures in absolute units, the picture is quite different. A much slower rate of increase in the developed world amounts to over 1100 units whereas the larger rate of increase in the developing world amounts to only 200 units. The amount of pollution is, to a large extent, related to the amount of energy consumed.

Carbon dioxide

Carbon dioxide emissions from fossil fuels are an important by-product of energy production. Carbon dioxide emissions for the largest contributing countries and for certain regions of the world are given in Figure 2.1. The United States is the largest single contributor of carbon dioxide, followed by the former Soviet Union. Contributions from the developing world are small, but the rate of increase here is much more rapid. Some projections suggest that China may be the largest single producer of carbon dioxide by the middle of the twenty-first century.

The direct measurement of carbon dioxide in air has been carried out systematically at Mauna Loa in the Hawaiian Island since 1958. During this period the concentration has risen from 315 p.p.m. to the current level of 351 p.p.m. A longer series of measurements has been possible by examining the carbon dioxide trapped in the Greenland ice cap (Neftel et al., 1985). Their data are plotted in Figure 2.2. The relationship between the increase of carbon dioxide and climate change, and its effect on agriculture, is discussed in Chapter 9.

Oxides of sulfur and nitrogen

The global input of SO_x from man-made sources is currently close to 100 million tons yr^{-1} and that of NO_x a little over 60 million tons. A major difference between the oxides of sulfur and nitrogen is the large difference in the relative importance of anthropogenic (man-made) and natural

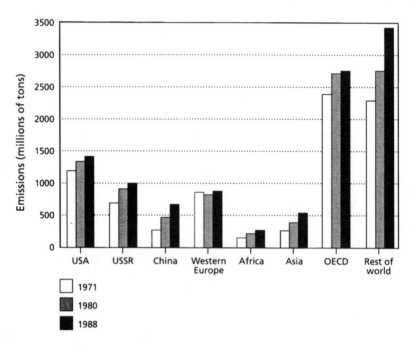

Figure 2.1 *Emissions of carbon dioxide, 1971, 1980 and 1988 (data from OECD,* The State of the Environment, *OECD, Paris, 1991).*

sources. SO_2 has only a small natural input, perhaps 5 million tons, largely from volcanic sources. In contrast, the natural inputs of nitrogen oxides are much larger than those from man-made sources. Bacterial action of the soil has been estimated (Henderson-Sellers, 1984) to liberate over 1000 million tons of NO and NO_2 and another 500 million tons of N_2O, which does not have any major anthropogenic sources, into the atmosphere. The impact of these acidic gases, and also ground-level ozone (which in some areas causes as many, or more, problems as SO_x and NO_x), on forests and agriculture is considered in Chapter 9.

The main sources of SO_x are shown in Figure 2.3. Coal used in power plants is the predominant source. In the case of NO_x, transportation is the largest single source, followed by utilities. While coal is the most environmentally unfriendly source of energy, it is also the most abundant (with reserves approximately 10 times that of oil) and is the cheapest of the fossil fuels. The cost of installing scrubbers is considerable. In new plants in Germany desulfurization of 80–95% has been achieved at an increase of 10–15% in the cost of producing electricity. To modify existing plants would be more expensive and such modifications have not yet been tackled on a large scale. Measures for the decrease of NO_x emissions have as yet

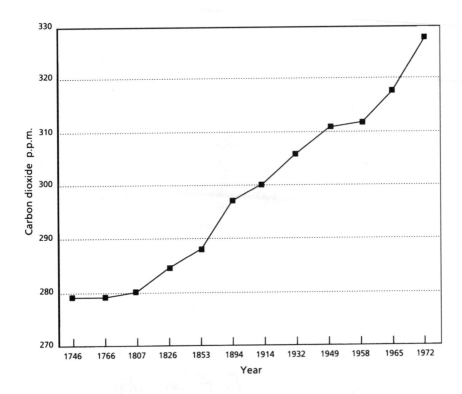

Figure 2.2 *Rise in carbon dioxide in the Greenland Ice Cap (data from Neftel* et al., Nature, Lond., **315**, 45–7, 1985).

received little attention, although stringent regulations for both NO_x and SO_x have been introduced in Japan. These regulations add roughly 25% to the cost of generating electricity, but give an air quality that again allows one to see Mt. Fuji from the skyscrapers of Tokyo.

Heavy metals

An organism's eye-view of the periodic table of the elements that comprise our world is given in Figure 2.4. Members of one important group are the main components of biochemical compounds. These are the elements hydrogen, carbon, nitrogen, oxygen, phosphorus and sulfur. Many other elements have known physiological functions. Some, such as iron (hemoglobin) and calcium (bone) have very vital functions. Several of these elements, although having physiological functions, can still cause toxic effects. Again, 'the dose is the poison'. Those identified as toxic and

(a)

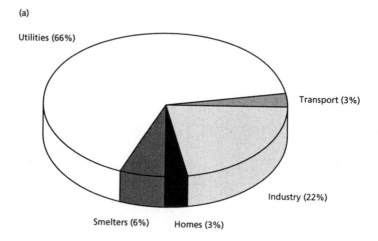

Utilities (66%)

Transport (3%)

Industry (22%)

Smelters (6%)   Homes (3%)

(b)

Transport (44%)

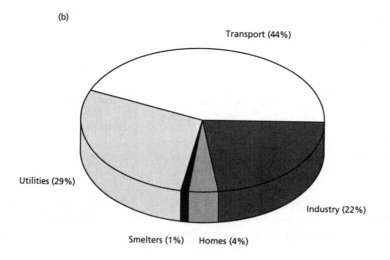

Utilities (29%)

Industry (22%)

Smelters (1%)   Homes (4%)

Figure 2.3 *Sources of (a) sulfur dioxide and (b) nitrogen oxide (data from Postel, S.,* Worldwatch Paper, No. 58, 1984).

nonessential have no known physiological function but are well known to cause toxic effects.

The term 'heavy metals' is used to cover the metals in the lower part of the periodic table. They are naturally occurring elements, but industrial usage has greatly increased the amounts in the environment. The relative importance of natural and man-made emissions is given in Figure 2.5, based on Nriagu (1989). The metal for which the ratio is highest for

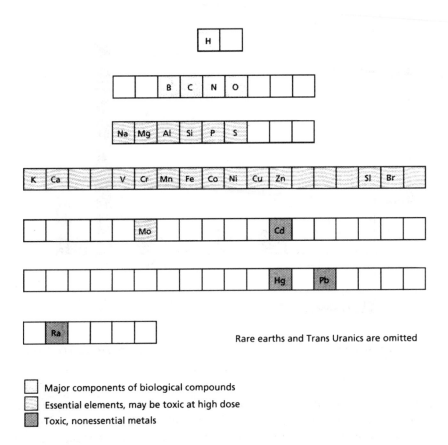

Figure 2.4 *Periodic table.*

man-made emissions is lead. This is largely because of the use of lead in gasoline; it is calculated that in the mid-1980s 80–90% of lead in ambient air came from the combustion of leaded gasoline. Considerable efforts have been made to switch over to lead-free fuel since this time. Total removal of this source would reduce anthropogenic emissions from 332 000 to 84 000 tons yr^{-1}. Lead production would then be reduced by 75% and it can reasonably be assumed that the emissions from the production process would be reduced by a similar amount. At this point mining operations for other metals would become the major sources of heavy metal pollution. Figure 2.6 shows the changes in the sources of lead. The contrast between the two diagrams is marked. The bottom right-hand quarter of the pie is now expanded to a full circle. If further improvements are considered

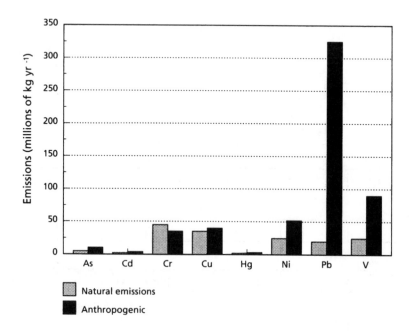

Figure 2.5 *Natural and anthropogenic emissions of metals (data from Nriagu, J.O., Nature, Lond., **338**, 47–9, 1989).*

necessary, then reducing mining/production operations would appear to be the best approach to reducing the pollution. These operations now account for over half the emissions and further removal of a small amount of lead from large quantities of coal, oil and wood before combustion is likely to be very difficult.

The changes in global contamination by lead have been dramatically demonstrated by examination of the levels of lead washed out of the atmosphere by rain and deposited as snow in Greenland. The levels started to increase in the 1700s, with a sharp rise in the 1950s, followed by a marked decline in the past 15 years. These changes since the 1950s correlate well with the increase and subsequent decrease of lead additives to gasoline. The measurement of the levels of lead is only one of the many measurements that can be made on the ice in Greenland; in fact the main driving force of the studies is to examine changes in climate. A major project has been set up to examine ice cores of this, the oldest ice in the world. Scientists from eight European countries are cooperating in the venture. Ice up to 100 000 years old has already been recovered from cores that are 2000 m in depth. It is estimated that the ice just above the bedrock is 300 000–500 000 years old.

(a)

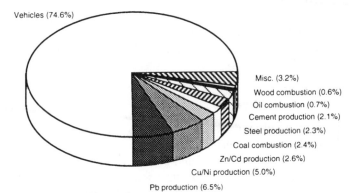

Vehicles (74.6%)

Misc. (3.2%)
Wood combustion (0.6%)
Oil combustion (0.7%)
Cement production (2.1%)
Steel production (2.3%)
Coal combustion (2.4%)
Zn/Cd production (2.6%)
Cu/Ni production (5.0%)
Pb production (6.5%)

Total emissions 332 000 tons yr^{-1}

(b)

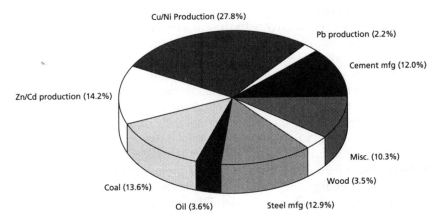

Cu/Ni Production (27.8%)

Pb production (2.2%)

Cement mfg (12.0%)

Zn/Cd production (14.2%)

Misc. (10.3%)

Wood (3.5%)

Coal (13.6%)

Oil (3.6%)   Steel mfg (12.9%)

Total emissions 78,000 tons yr^{-1}

Figure 2.6 *(a) Anthropogenic sources of lead; (b) sources excluding leaded petroleum (data from Nriagu, J.O. and Pacyna, J.M., Nature, Lond., **333**, 134–9, 1988).*

Highly stable compounds of environmental concern

There are two very different, highly stable synthetic compounds (or rather series of compounds), used for a variety of purposes, requiring the ability to withstand high temperatures and other adverse conditions, which have caused serious environmental problems. These are the PCBs and the chlorofluorocarbons (CFCs).

Polychlorinated biphenyls

The first commercial synthesis of PCBs was announced in 1930 in the American trade journal *Chemical and Engineering News*. The article that announced their synthesis predicted a wide variety of uses – in varnish, waterproofing, flameproofing and electrical insulation – based on their high chemical and physical stability. The PCBs were, indeed, a success story. The largest uses have been in the manufacture of capacitors, in plasticized products and in dielectric and hydraulic fluids. In short, they are used in almost every aspect of life in an industrialized society. PCBs (there are 210 possible compounds) are manufactured in various grades, containing different quantities of chlorine. By 1970 (the peak year before restrictions were introduced) the total global production was estimated by Risebrough and de Lappe (1972) at 100 million kg.

Despite the amounts used, PCBs generated little interest outside industrial circles until their discovery in the environment in the mid-1960s. A modest note in the *New Scientist*, reporting the work of the Swedish scientist Soren Jensen (1966), who discovered that these compounds had been found in several different species of wildlife in Sweden, generated enormous interest. Many scientists making measurements on the levels of environmental pollutants began to look for PCBs in their samples and it soon became apparent that PCBs were global contaminants.

The main uses of PCBs and their routes of entry into the environment in the USA, just before the voluntary restrictions were made in 1971, are shown in Figure 2.7. The routes of entry are complex. Leaks from the wide variety of equipment that used PCBs because of their stability were the largest source; leaching from landfills, which continues, is another major source. The effects of PCBs are considered in Chapter 8.

Chlorofluorocarbons

Like the PCBs, CFCs were first synthesized in 1930. Initially CFCs appeared to be both useful and harmless; they were chemically stable, almost completely inert and nontoxic. They were used as a propellant in spray

(a)

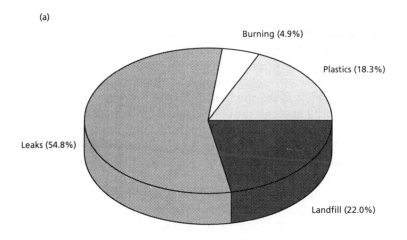

(b)

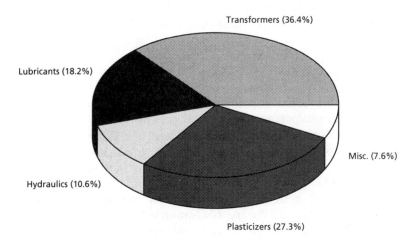

Figure 2.7 *(a) Routes of entry of PCBs into the environment, US 1970; (b) uses of PCBs, US 1970 (data from Nisbet, I.C.T. and Sarofim, A.F.,* Environ. Hlth Perspectives, **1**, *21–38, 1972).*

cans and fire extinguishers, as the compressor liquid in refrigerators and air-conditioners, as the blowing agent for the production of plastic foam and in other useful applications. The person credited with their discovery is Thomas Midgely, who also formulated tetraethyl lead. Whatever environmentalists may say about these inventions, he must have seemed to his employers, Chevron Corporation, to have had the Midas touch.

There are several different CFCs and the shorthand numbering system is widely used. The two most commonly used compounds are F-11 and F-12. The first digit is the number of fluorine atoms and the second the number of hydrogen atoms plus one. Thus F-11 contains one fluorine and no hydrogen atoms; its full chemical name is trichlorofluoromethane (CCl_3F). Similarly F-12 contains one fluorine and one hydrogen atom.

It was not until the 1970s that studies were undertaken to find out the fate of the CFCs. It was estimated that 800 000 tons yr^{-1} were entering the atmosphere. These studies showed that nothing happened to the CFCs in the lower atmosphere. They do not interact, they do not dissolve in the ocean, they do not get washed out of the air by rain; they just float around, slowly working their way upwards into the stratosphere. This focused the attention of scientists on what might be happening to the CFCs in the high reaches of the atmosphere 20–25 miles above the earth's surface. The possible effects of the interaction of CFCs with the ozone layer on primary productivity are considered briefly in Chapter 9. Major international efforts have been made to control CFCs (Montreal and London protocols) and these are outlined in Tolba's recent book *Saving Our Planet*.

Petroleum hydrocarbons

These are naturally occurring compounds, but man's extraction of them and subsequent transportation across the globe have vastly increased the environmental exposure to this group of chemicals. World production of oil totals 3 billion metric tons. In 1980, 1588 million tons of petroleum were transported by sea, and the offshore production was 658 million tons (NRC, 1985).

The total annual input of petroleum hydrocarbons to the oceans of the world has been estimated by the US National Research Council at 3.2 million tons. Figure 2.8 gives a diagrammatic representation of the relative importance of the major routes of entry to the environment. Waste disposal and transportation are the two largest sources, each of which contributes approximately a third of the total.

The single largest component of waste disposal is that from municipal wastes, which accounts for over half the total. In the case of transportation, normal operations account for the largest proportion. These figures are described in the report as 'best estimates'. Even allowing for the

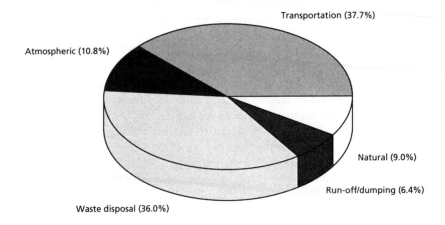

Transportation (37.7%)

Atmospheric (10.8%)

Natural (9.0%)

Run-off/dumping (6.4%)

Waste disposal (36.0%)

Figure 2.8 *Inputs of oil to the marine environment (data from NRC,* Oil in the Sea, *National Academy Press, Washington, DC, 1985).*

uncertainties in the details of the calculations, it is clear that most of the input of oil to the oceans occurs from a wide variety of small sources, both land-based and marine. Spills, despite the attention that they receive, account for only about one-seventh of the total.

Other high-volume chemicals

As mentioned earlier in this chapter, the OECD is now investigating all high-volume chemicals. The first group of high-volume chemicals assigned to member countries for evaluation did not divide readily into any particular classes of chemicals. The group included six chloro-compounds (four silanes and a substituted pyridine and butene), a number of pyridine compounds, several nitro-compounds and some pure hydrocarbons, such as pentadiene and cyclododecane.

Conclusions

The total volumes of chemicals liberated in the environment every year are staggering. There are over 6000 million tons of carbon dioxide, 170 million tons of carbon monoxide and 100 million tons of sulfur dioxide, to list three gaseous pollutants alone. Annual volumes of 2.5 million tons of pesticides are produced, 300 000 tons of lead and even 2000–3000 kg of that most toxic of compounds, tetrachlorodioxin.

It appears that any reasonably stable high-volume chemical will escape and can cause problems in the environment. The cases of PCBs and CFCs indicate the potential for 'sleeper' chemicals. Both were in widespread use for over 30 years before the problems were realized. The mode of discovery of their presence in the environment was quite different in each case. PCBs were discovered by identifying unknown peaks on a gas chromatographic printout of an extract of environmental samples. Once discovered in the environment, these compounds were looked for by other workers, and were soon shown to be a global problem. CFCs were discovered by Rowland and Molina, who set out to discover where CFCs were going. Once they had predicted that these compounds would move into the stratosphere – and potentially react with the ozone layer – they looked for CFCs high above the earth.

Have we identified all the serious chemical pollutants? This is a very difficult and, in an absolute sense, impossible question to answer positively. The two examples of 'sleeper' chemicals – PCBs and CFCs – already cited do not give confidence in our ability to predict environmental effects of chemicals produced on a large scale. A note in *Ambio* (Wesen, Carlberg and Martinsen, 1990) suggests that we have identified only 15% of organochlorine compounds present in the environment.

There is a strong case for an international program to discover the extent to which potentially pollutant chemicals are present in the environment. For high-volume chemicals, not only should calculations be made to see which compartment of the environment it is most likely that they will reach, but also that compartment should then be examined to see if the compound is present.

References

Bartle, H. (1991) Quiet sufferers of the silent spring. *New Scientist*, **18**, 30–5.

Bottrell, D.G. and Adkisson, P.L. (1977) Cotton insect pest management. *Ann. Rev. Entomol.*, **22**, 451–81.

Brydon, J.E., Morgenroth III, V.H., Smith, A. and Visser, R. (1990) OECD's work on investigation of high production volume chemicals. *Int. Environ. Reporter*, **June**, 263–70.

Demoute, J.-P. (1989) A brief review of the environmental fate and metabolism of pyrethroids. *Pestic. Sci.*, **27**, 375–85.

El-Sabae, A.H. (1989) Fate and undesirable effects of pesticides in Egypt. *Ecotoxicology and Climate, SCOPE*, **38**, 359–71.

Fleming, W.J. and Grue, C.E. (1981) Recovery of cholinesterase activity in five avian species exposed to Dictrophos, an organophosphorus pesticide. *Pestic. Biochem. Physiol.*, **16**, 129–35.

Gibbons, D.W., Reid, J.B. and Chapman, R.A. (1993) *The New Atlas of Breeding Birds of Britain and Ireland: 1988–1991*, Poyser, London.

Grue, C.E., Fleming, W.J., Busby, D.G. and Hill, E.F. (1983) Assessing hazards of organophosphate pesticides to wildlife. *Trans. N. Am. Wildl. Nat. Res. Conf.*, **48**, 200–20.

Henderson-Sellers, B. (1984) *Pollution of Our Atmosphere*, Adam Hilger, Bristol.

Hirano, M. (1989) Characteristics of pyrethroids for insect pest control in agriculture. *Pestic. Sci.*, **27**, 353–60.

Hudson, R.H., Tucker, R.K. Haegele, M.A. (1984) Handbook of toxicity of pesticides to wildlife. *U.S. Fish Wildl. Ser. Resource Publ.*, No. 153.

Jensen, S. (1966) Report of a new chemical hazard. *New Scientist*, **32**, 612.

Johnson, M.K. (1990) Organophosphates and delayed neuropathy – Is NTE alive and well? *Toxicol. Appl. Pharmacol.*, **102**, 385–99.

Mellanby, K. (1967) *Pesticides and Pollution*, Collins, London.

Neftel, A.E., Moor, E., Oeschger, E. and Stauffer, B. (1985) Evidence from polar ice cores for the increase in atmospheric carbon dioxide during the past two centuries. *Nature, Lond.*, **315**, 45–7.

NRC (1985) *Oil in the Sea*, National Academy Press, Washington, DC.

Nriagu, J.O. (1989) A global assessment of natural sources of atmospheric trace metals. *Nature, Lond.*, **338**, 47–9.

Passino, D.R.M. and Smith, S.B. (1987) Acute bioassays and hazard evaluation of representative contaminants detected in Great Lakes fish. *Environ. Toxicol. Chem.*, **6**, 901–7.

Peakall, D.B. and Bart, J.R. (1983) Impacts of aerial applications of insecticides on forest birds. *CRC Crit. Rev. Environ. Contr.*, **13**, 117–65.

Pearce P.A., Peakall, D.B. and Erskine, A.J. (1976) Impact on forest birds of the 1975 spruce budworm spray operations in New Brunswick. *CWS Prog. Notes*, No. 62.

Pimentel, D., Acquay, H., Biltonen, M. *et al.* (1993) Assessment of environmental and economic impacts of pesticide use, in *The Pesticide Question. Environment, Economics and Ethics*, (eds D. Pimentel and H. Lehman), Chapman & Hall, New York, pp. 47–84.

Potts, G.R. (1986) *The Partridge: Pesticides, Predation and Conservation*, Collins, London.

Pryde, P.R. (1991) *Environmental Management in the Soviet Union*, Cambridge University Press, Cambridge.

Ramesh, S., Tanabe, S., Tatsukawa, R. *et al.* (1993) Seasonal variation of organochlorine insecticide residues in air from Porto Novo, south India. *Environ. Pollut.*, **62**, 213–22.

Readman, J.W., Albanis, T.A., Barcelo, D. *et al.* (1993) Herbicide contamination of Mediterranean estuarine waters: results from a MED POL pilot survey. *Mar. Poll. Bull.*, **26**, 613–19.

Risebrough, R.W. and de Lappe, B. (1972) Accumulation of polychlorinated biphenyls in ecosystems. *Environ. Hlth Perspectives*, **1**, 39–45.

Sheehan, P.J., Baril, A., Mineau, P. *et al.* (1987) The impact of pesticides on the ecology of prairie nesting ducks, *CWS Techn. Rep. Series*, No. 19.

Tolba, M.K. (1992) *Saving Our Planet*, Chapman & Hall, London.

UNEP (1991) *Environmental Data Report*, Blackwell, Oxford.

Wesen, C., Carlberg, G. E. and Martinsen, K. (1990) On the identity of chlorinated organic substances in aquatic organisms and sediments. *Ambio*, **19**, 36–8.

CHAPTER 3

The integrated approach

The concepts of integrated pest management (IPM) and life-cycle assessment (LCA) of chemicals. Comparison of IPM today with the predictions in *Silent Spring*. The regulation of pesticide usage, especially in the developing world. Means of decreasing the impact of chemicals on human health and the environment, such as increasing biodegradability and/or recyclability, decreasing toxicity. The importance of devising processes that make long-lasting products with less energy and using a minimum of harmful materials is stressed.

Introduction

The word 'integrate' is not a technical word that one would normally put in a glossary of scientific terms. *The Oxford Dictionary* defines it as 'to put or bring together to form a whole' – simple enough, but in environmental terms difficult to do. One of the difficulties, and fascinations, of ecology is the complexity of its interactions. This book considers the management of both pesticidal and nonpesticidal chemicals, from the viewpoint of integration. Agriculturalists have now agreed on the term 'integrated pest management' (IPM), but different terms for the same principle are used in the environmental field. Although the importance of integration is clearly recognized, several different combinations of words have been used, such as 'integrated life-cycle management', 'integrated substances chain management' and more recently 'life-cycle assessment' (LCA). Although we would have liked to retain the word 'integrated', it seems that LCA has become the most widely used term and it will therefore be used in this book.

There is also a need for integration between concerns over food production and the use of chemicals, of which only some are pesticides. Modern developments in agriculture require increasing amounts of energy to produce the fertilizers and pesticides, to transport them to the farm, and for ever-larger farm equipment. The energy inputs in corn production in the United States rose threefold between 1945 and 1970 (Pimentel *et al.*, 1973), although this was more than compensated for by increased yields. Thus the yield per unit of energy increased over this period. Nevertheless agriculture, particularly Western agriculture, is a major user of energy and thus a contributor to such problems as acid rain and global warming. These problems can, in turn, affect the agriculture from which they stem. The effect of acid rain on forestry and the predictions of effects of global warming on agriculture are considered in Chapter 9.

Integrated pest management

The term 'integrated pest management' is not mentioned by Rachel Carson for the very good reason that it was not coined until nearly 15 years after the appearance of *Silent Spring*. Yet the seeds of the concept were sown before that book was published. In Peru, California and in greenhouses in the UK, overuse of insecticides caused pest crises in the 1950s. In each case they were solved by a combination of biological control with judicious use of insecticide. In a landmark paper in *Hilgardia* in 1959, the Californians Vernon Stern, Ray Smith, Robert van den Bosch and Kenneth Hagen described their work on lucerne in California, where they had solved the problem of aphids resistant to organophosphate insecticide by applying organophosphate insecticide at reduced dose! Some aphids had still been killed at this dose, but damage to biological control systems had been greatly reduced. In their seminal paper, the Californians defined 'integrated control' as 'applied pest control which combines and integrates biological and chemical control'. This integration was found immediately practical, and many scientists were able to devise similar integrations of chemical and biological control for other crops. Perhaps unfortunately for the implementation of integrated control, other scientists (particularly in Europe) proposed that integrated control should be an integration of pest control methods without the involvement of pesticides ('harmonious control'). The result was that in 1967, at a meeting of the UN Food and Agriculture Organization (FAO), 'integrated control' was redefined in terms very similar to 'harmonious control' (i.e. avoiding pesticides as far as possible).

The next stage in the evolution of IPM came in 1970. The term 'pest management', first proposed by Geier and Clark in Australia in 1961, was defined at a conference in Raleigh, North Carolina as 'the reduction of pest problems by actions selected after the life systems of the pests are understood and the ecological as well as the economic consequences of these actions have been predicted, as accurately as possible, to be in the best interests of mankind'. Thus single-component biological control would be as valid a pest management solution as an integrated approach. The term caught on and effectively replaced 'integrated control'; little distinction was made between the two terms after that.

Although the term 'integrated pest management' was first defined (by Lawrence Apple of the University of North Carolina and Ray Smith of the University of California) in 1976 merely to broaden pest management to cover all pests rather than all control methods, this definition of IPM has been totally ignored. Today 'integrated control' equals 'pest management' which equals 'integrated pest management' in most people's minds. Thus the definition of pest management given above is usually equally accepted as a definition of IPM.

The current status of IPM is given in detail in Chapters 5 and 6 of this book. Here we examine, briefly, the extent to which the visions put forward by Miss Carson in 1962 have been achieved. In contrast to the integration of methods in early 'integrated control' of the late 1950s, Rachel Carson saw the more biological alternatives to pesticides very much in isolation. The final chapter of *Silent Spring* discusses, in the following order, insect sterilization by radiation and chemicals, insect growth regulators, pheromones, attractants and repellents, insect pathogens and, finally, biological control. Each method is described with reference to specific pest targets in relation to which some success had already been achieved or might be anticipated. However, the chapter lacks critical evaluation of the already known disadvantages and specificity of the methods for particular target pests. *Silent Spring* is in many ways an astoundingly nonprophetic book! Rachel Carson's account of the alternative methods of pest control are given in an order very much the reverse of how they are viewed today. Here we discuss briefly how the various methods have progressed over the past 30 years; a more detailed account of the methods is given in Chapter 5.

Sterilization

Sterilization, Rachel Carson's number one suggestion, is hardly used in practice today. Although the chemical industry has searched actively for chemosterilants and has found several, very few have been marketed. Those that were, were generally not marketed for very long. They were usually not safe for release in the environment and had to be contained in traps or had other serious commercial drawbacks. Radiation sterilization was riding on a crest in 1962 with the successes at eliminating the cattle pest, screw-worm. Although the technique has been used against this pest ever since, most recently to prevent invasion of the Old World, attempts to use it commercially against other pests have not yet been reported. Screw-worm appears to have unique characteristics as a suitable target (p. 131).

Insect growth regulators

Insect growth regulators interfere with the internal hormone balances of insects and thus with growth and metamorphosis. At the time of *Silent Spring*, they were heralded as selective compounds to which no resistance would evolve. These claims, alas, proved false. Research is still in progress, and the early promise of this approach to pest control has not been fulfilled.

Sex attractants

Natural or synthesized sex attractants for male insects (sex pheromones) are used routinely in traps to estimate pest numbers to aid decisions about when to apply pesticides. For direct control, however, only isolated successes have been obtained with trapping out pests by combining pheromones with insecticides. More promise is shown by the 'confusion technique'. Here the release of artificial pheromone at many sources can confuse the orientation behavior of male pests. Clearly this ecologically desirable method can reduce or delay the need for insecticides. This is excellent IPM, but pheromones are still too expensive to be economically viable if they have to be used in combination with pesticides. Attractants (other than pheromones) and repellents have continued to be investigated over the past 3 decades, but until recently there have been few prospects of commercial use. The search for suitable chemicals has received the recent stimulus of industrial interest in the multiplicity of plant chemicals in tropical plants. New strategies for the deployment of such chemicals in pest control are emerging; it is a strong research and development area and may well assume practical importance in the near future.

Biological control

The method mentioned last in *Silent Spring*, biological control, has proved to be the most generally available method to substitute for at least some of the control previously achieved by total reliance on pesticides. Many successes have been achieved in the past 30 years with the introduction of biological control agents in systems where insecticides are not feasible or totally uneconomic. Such agents have also been introduced into systems (e.g. greenhouses) to supplement or replace pesticides. Also, IPM research and practice is increasingly targeted to maximizing the impact of natural enemies already present in systems, rather than (as in Rachel Carson's day) seeking to inoculate new species. Rachel Carson implied that we should be able to solve all pest problems with beneficial insects; today this view is totally outdated. Insect pathogens have many advantages over animal carnivores as biological control agents. In many ways their properties make them ideal for use in IPM, and in dealing with insecticide-resistant pests. However, they have limitations in efficacy and field viability; more serious, however, are registration and commercial difficulties. If these can be overcome, then the future for insect pathogens looks very encouraging.

It is astonishing that Rachel Carson does not even mention host-plant resistance. The method was well established for insect control from the 1950s onwards, and much earlier for plant diseases. More recently, particularly in the tropics with the establishment by the Consultative Group on International Agricultural Research (CGIAR) of a chain of international research stations, the breeding of cultivars that are less damaged by pests is seen as the principal aim of pest management research.This emphasis on host-plant resistance has spread world-wide. Potentially, plant resistance can provide a highly farmer-acceptable method of pest control that has little impact on the environment and requires little additional input.

Thus the visions of Rachel Carson have not been really fulfilled in any detail. She failed to appreciate the integrated approach to pest control that was gaining ground even as she was writing. Nor could she have realized how long it would take for world opinion to force moves to end routine reliance on pesticides as a sole pest control strategy. Her book was published in the early 1960s, yet a real pressure to change, especially in high-input agriculture, did not develop till the very late 1980s.

Role of pesticides in integrated pest management

An important philosophical difference between the views put forward in *Silent Spring* and those of today are that many proponents of IPM consider that pesticides must remain in the arsenal of weapons to be used against pests. Although it is stated early in *Silent Spring* (p. 12) that 'it is not my [Rachel Carson's] contention that chemical insecticides must never be used', the implications of the last chapter of *Silent Spring* are that pesticides can be phased out completely.

This view is clearly spelled out in the first seven words of the final chapter ('The other road') of *Silent Spring*: 'We stand now where two roads diverge.' These words, however, stand out as among the most misleading statements in the history of applied entomology. Rachel Carson presented the analogy of man having reached a crucial fork in the road. Now was the time to choose the narrow road to salvation and abandon pesticides in favor of biological control. The misleading nature of this analogy is only too evident when we consider the contemporary events that occurred in Peru, California and the UK referred to on p. 38. In Peru, the solution of pesticide resistant pests on cotton was based on restoring biological control under the protection of a selective insecticide (a stomach poison). As mentioned earlier, the solution in California was to supplement biological control with low-dose pesticide applications. In the UK all-year-round chrysanthemum crop, pesticide-resistant aphids were controlled by substituting a new insecticide, but one so selective that biological control of other pests could continue.

Far from representing two diverging roads, chemical and biological control were shown to be converging in those early days of integrated control. This convergence is still a key principle of IPM today. One of the most important developments stemming from the first definition of 'integrated control' was avoiding prophylactic pesticide treatments as far as possible by establishing 'economic thresholds' for pests. This principle has been widely practised from the late 1950s onwards, and has probably led to as great a reduction in pesticide use as all other components of IPM put together. The concept is not mentioned as part of 'The other road' in *Silent Spring*.

Regulation of pesticide usage

Silent Spring certainly played an important role in the development of the movement to ban the organochlorine pesticides. Legal action to phase out DDT started in the developed world in the late 1960s with the realization of its undesirable persistence and danger to wildlife. In the decision handed down by Ruckelshaus, the first administrator of the then newly formed USEPA, it was stated that there was compelling evidence for adverse impacts of DDT on fish and wildlife. The decision continued that, while there was no adequate epidemiological evidence for effects on man, DDT had been demonstrated to be a carcinogen in experimental animals (Dunlap, 1981). Although the environmental evidence was clear, whereas the evidence of carcinogenicity of DDT remained a matter for debate, it seems that it was the latter aspect that was vital in the final cancellation procedure. This cancellation of virtually all uses of DDT in the USA in 1972 was followed by cancellation of several other organochlorine pesticides within a few years. There are now over 40 pesticides on the suspended, cancelled or restricted list of the USEPA, while others are restricted to use by certified operators. It is important to distinguish between banned and nonregistered compounds. Nonregistration may not be the result of a known problem, but merely due to lack of a market large enough to justify the cost of re-registration. Therefore the fact that many pesticides not registered in the USA are registered in other countries is not in itself a matter for concern.

Some 10–15 years behind the industrialized nations, several African countries have recently banned DDT, aldrin and other chlorinated hydrocarbon insecticides (Brosten and Simmonds, 1989). Several developing countries still lack pesticide control legislation. In others that have such legislation, it is not strictly enforced. In Kenya, the Pest Control Products Act of 1982 was implemented in 1983 by the establishment of a Board which is responsible for managing pesticides in accordance with the FAO code of conduct. By 1985, 51 African countries had adopted the code (Brosten and Simmonds, 1989). Most countries in Central America have statutes or decrees to prevent problems associated with pesticides, but none are equipped for adequate enforcement.

An organization, Pesticide Action Network, campaigns to ban the use of 'undesirable' pesticides and documents the degree to which these materials are restricted throughout the world. In 1985 this organization launched a campaign to outlaw the use of the so-called dirty dozen pesticides, which has subsequently been increased to 18. The original 'dirty dozen' (aldrin, chlordane, chlorodimeform, dibromochloropropane (DBCP), DDT, dieldrin, ethylene dibromide (EDB), endrin, hexachlorocyclohexane (HCH), heptachlor, lindane and toxaphene) had been banned in 23–42 countries and severely restricted in 2–16 others by 1992. There are major differences even in neighboring countries. For example, Guatemala has banned all of them whereas Honduras has banned none and placed severe restrictions on only two.

Nevertheless, the trade in restricted pesticides remains considerable. In 1990 over US$12 million of pesticides that were banned or had their registration canceled were shipped from US ports (Smith and Beckmann, 1991). This includes over US$1.5 million of chlordane and heptachlor. Additionally, US$7 million of unregistered pesticides and US$60 million of pesticides whose use is restricted were exported. This total of US$68.5 million of restricted pesticides is out of a total of US$400–600 million worth of pesticides exported from the United States.

FAO/WHO established the Codex Alimentarius Commission in 1963, which set acceptable maximum residue levels for food coming into international trade. These levels are then offered for acceptance to countries that wish to participate and the code is being taken up, even if slowly. Quarantine provides the policing and helps to avoid the 'circle of poisoning' phenomenon that involves the shipping of food contaminated by unregistered products into foreign countries.

An International Code of Conduct on the Distribution and Use of Pesticides was developed by FAO in 1985, and amended to include prior informed consent in 1989. The voluntary code is intended to serve as a point of reference, particularly for countries that have not yet established their own adequate regulatory infrastructures.

The environmental procedures of the US Agency for International Development (USAID) require that agency to assess the public health and environmental impact of any pesticides before funding for their use and supply is approved. Although USAID missions are in a country only by invitation and cannot enforce procedures, the desire for aid can influence acceptance of reforms significantly. Many countries still consider environmental management to be a luxury and view the regulations as too strict, but changes have nevertheless resulted from the implementation of USAID regulations. These changes include reductions in the amounts of pesticides used, restrictions on use to protect public health and the environment, reconsideration or revision by countries of their pesticide use policies and the initiation of projects that address environmental concerns directly.

Most pesticides used in the tropics originate in developed countries and, under current legislation, may be exported from the country even if prohibited or severely restricted there (Bottrell, 1983). Taking the USA as an example, some $400–600 million worth of pesticides is exported annually, with most going to Asia and Latin America. The USA export trade is, however, only 16.5% of such exports world-wide. The USA, along with other developed countries, has been accused of dumping their banned pesticides on developing countries. These exported pesticides include highly toxic insecticides such as aldicarb, methyl parathion and carbofuran as well as persistent hydrocarbons such as DDT (Bottrell, 1984). US exporters must now inform foreign buyers of known hazards of the materials as well as of any changes in the regulatory status of the chemical in the USA (USAID, 1990a,b). Further, the FAO code recommends that 'prior informed consent' should be obtained from the governments of destination before shipping of the pesticides. However, it is important to realize that some chemicals, banned in the country of manufacture, may still have virtues for the receiving country. Nonvolatile insecticides such as DDT may be the only safe option for farmers without mechanized application methods and protective clothing, or unable to afford safer but much more expensive compounds such as the synthetic pyrethroids. Some developing tropical countries, such as Sri Lanka, which have restricted the use of DDT, are now at the top of the league table of occupational deaths from insecticide poisoning. In 1974, it was estimated that there were 500 000 acute pesticide poisonings resulting in over 900 deaths of which 99% were in the developing world (Wasilewski, 1987). A more recent estimate by WHO in 1991 is 1.1 million cases per year of which 20 000 are fatalities.

Mellanby (1992) in his recent book *The DDT Story* suggested that DDT still had a future. He considered that if we accept the advice offered by WHO 'some years ago', that the indoor spraying of DDT in routine antimalarial operations does not involve significant risk to man or to wildlife and that the outdoor use of DDT should be avoided if possible, then DDT could, and should, continue to be used in antimalarial campaigns.

Regulation of nonpesticidal chemicals

Although bans have also been used to control nonpesticidal chemicals, none has been as confrontational as the banning of DDT in the United States. The hearing that led to the ban on DDT lasted for 7 months. It involved 125 expert witnesses and generated over 9000 pages of testimony (Dunlap, 1981). When it became obvious that PCBs were a widespread environmental contaminant, Monsanto (the sole US producer) put into place voluntary restrictions (Chapter 8). Initially 'closed-circuit' uses, such as the use of PCBs in electrical generators, continued. However, not even the best generator lasts for ever, and eventually the problem of what to do with the PCBs has to be tackled. This 'cradle to grave' or the LCA approach has been used by agencies such as the USEPA to ensure the safe use of chemicals.

However, sustainable development calls for more than the LCA approach to ensure that harm is not caused by chemicals. Continued economic development will require drastic adjustments of the use of energy and materials to produce products and services. Two major factors are involved in the need for drastic action. First, there has been a call for major decreases in emissions to stabilize the atmosphere. For example, the Intergovernmental Panel on Climate Change (IPCC) (cited by Schmidt-Bleek, 1992) has recommended the following reductions: CO_2 by 50%, CFCs 80–100%, methane 20% and NO_x 70–80%. Secondly, the aspirations of the developing world to increase their standard of living, with their still rapidly increasing populations, have to be accommodated.

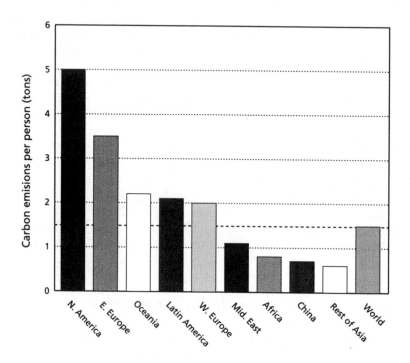

Figure 3.1 *Carbon emissions per person, 1988 (data from Flavin, C., State of the World, 1992, Norton, New York, 1992).*

Regulation of carbon emissions

The difficulties can be illustrated by an examination of the implications of reducing carbon emissions by 50%. The emissions in 1988 on a per person basis are given in Figure 3.1 for various regions of the world. The North American figure of 5.07 tons per person is 3.5 times, and that of western Europe 1.4 times, the world average, whereas Asia and Africa are well below the average. The high figure for Latin America, roughly comparable to western Europe, is due more to forest destruction than to high energy use (Figure 3.2). If, globally, we were to meet a 50% reduction in carbon emissions but also to give each person an equal share, this would call for drastic changes. Emissions in North America would have to be cut to only 14% of current usage and even Africa would have to make a modest reduction (Figure 3.3).

While a 50% reduction in carbon emissions and equal shares for all may be dismissed as utopian, it does illustrate the difficulties ahead as we try to control the 'greenhouse' gases. At the moment, three-quarters of

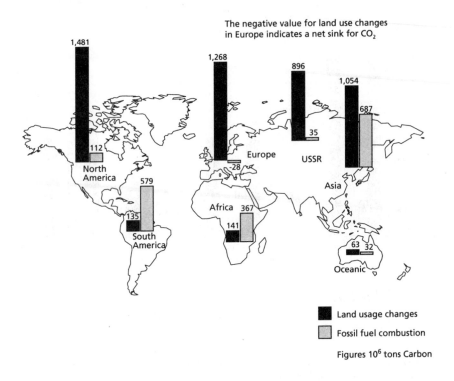

Figure 3.2 *Emissions of CO2 from human activities (from UNEP,* Chemical Pollution: a Global Overview, *UNEP, Geneva, 1992, with permission).*

the world's energy is produced by fossil fuel. Obviously achieving decreased dependence is going to take a considerable amount of time and investment. The best solution in the short term is likely to be to decrease as rapidly as possible the use of coal, which contributes more carbon per unit of energy produced, as well as other major pollutants, than any other source, and to increase the use of natural gas, which produces 43% less carbon per unit of energy produced (Flavin, 1992). Oil is intermediate between coal and natural gas both in carbon (17% less than coal) and in sulfur content. The expanded use of natural gas is a stopgap measure, since world reserves are small compared to coal. Beyond this stage it will be essential to move into an era of dependence on renewable energy sources.

The nuclear industry is in difficulties world-wide, because of high maintenance costs and problems over waste disposal (considered later). Commercial nuclear fusion – the dream of an artificial sun to provide power on earth – seems as far away as it has been for the past 30 years. However, several renewable sources of energy (wind power, geothermal

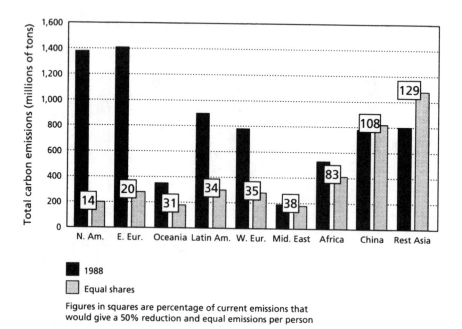

Figure 3.3 *Total carbon emissiions for various regions of the world. Figures in squares are percenatge of current emissions that would give a 50% reduction and equal emissions per person (data from Flavin, C.,* Worldwatch Paper, *No. 91, 1989).*

and solar) are breaking into the commercial market, and solar photovoltaic technology (which converts the sun's radiation directly to electricity) seems likely to follow (Shea, 1988). The transfer from fossil fuels to these forms of energy production will, however, require a good deal of realignment of capital. This is particularly a problem in the developing world already saddled with huge external debts. One proposal has been a 'carbon tax' on fossil fuels. A tax of US$50 per ton of carbon would, world-wide, generate US$280 billion per year. In the United States this would amount to US$240 per person and in India to US$9 per person. It has been calculated that a tax at this level would provide the stimulus for investment, both in efficiency and in the transfer to renewable sources of energy production (Flavin, 1989).

The other side of the carbon dioxide equation is the amount fixed by plants. In all continents except Europe, deforestation is contributing more carbon dioxide to the atmosphere than reforestation is removing. The relative importance of fossil fuel emissions and land use changes is shown in Figure 3.2. While reforestation is not a strategy that can offset all, or even most, of the carbon emissions from power plants, it could play a role.

In the United States, for example, converting 12 million hectares of poor cropland to trees would absorb 60 million tons of carbon annually. This equals a reduction of about 4% in emissions. The carbon balance due to land use is negative in all continents except Europe, and a reversal of this would not only help this balance, but also reduce the loss of biodiversity. Maintenance of biodiversity is critical to one of the two major themes of this book, integrated pest management.

Regulation of oxides of sulfur

Reducing SO_x emissions is going to require either a substantial move away from coal as an energy source or a substantial investment into devices to remove sulfur. So far there has been little movement in either direction. That the industrialization of China is based almost entirely on coal has serious implications for the acidification of large areas of Asia (Chapter 9), as well as for the global carbon balance.

Regulation of CFCs

The control of CFCs has been one of the success stories of international cooperation. The Montreal Protocol, which called for first holding the level of consumption and within 10 years reducing consumption to half existing levels, was adopted in September 1987. This protocol came into force in January 1989 and has been signed by 73 countries and the EEC. As information became available on the 'ozone hole' this protocol was amended in 1990 to call for complete elimination of the use of CFCs by the year 2000.

So far we have been considering chemicals that end up in the atmosphere. Currently these are of great concern because of the impact that they have. The problems for agricultural production of climate change, acid rain and ozone depletion are considered in Chapter 9. Although, in the broadest sense, the liberation of carbon dioxide and other gases into the atmosphere is waste disposal, the term 'waste disposal' is usually reserved for putting liquid and solid chemicals into landfills, incinerations or discharging them in water.

Waste disposal

The most difficult and intractable case is the disposal of nuclear waste. This is because many disposal methods – incineration, conversion to other materials – are not possible, and also because of the long time these materials remain hazardous. The half-life (the time it takes for 50% of the original activity to decay) of plutonium is 24 400 years, which means that this isotope can be a problem for a quarter of a million years. The world's 400 commercial nuclear reactors, producing about 5% of the world's energy, created some 9500 tons of radioactive waste in 1990. This brings the accumulated total to 84 000 tons, double the 1985 level (Lenssen, 1992). From the beginning of the nuclear industry, this problem has not been adequately addressed. The favored option is deep (half a kilometre or more) burial in specially constructed facilities. Technically this poses no insuperable problems, but the time-frame for which such facilities would have to be secure is tens to hundreds of thousands of years. There is no certainty that volcanic action or another geological force would not later release the radioactive material. Certainly, major changes to the earth's surface have occurred within this time-frame. For example, 10 000 years ago the English Channel did not exist.

With these problems it is not surprising that the NIMBY (not in my backyard) syndrome has made it virtually impossible to agree sites for the disposal of nuclear waste. In the United States, the site selected is in the Yucca Mountains of Nevada. However, faced with technical difficulties and firm opposition, the earliest possible date – if it is approved at all – for the opening of the site would be 2010. In France, which generates a higher proportion of its energy from nuclear power than any other country, the situation is similar. Sites are being investigated, but final selection is at least 15 years away. The situation is no better elsewhere. No waste disposal sites for high-level nuclear waste have been constructed any-where in the world. At the moment the ever-growing amount of nuclear waste is being stored in surface facilities near to the nuclear plants. The problems of waste disposal, compounded by the accident at Chernobyl, have caused the massive and probably irreversible decline of the nuclear industry. There is considerable international concern about the safety of the obsolete nuclear power plants in eastern Europe. The cost of upgrading or decommissioning them will be enormous. Also, the loss of their production will be serious and likely to lead to increased pollution. Virtually no new nuclear plants have been started in the past 5 years anywhere in the world. For the first, but perhaps not the last time, an industry has been killed by its failure to deal with waste disposal.

Within the chemical industry as a whole there are much broader opportunities for safe disposal. Just over half the total industrial waste of Japan was recycled and reused in 1983 (Postel, 1987), with a further 31% reduced through treatment (incineration, dewatering, etc.). This left only 18% to be disposed off. In contrast, two-thirds of the hazardous waste in the United States was placed in land fillings and a further 22% discharged in sewers, rivers or streams. Tolba (1992) gives similar data for chemical waste in the United Kingdom in 1985, where the vast majority of the material was disposed of in landfills. These figures are not strictly comparable, as the Japanese figures refer to all industrial waste, the American ones only to hazardous waste and the UK ones to all chemical waste. Differences in definitions presumably also account for the huge percentage of the global total of hazardous waste being generated by the United States (81%, some 10 times more than OECD Europe) cited in Tolba's (1992) recent book *Saving Our Planet.*

While most countries rely on land disposal methods for their hazardous waste, advanced systems have sometimes been put in place elsewhere. Denmark has introduced a system that treats, detoxifies or destroys most of the nation's toxic waste generated since the mid-1970s. In outline the system consists of a central disposal facility capable of a wide range of detoxification systems and high-temperature incinerations. This is supported by 21 transfer stations throughout the country for industrial wastes and 300 smaller collection points for household wastes such as paints and solvents.

The problem of transboundary transfer of hazardous waste began to be tackled seriously in the early 1980s. At this time some 2 million tons of hazardous waste crossed OECD European frontiers each year; in addition some 250 000 tons were transported annually to eastern Europe. It has been estimated that 120 000 tons of hazardous waste were exported annually from North America and Europe to the developing world. The hazards to both human and environmental health of disposing of highly toxic waste by shipping it to developing countries are obvious. The recent attempt by an Italian company to site an incineration plant off the shores of war-torn Somalia indicates the need for international controls. Principle 14 of the Rio Declaration deals directly with this issue: 'States should effectively cooperate to discourage or prevent the relocation and transfer to other States of any activities and substances that cause severe environmental degradation or are found to be harmful to human health.' There is a need for Third World governments and leaders not to give way to the temptation of short-term gain in exchange for long-term problems. Hopefully the Basel Convention on the Control of Transboundary Movements of Hazardous Wastes and their Disposal, which was adopted by 116 governments and the European Community in March 1989 but so far has been ratified by only a handful of countries, will tackle this problem.

Life-cycle management of chemicals

A recent review (VNCI, 1991) on integrated life-cycle management of several chemicals has been carried out by the Dutch chemical industry at the request of their government. A diagram of the life cycle of a chemical, consisting of six stages and four valves (Figure 3.4), is taken from this. The review states that 'the art of integrated management is to control the outflow of the substances into the environment by manipulating the valves

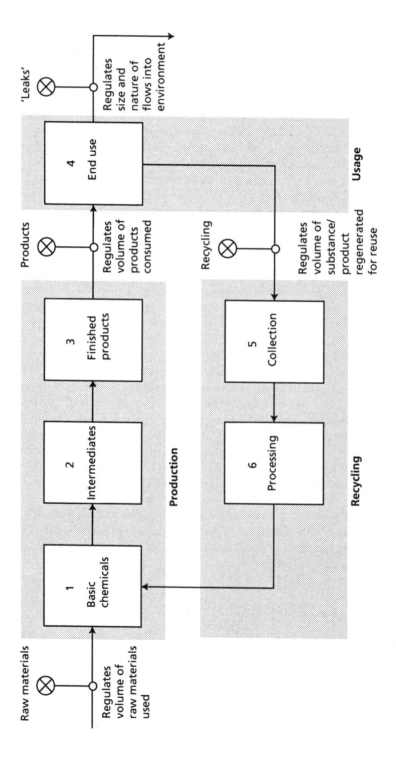

Figure 3.4 *Integrated life-cycle management of chemicals (from VNCI, Report, Association of the Dutch Chemical Industry, Leidschendam, 1991, with permission).*

as effectively and efficiently as possible.' One of the case studies published was the life cycle of an organochlorine compound referred to only as 'HCFC-22'. HCFC-22 is currently in the process of replacing CFCs in a wide range of applications. Although HCFC-22 contributes substantially less to ozone depletion than other CFCs, it does have some impact. When the life cycle of HCFC-22 was examined, it was found that the two most critical areas were reduction of leaks by improving maintenance practices and establishing a collection and recycling process. A review of the process established that these improvements could be made without serious economic cost.

LCA can be, and usually is, carried out at the level of the individual chemical. While the objectives of the Dutch review are laudable, they could usefully be extended to include the concept of the reduction of the amount of raw material used and the energy used during the process. Schmidt-Bleek (1992) has recently stated that 'de-materialization by technology is the only real hope, particularly in the Third World, to make room for economic growth under condition of not forcing the ecosphere to additional catastrophic reactions'. His diagram on the changes necessary in energy consumption and material usage is shown in Figure 3.5. He emphasizes that it is vital that products and services should be much less material and energy intensive than they are today. Such changes are going to require increased prices for natural resources as the market itself will not internalize environmental costs (OECD, 1991). One possibility to

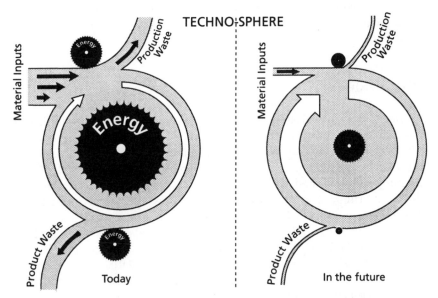

Figure 3.5 *Energy consumption and open material streams (from Schimdt-Bleek, F., Fresenius Environ. Bull., 1, 417–22, 1992, with permission).*

avoid increased strains on budgets would be to transfer some part of personal income taxation to taxes on natural resources.

The concept of LCA can also be used to decide which chemicals should be used. Take for example the plastic industry. Plastics are widely used and, for many purposes, have advantages over glass or metals. The problems of recycling are, however, greater than for either alternative. There are difficulties over collection, and the separation of plastics and breaking them down to the original chemical constituents are difficult and often economically unattractive. With polyvinyl chloride (PVC), there is an additional difficulty in disposal. Although PVC contains roughly as much energy as wood, its chlorine content poses problems. Incinerators that burn PVC need scrubbers to prevent emissions of hydrochloric acid. Also, the possibility of forming dioxins, considered powerful carcinogens, has to be addressed. As a result the incineration of PVC is discouraged. Recycling and disposal problems are less serious with other plastics such as polyethylene, suggesting that a move away from PVC would be, from an environmental point-of-view, advantageous.

The ideal would be for industrial processes to use waste from one process as the raw material for the next, analogous to biological ecosystems. While it is unlikely that industrial processes could match the biological system by which plants synthesize nutrients that feed herbivores, which in turn feed carnivores whose wastes and bodies eventually support further generations of plants, we need to move in this direction. Even the biological system is, of course, not a closed system since energy from the sun has to be supplied. However, unlike most of the energy we consume, this is renewable.

Conclusion

We need to think broadly about the relationship between industrial production and environmental concerns. Too often the positions of industrialists and environmentalists are polarized. To the former, effluents and other waste are things to be dealt with as cheaply as possible; to the latter any effluent is bad. We need 'industrial ecologists', a breed of persons who can think through the processes involved and reduce the environmental impact of industrial processes to a minimum.

The use of pesticides is, in most countries, covered by specific legislation. In addition to national laws, which vary greatly in their strength both in the actual legislation itself and even more on the degree of enforcement, there are legislative measures and also pressures at the international level. These are considered in Chapter 4.

It is clear that the use of both pesticidal and nonpesticidal chemicals requires an integrated approach. In the case of pesticides, there are concerns for both the safety of the operator and for unwanted environmental side-effects; these concerns may often be in conflict. The choice of a pesticide to control a pest needs to include consideration of the problems of future resistance of the pest species, and alterations to the community structure that can affect natural controls need to be considered when the efficacy of the pesticide to control the pest is concerned. The IPM philosophy makes it possible to envisage achieving worthwhile increases in yield by introducing minimal insecticide use in that third of the agricultural hectarage that currently receives no pesticides, largely for economic reasons. Full protection with pesticides would be prohibitively expensive, besides making it likely that problems of resistance and environment damage would occur. In contrast, the increase of yield with IPM using limited pesticide input could be expected to have no serious side-effects.

Equally for nonpesticidal chemicals, the 'broad canvas' approach should be used. The concept of LCA can be applied from the individual chemical to complex processes involving the use of many chemicals. It is essential to apply these principles in order to continue our industrialized society, let alone to allow for increasing industrialization of the developing world.

References

Apple, J.L. and Smith, R.F. (eds) (1976) *Integrated Pest Management*, Plenum, New York.

Bottrell, D.G. (1983) Social problems in pest management in the tropics. *Insect Sci. Applic.*, **4**, 179–84.

Bottrell, D.G. (1984) Government influence on pesticide use in developing countries. *Insect Sci. Applic.*, **5**, 151–5.

Brosten, D. and Simmonds, B. (1989) Inputs for the starving continent. *Agrochem. Age.*, **33**, 6–7, 26–7.

Dunlap, T.R. (1981) *DDT; Scientists, Citizens and Public Policy*, Princeton University Press, Princeton.

Flavin, C. (1989) Slowing global warming: a worldwide strategy. *Worldwatch Paper, No. 91*.

Flavin, C. (1992) Building a bridge to sustainable energy, in *State of the World 1992*, Norton, New York, pp. 27–45.

Geier, P.W. and Clark, L.R. (1961) An ecological approach to pest control. *Proc. 8th Technical Meeting, International Union for Conservation of Nature and Natural Resources, Warsaw, 1960*, pp. 10–18.

Lenssen, N. (1992) Confronting nuclear waste, in *State of the World 1992*, Norton, New York, pp. 46–65.

Mellanby, K. (1992) *The DDT Story*, British Crop Protection Council, Farnham.

OECD (1991) *The State of the Environment*, OECD, Paris.

Pimentel, D., Hurd, L.E., Bellotti, A.C. *et al.* (1973) Food production and the energy crisis, *Science, NY*, **182**, 443–9.

Postel, S. (1987) Defusing the toxic threat: controlling pesticides and industrial waste. *Worldwatch Paper, No. 79*.

Schmidt-Bleek, F. (1992) Will Germany remain a good place for industry? The ecological side of the coin. *Fresenius Environ. Bull.*, **1**, 417–22.

Shea, C.P. (1988) Renewable energy: today's contribution, tomorrow's promise. *Worldwatch Paper, No. 81.*

Smith, C. and Beckmann, S.L. (1991) Export of Pesticides from U. S. Ports in 1990, Report to the Committee on Agricultural, Nutrition and Forestry of the United States Senate, Washington, DC.

Stern, V.M., Smith, R.F., van den Bosch, R. and Hagen, K.S. (1959) The integrated control concept. *Hilgardia*, **29**, 81–101.

Tolba, M.K. (1992) *Saving Our Planet*, Chapman & Hall, London.

UNEP (1992) *Chemical Pollution: a Global Overview*, UNEP, Geneva.

USAID (1990a) Integrated Pest Management: Aid policy and Implementation, Report to the United States Congress, September 1990, Washington, DC.

USAID (1990b) Pesticide Use and Poisoning: a Global Review, Report to the United States Congress, September 1990, Washington, DC.

VNCI (1991) Integrated substance chain management, Report, Association of the Dutch Chemical Industry, Leidschendam.

Wasilewski, A. (1987) The quiet epidemic. Pesticide poisonings in Asia. *IDRC Report, No. 16*, 18–19.

CHAPTER 4

Pressures on pesticides in IPM

Pesticides still have an important place in sustainable pest control. Their role is changing, however. This is influenced by international pressures from world agencies, and by local national legislation and policy. Public pressure arising from real and perceived health and environmental risks is also important. Industry is under pressure to produce more IPM-compatible products. However, the greatest pressures come from the development of resistance to pesticides in the target pests. While there are pesticide-based solutions for this problem, most are still controversial. IPM is probably the ideal solution, but this depends on the availability of suitable programs involving alternative control methods.

Introduction

It was pointed out in the previous chapter that the role of pesticides in IPM in the past 30 years has been very different from the future envisaged by Rachel Carson in *Silent Spring*, and that real public and government pressure to reduce pesticide use in agriculture is a comparatively recent phenomenon. Partly, the slow progress of IPM is due to low financial investment for the development of alternative approaches, particularly in the ability to match the level of control provided by pesticides. Partly, it reflects the possibilities for coping with pesticide resistance by switching to alternative compounds. Moreover, much of any evolution from reliance on chemicals to IPM has been controlled by attitudes of governments and international agencies.

The evolution of concepts

Pesticide use has moved a long way since *Silent Spring*, when most pesticides were highly persistent (e.g. the organochlorines) and were applied either according to the calendar or when the pest was observed. The first major change was, as described earlier, the introduction of 'economic thresholds', when treatment was only applied when warranted by established relationships between pest densities and crop yield. The early introduction of this concept meant that, for a while, much of the practice of what we now call IPM was little more than modified spray programs involving the intelligent use of pesticide. There is still probably an over-reliance on the use of pesticides in IPM programs; this is because the third phase, the maximum use being made of more biological tactics, is still only in late infancy. Pesticides can often cope with pest problems as a sole control measure, but at an environmental cost and with the danger of the development of pesticide-resistant pest strains. Although there are many examples where other control measures (e.g. host-plant resistance, biological control, cultural control) have, on their own, proved as, or even more, effective than pesticides, such other measures more often fail to provide an acceptable level of pest reduction. It is a pity to ignore their contribution and replace it with sole reliance on pesticides. The remaining gap can often be filled with minimum pesticide use. Thus pesticides are still, and are likely to remain, a component in many, if not most, IPM programs where the economic threshold for pest damage is low. The key to the use of toxic chemicals in IPM is that the chemical input must be designed so as not to prevent the operation of the more biological control components. As pointed out above, it then ceases to be necessary for the chemical to provide the total required control on its own. How this may be achieved will be discussed in Chapter 5. Less toxic and more environmentally friendly chemicals are likely to be effective (even when used in smaller amounts), and the accent is on selective compounds and on ingenious solutions for applying broad-spectrum compounds selectively. The shift away from emphasis on chemicals in IPM does not detract from the logic and appropriateness of their use in a multiple, all-suitable-techniques strategy for sustainable pest control.

Factors influencing the changing role of pesticides in IPM

Besides the conceptual shifts described above, the selection and extent of pesticide use in IPM has been strongly influenced by several outside factors. These have exerted often unseen or unrecognized, positive or negative pressures on the need for pesticides in agricultural and public health situations. They have also affected the availability and selection of materials, and the extent to which they have been used. These factors have had an importance equal to that of the evolution of concepts in changing the use of pesticides since *Silent Spring*, particularly in the tropics.

Pesticide policy and legislation at the international level

The use of pesticide in developing countries has been affected by several international actions aimed at avoiding misuse of pesticides and the loss of their effectiveness from resistance.

USAID and FAO/WHO have been in the forefront of international influence on the use of pesticide in developing countries. The linking of aid to the use of less-damaging pesticides by USAID and FAO's code of conduct on the distribution and use of pesticides has already been described in Chapter 3.

Pesticide legislation in the USA has grown in complexity from the Federal Insecticide Act of 1910 to the Federal Environmental Pesticides Control Act of 1972. Each proliferation of legislation has increased the stringency for registration or re-registration of pest control products. This has had a significant world-wide impact on the availability, selection and use of crop protection chemicals and has increasingly raised the economic incentives for farmers to use nonchemical methods. Certainly the lowering of tolerances for pesticide residues has forced changes in the kinds and quantity of pesticides used.

National efforts and legislation governing import and use of pesticides

Reference was made in Chapter 3 to the increasing legislation against chlorinated hydrocarbons in developing countries and the problems of enforcing such legislation in many such countries that have adopted the FAO code of conduct.

Economic conditions and the need to stimulate crop production often lead to government pricing controls and subsidies for pesticide inputs. Senegal, for example, subsidizes pesticide use on rice to avoid losing foreign exchange on rice imports. Such practices can, of course, lead to increased use (and misuse) of pesticides. In a study of nine countries (Repetto, 1985), the average subsidy of total retail cost of crops was 44%, with Honduras having the highest *per capita* subsidy. This resulted from giving pesticide importers a favorable rate of exchange and exempting pesticides from sales taxation. The use of alternative methods was thus discouraged, and the use of economic thresholds undermined. Often such policies benefit larger farmers over smaller ones, and mostly on cash crops (Repetto, 1985). Similar consequences result from subsidies in vector control programs.

The introduction of structural adjustment programs in some West African countries has forced the removal of government subsidies on pesticides and their replacement with the incentive of high producer prices. Farmers are then able to purchase pesticides, even at the higher prices. Indeed, pesticide consumption has continued to rise. The GATT agreement (Chapter 1) may well affect the issue of subsidies.

Excessive pesticide use can also stem from the economic pressures that cause governments to give more funding and support to ministries of agriculture and commerce than to environmental ministries. Control measures that are perceived to require little extension support are preferred by policy makers. Thus pesticides included in simplified technology packages (incorporating resistant varieties and biological control releases) remain under centralized government control. However, pesticide management is not so simple, and requires high levels of extension if pesticides are to be used appropriately in IPM.

Public perception of pesticide-associated risks

Despite the greatly increased stringency of registration procedures, agricultural pesticides are still perceived as intrinsically hazardous to health and the environment. Such perceptions take no account of the amount used or of the care taken. It has been argued by a journalist that this perception is based on three false premises (Tucker, 1978):

> First is the myth, which environmentalists have fashioned, of an ideal, pre-pesticide past, when crops were good, living was easy, and insecticides were few. This is a complete fantasy. Second, there is the false distinction between natural and unnatural chemicals and implicit assumptions that chemicals like pesticides never occur in nature. Third, there is the myth that these unnatural chemicals are causing an equally mythical, epidemic increase in cancer. Unfortunately the genesis of all three of these ideas can be traced directly to *Silent Spring*.

On the other hand, neither have pesticides proved to be a panacea. Increases in pesticide usage have often not been associated with decreased losses from pests (p. 65). Also, insecticides do carry both environmental and health risks, especially where they are misused or overused (Chapter 8). We would, however, agree with Tucker that the risks, especially of modern pesticides used appropriately, tend to be highly overstated.

The development of 'organic farming' in developed countries has been partly motivated by a minority of consumers who have a pronounced distaste for pesticides. The market is still very small; it is less than 1% of vegetable and fruit sales in the UK, for example. More important, perhaps, is pressure being put on farmers in some countries to adopt IPM practices by offering certification or other inducements likely to lead to participating farmers increasing their market share. In Europe, the International Organization for Biological Control (IOBC) is offering certification labels to growers (especially in viticulture) producing crops to their IPM standards (which include pesticide recommendations). In the UK, a number of large supermarket retailers have combined with the Ministry of Agriculture to provide production protocols, which specify crop protection by IPM so long as suitable methods can be suggested, for several vegetable crops. The aim is gradually to expand the range of crops covered, and for the supermarket chains concerned to place their purchasing contracts preferentially with growers complying with the protocols. This seems a particularly promising approach for increasing the uptake of IPM by farmers and growers.

IPM itself has a credibility problem with farmers. It is often seen as 'pest control without chemicals' or just another version of 'biological control'; moreover, it does not carry the security provided by pesticides that the farmer may be able to seek compensation from the agrochemical company for a control failure. Thus farmers are reluctant to relinquish pesticides as dependable pest-control tools.

Industrial pesticide development and production

Older chemicals are often used in developing countries. Such chemicals are relatively inexpensive; they are no longer patent protected and may be manufactured in the developing country, often without the safety standards of the larger companies, as was illustrated dramatically at Bhopal.

The larger international companies have redirected their efforts towards 'more IPM compatible products', i.e. chemicals with increased target specificity, such as cuticle inhibitors (p. 155). There are several benefits of such compounds to industry: better resistance management and therefore longer product life, improved grower satisfaction, support from the public and policy makers, and compliance with the FAO code of conduct. However, these compounds are more expensive and chiefly directed to markets that can afford them. The costs of developing and marketing a new insecticide are now around US$75 million. Escalation of these costs is linked with the more sophisticated structures of newer compounds, general inflation and the more stringent requirements for registration. Based on increases observed in the 1970s, a doubling time in the cost of developing pesticides of about 3.2 years has been calculated (Metcalf, 1980). In view of likely future resistance problems, this outlay needs to be recouped by the pesticide industry with the minimum of delay.

In addition, in developed countries, re-registration requirements for older compounds have resulted in the voluntary withdrawal of compounds by agrochemical companies, who often feel the costs of re-registration are not economic for many older products. Even those submitted for re-registration face a considerable delay and are, meanwhile, unavailable to growers. Where growers are left with a severely limited range of products, they can see the danger of resistance developing, and many turn (for the first time) to IPM.

Resistance to pesticides

The first case of resistance to pesticides was detected in 1914. By 1990, nearly 500 cases of resistance of insects and mites could be listed (Schulten, 1990). To these cases can be added ever-increasing numbers of plant diseases, weeds, nematodes and rodents which are also showing pesticide resistance. Among the arthropods listed as resistant in 1986 (Georghiou, 1986), 59% were of agricultural importance, 38% of veterinary or medical importance and only 3% were beneficial insects.

The development of resistance affects IPM by reducing the number of pesticides that remain effective, applies pressure for application at higher dosages and frequencies (with potentially devastating implications for biological control), and decreases the incentive for industry to market more highly specific compounds. This, coupled with the escalating costs of pesticide development, suggests that there will be a decrease in the number of materials available for use in new IPM programs. Metcalf (1980) claims that resistance to insecticides is the greatest single problem facing applied entomology.

How to manage pesticides to prolong their useful life is still the subject of controversy. Several strategies have been proposed; these are discussed briefly below.

High v. low kill strategy

There is still argument about even this simple dichotomy of possible strategies. Georghiou and Taylor (1976) considered biological and operational factors accelerating the development of resistance based on a population genetics model. They concluded that a low kill strategy would delay resistance in comparison with aiming for maximum kill. Important elements in the argument were that a high kill strategy must inevitably lead to resistance, whereas a low kill strategy would maintain susceptibility in the population. Breeding could then dilute any resistance that developed.

The principal opponents of this conclusion have come from the agrochemical industry. Not surprisingly, the industry is worried about reduced pesticide use. The main argument against the low kill strategy is that heterozygote pests with partial resistance would survive the pesticide application and pass on higher resistance levels through offspring homozygous for resistance. The counter-argument of proponents of IPM is, of course, that other control components would then provide mortality for most of the heterozygotes surviving the pesticide.

Sequential v. rotational use of pesticides

It is surprising that, in view of man's long experience of pesticides and resistance development, this second dichotomy of strategies has also not been resolved. Will a number of compounds in total last longer if used one at a time until resistance forces a switch, or is it better not to wait for resistance to develop and rotate their use? The latter approach carries the danger of multiple resistance to all the compounds in the rotation. Again, there are disciples of both viewpoints. The history of resistances in the field, however, suggests the following conclusion. Resistance in insects appears to be delayed by sequential use, whereas the rotational strategy seems the more effective with mites.

Switching the life-stage target

Georghiou and Taylor (1976) concluded that resistance to insecticides would develop more rapidly if applications were directed against larvae rather than adults, and more rapidly still if targeted against both stages.

Most insecticides are applied to control insects in the larval stage. Georghiou and Taylor's conclusion immediately suggests that a switch to attacking the adult when resistance develops in the larva should prolong the useful life of an insecticide. It is relevant that rotations of acaricides against mites usually involve compounds with differential activity against eggs, immature stages and adults.

Windows of pesticide use

Once resistance to the older insecticides has appeared in a pest population, farmers often switch to a synthetic pyrethroid and continue with it from then on. This has often led to selection for rapid resistance to the pyrethroids. Many pests are migratory, and there is no reason why the arriving pests should have the degree of pesticide resistance that was present in the farmer's crop at the end of the previous season. Thus the usefulness of pyrethroids can be extended by reverting to older insecticides early in the season and restricting pyrethroid use to a window towards the end of the season, if or when resistance to the older compounds again appears.

Restricted spray windows are used in a number of cotton systems. The best-known example comes from Australia, where bollworms rapidly developed resistance to the synthetic pyrethroids. It was decided to restrict the spraying of pyrethroids to a window between January 10 and February 20 (in the middle of the crop season). The aim was that only one generation of bollworm would be exposed to this group of insecticides. However, after initial apparent success of the strategy, the level of pyrethroid resistance in bollworm appears to be rising again (Matthews, 1989).

In Zimbabwe, the pyrethroid window is 9 weeks during the main flowering and boll development period, when bollworm is most likely to infest the crop. In the Philippines, it is between 84 and 112 days after planting.

Pesticide rejuvenation

If the biochemical mechanism of the pesticide resistance is known, it may be possible to add to the formulation a compound (synergist) which blocks the mechanism.

Such rejuvenation has been attempted with synthetic pyrethroids by adding the synergist piperonyl butoxide. Adding a synergist unfortunately does not change the mode of action by which the organism is killed. It merely means the organism will need the same mechanism considerably strengthened to develop resistance to the synergized pesticide. This raises the danger of 'super-resistance', which could render a pest immune to a wide range of pesticides with a similar mode of action.

Diversification of control methods (IPM)

The selection pressures on pests can be diversified by increasing the contribution of other sources of control. This is probably the ideal way of delaying or preventing resistance to pesticides developing.

In a few of the organochlorine pesticides (e.g. endosulfan), in the organophosphates and in the systemic carbamate insecticides, we have an arsenal of considerable flexibility in IPM programs, with a variety of routes to the target. This variety is not shown in the newer insecticides (e.g. the synthetic pyrethroids). Thus, to be able to do IPM at all, it is vital that the current pesticides are not allowed to be lost through resistance. This danger is perhaps the most cogent reason of all for using IPM methods to reduce the quantity and frequency of current insecticide applications.

Availability of practical IPM programs with alternative control tactics

A follow-up survey after a pest-monitoring training session in the Philippines revealed that adoption was low because of the tedium, time taken, marginal benefits and requirement for new skills (Medina, Velasco and Soriano, 1992). The count and record system proved too advanced for the farmers' capabilities and needs. The complexities of impractical IPM systems, particularly monitoring and the determination of crop loss and economic thresholds, will discourage the adoption of the IPM approach and encourage reliance on the single tactic of pesticide application.

The use of economic thresholds alone usually results in a worthwhile reduction in pesticide use. It is nevertheless only a first stage towards IPM. There are still few pest- and disease-resistant varieties being released. They tend to be designed for monocultures rather than for the less profitable mixed cropping systems of small farms in the tropics. There are also many scenarios where no proposals are available for increasing the impact of biological control agents. Cultural controls were the cornerstone of pest control in pre-insecticide times and could still reduce dependence on pesticides. However, without an IPM system in which cultural controls can find inclusion, the effect of cultural control seems unspectacular and is little appreciated. Unless research has produced, for example, a successful biological control measure that requires no manipulation by the farmer, cultural controls are often replaced by the insecticide tactic to save the time and labor they involve.

Conclusions

The role of pesticides, particularly in agriculture, has therefore changed dramatically in the 30 years since *Silent Spring*, though in different ways in different parts of the world.

In developed countries, pesticide use continued to rise steeply during the 1960s and 1970s. For example in Japan, there was a 33-fold increase of insecticide inputs between 1950 and 1974; that this was a 'pesticide treadmill' phenomenon is evident from the fact that little yield increase was associated with the rise (Kiritani, 1979). Similarly, Pimentel *et al.* (1978) estimated that crop losses from pests in the USA had doubled in the previous 35 years despite a tenfold increase in pesticide use. However,

Pimentel *et al.* conceded that the return per dollar invested in control with pesticides was US$4. Resistance to pesticides has usually been the driving force behind the introduction of IPM methods in the developed countries. Cotton and glasshouse and orchard crops have been among the first arenas where resistance to insecticides has forced the abandonment of routine reliance on pesticides. More recently, the banning of a number of compounds has removed the possibility of routine chemical control of some pests for growers. Increasingly, public pressure stemming from environmental and health concerns is influencing both pesticide legislation and grower attitudes in favor of IPM. Long delays are now occurring in pesticide registration and re-registration, to the point where many growers are severely limited in the pesticides they may legally apply. The result of all these changes has been an increasing implementation of IPM and a decrease in pesticide use in developed and developing countries. Examples of practical IPM remain relatively few and it is clearly still early days. One has the feeling, however, that at long last IPM is taking off. Pesticides are still very much a part of IPM programs, but increasingly used as a stiletto rather than as a scythe. As pointed out earlier, there are excellent pesticides if used judiciously and IPM will develop more rapidly and successfully if they are accorded a proper role among the IPM strategies that are available for integration than if attempts are made to 'go it alone' with more biological methods. Indeed, the appropriate use of insecticides remains the key to translating the incomplete control often available from other methods into a grower-acceptable IPM package.

Thus IPM, including some pesticide use, is the route whereby it has been possible to reverse the pesticide treadmill in developed countries.

The same strategy is also appropriate to preventing the pressures for greater food production in developing countries leading inexorably to the same pesticide problems that have been experienced elsewhere, at least in countries or crops where it is not already too late (Chapter 6). In developing countries, as described above, international agencies are being at least partially successful at introducing and encouraging the use of insecticides in an IPM context. Without such influences, pressures for yield increases and the lack of regulation and lack of alternative control measures would be likely to lead to routine pesticide applications as the easy option.

References

Georghiou, G.P. (1986) The magnitude of the resistance problem, in *Pesticide Resistance: Strategies and Tactics*, National Academy Press, Washington, DC, pp. 14–43.
Georghiou, G.P. and Taylor, C.E. (1976) Pesticide resistance as an evolutionary phenomenon. *Proc. 15th Int. Congr. Ent., Washington, DC, August, 1976*, pp. 759–85.
Kiritani, K. (1979) Pest management in rice. *Ann. Rev. Entomol.*, **24**, 279–312.

Matthews, G.A. (1989) *Cotton Insect Pests and their Management*, Longmans, Harlow.

Medina, C.D., Velasco, L.R.I. and Soriano, J.S J. (1992) Developing an insect monitoring system for rice farmers in the Philippines. *Abstracts 19th Int. Congr. Ent., Beijing, July, 1992*, p. 367.

Metcalf, R.L. (1980) Changing role of insecticides in crop protection. *Ann. Rev. Entomol.*, **25**, 219–56.

Pimentel, D., Krummel, J., Gallahan, D. *et al.* (1978) Benefits and costs of pesticide use in U.S. food production. *BioScience*, **28**, 778–84.

Repetto, R. (1985) *Paying the Price: Pesticide Subsidies in Developing Countries*, World Resources Institute, Holmes.

Schulten, G.G.M. (1990) Needs and constraints of integrated pest management in developing countries. *Med. Fac. Landbouw. Rijksuniv. Gent*, **55**, 2–216.

Tucker, W. (1978) Of mites and men. *Harper's Mag.*, **August**, 43–58.

CHAPTER 5

Principles of IPM

The causes of pest mismanagement, and how these translate to components of IPM. The principles of economic thresholds, how increased selectivity of pesticides may be achieved. Insecticide application in relation to IPM. The principles and techniques of biological control, the role of biodiversity in IPM both within and outside the crop. Cultural control methods. Semiochemicals, genetic controls. The classification, sources, manipulation, mechanisms and problems of host-plant resistance. Integration of host-plant resistance in IPM. Insect growth regulators (IGRs). Different approaches to putting the technology together as IPM systems.

Introduction

The last chapter of *Silent Spring*, entitled 'The other road', describes Rachel Carson's view of what today would be described as the principles of IPM. The development of IPM has been quite different from Rachel Carson's vision of the future of pest control, and is better described as a road network linking her two distinct and opposing roads. This is because several new approaches to pest control have emerged since *Silent Spring*, and because modern pesticides are more environmentally friendly than those discussed by Rachel Carson. Thus modern pesticides can be used effectively in IPM, and most proponents of IPM consider pesticides as part of their armory.

The fundamental principle of IPM is relatively simple. It is to prevent, or at least delay, counter-adaptation by pests to control measures by diversifying the latter. This should be accomplished by maximizing the use of nonpesticidal methods; pesticides may then have to be used to fill the gap between what is attainable without them and the level of control required by the grower. It is then important that the pesticides be used with minimum disruption to the other measures, particularly biological control.

As the need for IPM arises from pest mismanagement in the past, the causes of this mismanagement are highly relevant to how IPM might proceed. These causes are generally considered to include:

1. Overdosing with pesticides, leading to development of resistance.
2. Loss of biological control. Pesticide use is an important cause, leading to increased severity of pest outbreaks and the appearance of new pests. A second component is the reduction of biodiversity in agroecosystems.
3. The introduction of high-yielding, genetically uniform and pest-susceptible varieties to large areas of monoculture.
4. Agronomic changes such as the abandonment of cultural controls such as crop rotations and the move from mixed cropping to monocultures.

These causes of pest mismanagement translate into the following aspects of IPM:

1. Use of economic thresholds to guide spraying decisions.
2. When pesticides are needed, they are used in a way that is least damaging to biological control. Biological control agents can also be purposefully introduced and/or promoted by habitat modification, including the planned diversification of the agroecosytem.
3. Introduction of host-plant resistance to pests, as far as possible involving several mechanisms.
4. Introduction of cultural controls if these can be compatible with farm management systems.

Since the recognition of the causes of pest mismanagement, several other techniques (e.g. pheromones, male sterilization) have been added to the IPM armory.

The purpose of this chapter is to discuss the development of the diverse approaches to pest control listed above. Their potential role in pest management, particularly in the integrated approach represented by IPM, will be emphasized.

Economic thresholds

Economic thresholds are used to decide when a pest has reached levels requiring a control intervention. They have now been evaluated for many pests of many crops to replace reliance on routine prophylactic treatments. This change of emphasis to applying pesticides only when necessary, has been a major development since *Silent Spring*. Reductions in the amount of pesticide used on a crop of at least 20–30% can normally be achieved.

Stern *et al.* (1959) defined the economic threshold as the 'density at which control measures should be determined to prevent an increasing pest population from reaching the economic injury level.' They defined the economic injury level as 'the lowest population level that will cause economic damage,' i.e. the cost/benefit ratio of the control measure rises above unity. Therefore the economic threshold is normally lower than the economic injury level to give time for control measures to be mobilized and for them to take effect.

The cost/benefit concept has, of course, the two elements of 'cost' and 'benefit'. 'Costs' are now recognized as considerably greater than the direct ones of pesticide and the labor and fuel charges involved in their application. Pesticide application may cause crop damage by phytotoxicity (e.g. reduced yield), stem or leaf breakage, soil compaction, etc. Increasingly, environmental pollution (Chapter 8) and damage to biological control and wildlife are seen as additional costs. A further cost may be a reduction in the value of the crop if customers are prepared to pay a premium for reduced or zero pesticide use. This may be either for reasons of conscience or because of a real or perceived benefit in crop quality. Another example was the premium customers in the UK were prepared to pay during the 1970s for the more natural and pleasing appearance of the foliage of chrysanthemums grown under a biological control (as opposed to a pesticide) regime.

'Benefit' also has several aspects. At very low densities, pests may actually increase the quantity or quality of the yield. For example, some early thinning of fruits or buds by pests may reduce the competition between them and lead to increased size and quality of those that remain. Some removal of leaf area by pests, especially on the lower leaves, may reduce plant respiration while hardly reducing photosynthesis. There are also examples where a little wounding by insects provides a stimulus to plant growth.

A second aspect of 'benefit' is that no benefit is obtained by controlling pests at densities at which the plant can compensate fully for the degree of attack involved. Here cotton is a particularly good example. Provided the crop is not stressed by lack of nutrients or water, the harvested crop will be formed on only about 12 of the 90 or so flower buds that each plant produces. The extra buds are shed naturally even if they are not damaged early by pests. Cereals are an example where interplant competition in a densely sown crop results in yield ha^{-1} remaining unchanged over a wide range of tillers m^{-2}. Even total loss of some individual plants from pest attack will not result in a yield loss. Reduction in leaf area on individual plants can also be compensated; it has been claimed that sugar beet can fully compensate for up to 60% defoliation, but these experiments were carried out with manual and not pest damage. Pests usually reduce the yield far more than the equivalent simulated damage (see below). Indeterminate crop varieties will continue to produce new fruiting/podding points after damage if nutrients and water remain available; thus early damage can be compensated, although there is a penalty of a delay in harvest. The potential for compensation in leafy indeterminate crop varieties is high, but both leafiness and the potential for an extended period of fruiting and podding are undesirable characters in modern crop improvement breeding programs. Newer high-yielding varieties tend to be selected under full insecticide protection and then unfortunately usually require such protection to be maintained if they are to produce good yields. Without the protection, they may well yield less than the older varieties.

Economic thresholds are based on economic injury levels, which can be determined by several methods (van Emden, 1978). Surprisingly enough, the most obvious approach of comparing the yield of plants with artificially imposed different insect burdens has frequently proved unsatisfactory. Many insects are mobile between plants, so that original differences in numbers become evened out. The effects of cages on plants and on the reproduction of the contained insects are so drastic, that attempts to confine the insects can lead to misleading results. More satisfactory, in spite of possible direct effects on yield, is to use insecticides to cut off pest population increase at various points in time. Many workers have attempted to simulate the effects of pests on yield by artificial damage, especially in relation to chewing insects and the defoliation they cause. However, much of the yield loss of insect-chewed plants results from insect saliva rather than removal of plant tissue. The damage to the plant of leaf area removed by a multiplicity of small bites is very different from the effects of a single cut with a pair of scissors. Thus, as pointed out earlier, the equivalent defoliation by insects results in much higher yield loss than when the defoliation is artificial. Yet another approach has been to seek a relationship between pest density and yield loss from figures obtained in farm surveys. Such relationships may also be very misleading. That a range of pest densities can be obtained may be due to latitudinal, environmental (e.g. soil, temperature, rainfall), disease and even cultivar variations that themselves directly influence crop yield. A further approach used in apple orchards in the UK has been to persuade farmers to compare two pesticide regimes in different blocks of the crop. Routine spraying was continued on one block, but the other block was monitored for pest occurrence. Season by season, originally very low pest thresholds for spraying were raised until the grower complained of economic damage. Whether the complaint was justified or not, it enabled the really practical threshold, the 'grower worry' threshold, to be determined.

Finding out whether a threshold is exceeded or not can also take a variety of forms, partly representing an evolution in sophistication. The first form of threshold monitoring introduced for a pest usually requires the farmer to count the insects on a small sample of his crop. This form of monitoring is often called 'crop inspection' or 'scouting', and would normally be carried out by the farmer himself or a crop consultant employed by him. An example is monitoring for red bollworm in cotton in Central Africa, where it is recommended that crops are examined twice weekly from 6 weeks after germination. Twelve plants are sampled for bollworm eggs along each diagonal of the field. The economic threshold for such a sample is an average of 0.25 eggs plant^{-1} (Matthews, 1989).

Such a basic threshold can be refined by, for example, scoring only caterpillars over a certain size, as recommended in soybeans in the USA. Alternatively, different size categories of a pest can be given a different weight in determining whether the threshold has been reached. Increasingly, monitoring is becoming more detailed. For example, thresholds for phytophagous mites in apple orchards are varied according to the densities of predatory mites that are monitored simultaneously.

Farmers can find such self-operated crop inspection too complicated or time-consuming. The UK threshold for cereal aphids referred to above requires the farmer to count small insects on 20 tillers, and specifically well away from the field edges. Few farmers therefore carry out such inspections, though they will readily claim to do so to government advisers. It is therefore often hard to establish whether recommended thresholds are or are not widely used.

It is therefore better if crop monitoring can be provided as a service to the farmer, or to replace it by procedures such as trapping where the farmer has only few locations to monitor with a simple check.

The former approach may be possible where only a single sampling occasion is involved and where the resulting recommendation can be applied over a large area. As an example, advice in the UK on the need to spray against black bean aphid (*Aphis fabae*) has been based on counts by the Advisory Service of winter eggs on selected bushes of the over-wintering host, the spindle tree. A separate forecast is given for each of 16 areas in southern England (Way *et al.*, 1981). Farmers are then advised whether economic damage to spring-sown field beans is unlikely, possible or probable. The 'unlikely' and 'probable' forecasts have proved very reliable. A second sampling of spindle in mid to late May enables advice to be given on the timing of any necessary chemical treatments.

Monitoring by trapping may also be a service rather than a farmer-operated tool. The UK Advisory Service monitors adult *Agrotis segetum* activity by light and pheromone traps (i.e. traps baited with the female moth sex attractant). The monitoring is carried out in early summer and uses temperature and rainfall data to estimate egg development and larval (cutworm) survival. Each 0.1 mm of daily rainfall is estimated to kill 1% of surviving small larvae and any daily fall of 10 mm or more effectively destroys all pre-third-instar larvae. Based on all these data, warnings are issued in relation to particular types of crops (Emmett, 1984).

The UK pea moth monitoring scheme combines the operation of pheromone traps by farmers themselves with further calculations by the Advisory Service (Emmett, 1984). The farmer sets up two pheromone traps at right angles in a corner of the pea field and inspects them three times a week. The threshold is reached when 10 or more moths have been caught in either trap on each of two consecutive sampling occasions. However, this is not a threshold for commencing spraying, but a warning that damage is likely. It provides a date from which the Advisory Service can predict the appropriate spray date for the individual farmer. This prediction uses forecast temperatures to estimate when females will lay eggs and the subsequent development time of those eggs. This is computer predicted, and is updated with actual weather records as the days pass.

Often, however, the decision if and when to spray is based entirely on the farmer's own operation of his traps. Pheromone traps catch only males, since the synthesized attractant is based on the female-produced phero-mone. The use of pheromone traps for monitoring is therefore only possible if there is a predictable relationship, not only between male and female numbers, but also between male numbers and the timing of female oviposition on the crop. Where such relationships can be established, pheromone traps provide a very timesaving monitoring method. Codling moth is an example of a simple and satisfactory system. Only one trap is needed per hectare, and the action threshold for applying sprays is reached if five codling males are found per trap in two consecutive weeks (Emmett, 1984).

The ultimate step is not to base spray decisions on monitoring or trapping in the field, but on a centrally managed computer simulation of the field situation driven primarily by weather variables, and incorporating models for crop growth and development. Such a level of sophistication is still largely for the future, but impressive strides have been made in developing such a model for cereal aphids in Europe, both in the Netherlands and in the UK. The models still need to be 'synchronized' with the situation in crops by periodic reference to field data, but can already produce realistic forecasts.

The integration of pesticides into IPM

The role of pesticides in IPM has already been described in general terms (Chapter 3). Here, it is relevant to consider the 'proactive' (i.e. positive) use of pesticides to improve the IPM package. This is in total contrast to regarding pesticides as a 'necessary evil', and merely considering how to use them to cause as little damage as possible to biological control and the environment.

The principle of the proactive use of pesticides in IPM is that they can be used to improve biological control if their application results in an

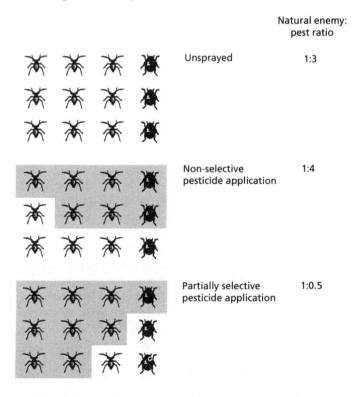

Natural enemy: pest ratio

Unsprayed   1:3

Non-selective pesticide application   1:4

Partially selective pesticide application   1:0.5

Figure 5.1 *The importance of natural enemy:pest ratios after pesticide applications.*

improvement of the natural enemy:pest ratio. In other words, we accept that kill of both beneficials and pests will occur. If, however, there is even partial selectivity in favor of natural enemies, then biological control of the pest will have been improved rather than damaged by the application of a pesticide (Figure 5.1).

Use of a selective pesticide

This is the obvious approach when possible. Some products such as sprays of insect pathogens have considerable selectivity by the very nature of insect–pathogen relationships (p. 86). Other 'novel' insecticides, such as insect growth regulators (p. 153), tend to show similar selectivity, at least to a definable type of pest (e.g. caterpillars).

Selectivity is harder to find among the traditional toxic insecticides, since they are nearly all targeted at a site of action, the nervous system, which is biochemically similar in insects and most other animals, including man. Moreover, at least until recently, there was little incentive for agrochemical companies to market compounds that were not broad spectrum, since such a property narrowed the potential market. None the less, selectivity is known for some insecticides and, for any particular crop and pest situation, experiment may reveal unsuspected selectivity. This is sometimes the consequence of behavioral differences between the pest and its principal natural enemies. The flexibility of route to the target (residual, fumigant, translaminar, systemic, etc.) offered by the organophosphate insecticides and some carbamates is particularly valuable in this context. Even the uniformly residual and contact synthetic pyrethroids possess a contact irritancy that varies between insect species, and could be a source of partial selectivity.

A few insecticides show true selectivity for biochemical reasons. Most striking is the carbamate insecticide pirimicarb, which is toxic to aphids and Diptera, yet not to other insects at equivalent doses. This is because acetylcholinesterase, the enzyme in the insect nervous system that is inhibited by pirimicarb, varies in insects and in most is insensitive to the pesticide. The organochlorine endosulfan is also biochemically selective and Hymenoptera, an order containing many species valuable in biological control, are far less affected than other insect groups.

Selective use of a pesticide is also, of course, possible where the biological control agent at risk is resistant to that pesticide. Mite predators of orchard spider mites have developed resistance to organophosphates, and were first reported in 1958 from British Columbia. Such resistant predators have been transferred or discovered elsewhere and form an important component of orchard IPM in many countries. Pesticide resistance has also appeared of its own accord in at least eight other beneficial species, as listed by Croft and Brown (1975).

In America, the natural organophosphate-resistant predatory mites found in both orchards and vineyards have been artificially selected in the laboratory for resistance to additional pesticides, including pyrethroids. The field population of predatory mites has to be reduced to a very low level with pyrethroid insecticide before release of the multi-resistant mites. Otherwise interbreeding might occur and dilute the resistance. Under these circumstances the resistant mites have maintained their resistance with time, allowing simultaneous use of insecticides. The system has been adopted by about 60% of Californian almond growers, and savings due to decreased pesticide use are estimated at US$60–100 per hectare (Headley and Hoy, 1987).

That resistance occurs naturally and can be supplemented or accelerated by selection in the laboratory suggests great potential for this source of selectivity. Molecular biological techniques now make the transfer of genes between species a real possibility. It is attractive to think that the source of resistance for a genetically engineered natural enemy may well be its pest prey! However, at present entomologists lack the kind of efficient techniques for DNA transfer possible with plants (p. 140). Also, worries have been expressed that availability of pesticide-resistant natural enemies would encourage continued pesticide use. A further constraint is that the economic feasibility of the whole approach is limited by the need to develop new strains each time a new pesticide is adopted by growers.

In IPM, one is more usually faced with the challenge of how a broad-spectrum pesticide can be made at least partially selective by the way it is used.

Formulation

Formulation (the form in which the active ingredient of a pesticide is marketed) can impart some selectivity. Wettable powders, emulsions, flowables, etc. have different additives such as wetters and emulsifiers that can affect persistence on the leaf and contact with target and nontarget organisms. Systemic insecticides applied to the soil as granules are taken up by the plant to poison sap suckers such as aphids, without leaving a residue on the leaf to be toxic to other insects. Many insecticides can be encapsulated so that the 'droplets' are contained in a sticky polymer shell that adheres to the plant. This imparts the properties of a 'stomach poison' to an otherwise broad-spectrum compound (i.e. only insects chewing the leaves and ingesting the capsules will be poisoned).

Reduced dose

The variability in insecticide tolerance between individuals of a plant-feeding insect species tends to be greater than for carnivorous insects such as biological control agents. It is usually suggested that this is because the plant feeder has had to evolve an armory of enzymes to detoxify plant secondary metabolites; such enzymes can equally be turned against insecticides. Possession of the armory inevitably results in quantitative variation within the population, and there will be a wide pesticide concentration range between 1% and 100% mortality. Carnivorous insects, on the other hand, do not have to detoxify plant secondary compounds to the same degree. They may therefore show a much smaller variation in insecticide susceptibility between individuals. This explanation accounts for the fact that mortality of many carnivores rises more steeply than that of herbivores as pesticide concentration increases (Figure 5.2).

Figure 5.2 illustrates a situation where the carnivore is more susceptible to the pesticide than its herbivore prey. This is probably the norm where parasitoids and their prey are compared by laboratory bioassay. Croft and Brown (1975) reported this relationship for 11 out of 15 parasitoid (see p. 86 for definition) and prey combinations. However, it should not be assumed, as often it is, that carnivores are invariably more sensitive to pesticides than their prey. Even in laboratory assays, many predators show great tolerance to pesticides, possibly because their thick cuticles are a barrier to penetration. Thus Croft and Brown found that in 77 cases of predator–prey combinations, the predator showed greater pesticide tolerance in 63, often by a large factor (in 34 cases the predator was a coccinellid beetle). Thus, particularly in the field where differences in herbivore and carnivore behavior may increase pesticide selectivity towards the carnivore, it is far too glib to generalize that pesticides will destroy biological control.

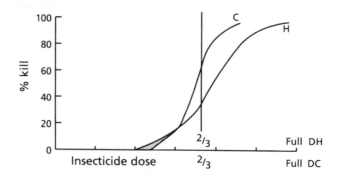

Figure 5.2 *Theoretical selectivity inherent in pesticide application. H, dose mortality curve of a herbivore; C, dose mortality curve of predator or parasitoid; DC, dose scale for carnivore; DH, dose scale for herbivore. (From van Emden, H.F., Proc. Br. Crop Prot. Conf., Pests and Diseases, Brighton, 1990, p. 945, with permission.)*

Whatever the relationships in detail of the two mortality curves shown in Figure 5.2, clearly dose reduction will progressively benefit the carnivore till, in the theoretical example shown, a pesticide selective against the carnivore at high concentrations becomes equitoxic to both carnivore and herbivore; at still lower concentrations a 'selectivity window' opens, in which a higher percentage of the herbivore than the carnivore is killed.

Application of pesticide in time

The arrival or emergence in the crop of natural enemies is usually later than that of pests. This can provide the opportunity for an early pesticide application of an ephemeral compound without risk to biological control.

A second possibility is that pesticides with a short residual life can be applied selectively when most of the natural enemy population on the crop is protected in some way. For example, the first generation of aphid parasitoids is sufficiently synchronized that there is a possibility for applying an aphicide while the parasitoids are pupated within the protection of the skin of the dead aphid (the 'mummy'). A short-lived compound is necessary, as adult parasitoids can pick up a toxic dose from the surface of the mummy while emerging. A similar philosophy has been used for coccinellid predators of cowpea aphid in Nigeria (Morse, 1989). Here, aphid populations were not sprayed before ladybirds had oviposited on the crop. The rationale was that oviposition by ladybirds tends to be quantitatively dependent on the density of the aphid prey and the maximum number of eggs was obviously desired. It was found that dimethoate could be used safely while the coccinellids were still in the egg stage and yet it greatly reduced aphid numbers. The result was a very favorable ratio of ladybird larvae hatching from eggs to the number of aphids remaining.

Many pest and beneficial insects move about the plant diurnally, but the two types of insect do not necessarily show the same or a synchronized pattern. Thus leafhoppers in cotton and grain legumes are difficult daytime targets for sprays since they sit on the underside of the lower leaves. In the evening, however, they move up the plant to sit on the upper surfaces of the higher leaves. They then become as easy target, and good control may be achieved with a little ephemeral insecticide with little harm to their natural enemies.

Apart from selectivity between a pest and its own natural enemies, selectivity between a pest and the natural enemies of other pests on the crop needs to be considered. Here a fundamental principle is to use pesticides against a pest at a time when natural enemies of other pests are either absent or protected. A typical example comes from Pennsylvania apple orchards, where the peaks of adult leaf roller moths and their oviposition occur in June and September. Spider mites and their ladybird predators do not build up until July and August, and this makes it possible to time leaf roller control not to interfere with the biological control of the mites.

Application of pesticide in space

Localizing the application of pesticide is an extremely effective and valuable technique in IPM for obtaining contact of the pest with the pesticide while avoiding or minimizing contact by natural enemies. Insecticide-treated traps baited with a specific attractant for the pest are an example. Such baited traps were used some years ago in the control of fruit flies. The females respond to a cheap bait of molasses or protein hydrolyzate which mimics the odor stemming from microbial activity on fruit surfaces. More recently, screens treated with a pyrethroid insecticide and baited with octanol have been used in control of the tsetse fly. Similarly, pheromones have been used as baits; for example, in Egypt there has been work on applying bollworm pheromone plus pyrethroid as a small smear on cotton leaves at intervals along the row. In Nicaragua, the aggregation pheromone of boll weevil has been used to concentrate weevils emerging from hibernation on to small, early planted areas of cotton, where the pest is then controlled with a pesticide. The problem of the approach is, of course, that natural sources (the crop to be protected or insects there) give off a similar odor in competition with the traps. The specific use of pheromones for 'trapping-out' is further discussed on p. 125.

A very general approach to limited application in space is 'band spraying', where swathes covered by the pesticide are purposefully not adjacent, leaving unsprayed crop areas between them. This concept capitalizes on the greater mobility of adult natural enemies than of most pests, and therefore their ability to avoid the spray. The unsprayed 'refuge' crop areas thus tend to have a higher natural enemy:pest ratio after than before spraying. If necessary, the areas sprayed and unsprayed are reversed at a second pesticide application. The band-spraying concept has been ingeniously combined with the potential selectivity of reduced doses (see above) in orchards in Pennsylvania. In this 'alternate-row middle' technique (Lewis and Hickey, 1964), half-dose insecticide is sprayed down alternate rows. Each tree is thus only sprayed from one side on each spraying occasion, but receives spray from the other side at the next application. Predators build up with this spray regime to a point where the frequency of spraying can be greatly reduced.

Where resistance to pesticides has developed in pests, it may become essential to modify spraying practices to leave unsprayed refuges for natural enemies. In USA citrus orchards, sprays against fruit flies were directed as far as possible to the bottom half of the tree, where the pest is more abundant than in the top half. Some natural enemies then survived higher up. The system was improved in Australia by adding to the insecticide a bait which attracts the fly. Thus fruit flies in the upper part of the tree were lured down to contact the insecticide.

A classic success with pesticide selectivity in space was achieved on coffee in Kenya as long ago as the 1950s (Wheatley, 1963). Remarkably enough, the 'selective' insecticide was the notoriously broad-spectrum DDT. This was painted around the trunks of the coffee trees early each season. Whenever pests, particularly large looper larvae, became too numerous in the coffee canopy, the tree was sprayed with natural pyrethrum at a sublethal dose, yet sufficient to 'knock down' both pests and beneficials. These then fell out of the tree on to the ground. Many beneficials flew back to the tree after recovery, but the caterpillars could only regain the leaves by crossing the DDT band and picking up a lethal dose. Each use of the DDT/pyrethrum system therefore greatly improved the natural enemy:pest ratio in the canopy.

Insecticide application technique in relation to IPM

Since *Silent Spring*, pesticides have largely continued to be applied by conventional hydraulic sprayers with nozzles that have a simple orifice. These systems produce a wide range of droplet sizes that are dispersed into the environment. The destinations of these drops are determined by an interaction between their size, air movement and the collection efficiency of surfaces in their path. With any air movement, winnowing of the droplets occurs with gravity. This affects the larger drops; those smaller than 50 μm remain airborne. In the field these small drops will be lifted by convection and travel considerable distances from their source. Better control of the fate of droplets can reduce the environmental hazard of pesticides as well as contribute to the provision of 'refugia' for natural enemies inside and outside the crop.

Droplet spectra can be measured directly in the air using the diffraction of laser light and computer interpretation. As hydraulic sprays are the principal method of application world-wide, reference fan nozzles have been used to define categories of spray quality in the UK (Doble *et al.*, 1985). The main categories are 'coarse', 'medium' and 'fine'. Sprays with higher numbers of droplets below 1100 μm in diameter than in the 'fine' sprays are categorized as 'very fine'. These are not recommended for arable crops due to the drift hazard.

Most public perception of hazards due to spray applications focuses on this downwind drift. Although in practice only a small proportion of the spray volume drifts outside a treated field, the effects of the many very small airborne droplets of some pesticides can have significant impact several meters downwind, such as herbicide damage and the death of bees and other beneficial organisms. The smaller the droplet, the greater is its movement in natural air currents. Thus the shrinkage of drops by evaporation of water increases drift in hot, dry conditions. The smallest droplets remain suspended in the air and do not readily deposit on surfaces. They are therefore likely to remain carried around in the air and finally remain airborne over long periods as concentrated particles of pesticide. Paradoxically, wind may reduce drift by increasing the impaction of drops on leaves or other surfaces.

Habitats, including other crops, hedgerows or woodland, next to treated crops provide refuges for natural enemies (including when overwintering). Contamination with pesticide may affect the survival of the beneficial fauna there. Increasing droplet size by, for example, using a larger orifice nozzle may reduce aerial drift, but merely results in drift of a different kind. Only a small window in droplet size for any combination of environmental conditions (including the plant surface) contains droplets likely both to impinge and to be retained on the plant leaves. The coalescence of large droplets on a leaf will cause runoff to the ground, and very large drops bounce from leaves instead of depositing on them. Most pesticide in larger drops directly contaminates the soil; thus a broad-spectrum insecticide and some other pesticides can directly affect soil-inhabiting natural enemies, including the generally prevalent ground beetles and spiders. That part of the spray spectrum which reaches the soil also provides the potential for groundwater contamination, though many pesticides remain bound in the surface soil layers and are then degraded by microbes. Registration authorities tend to favor the use of larger droplets to minimize drift, but it is droplets around $100 \mu m$ that deposit on plants. Even smaller droplets are caught by the hairs on leaves and insects. In terms of biological efficiency, therefore, there is a need for spray application techniques that can deposit small droplets with least risk of drift.

Today we have more selective chemicals and more active molecules, such as the synthetic pyrethroids, requiring only a few grams per hectare. In contrast to these significant improvements in the chemistry of pesticides since *Silent Spring*, investment in improving application technique has been small. A few specialized techniques such as granule application, seed treatment and weed wipers enable treatments to be more localized and so appropriate for IPM. However, the number of pesticides available for these techniques is small. Most pesticides are still applied as aqueous sprays. Although volumes of spray applied have decreased over the past 3 decades, 100 liters of water or more per hectare are still the norm. In the semi-arid tropics, for example many countries in Africa, ultra-low-volume application using hand-operated battery-powered spinning-disc applicators have been introduced. However, this technique, originally conceived for locust control, does rely on air movement to disperse the $50–100 \mu m$ droplets downwind from the nozzle into the crop canopy. Drift beyond the crop boundary is small in large areas of a crop, but in many multicrop situations the height of plants and the resultant air turbulence can cause loss of the small droplets to drift. Slower disc speeds increase droplet size, but increase deposition on the soil.

An alternative approach to increase the proportion of a spray deposited on a crop is the electrostatic charging of spray droplets. When charged sprays are applied to foliage, most droplets are deposited on the nearest earthed surface, i.e. the outer and upper foliage. Mutual repulsion of droplets of the same polarity inevitably means that the spray cloud not only expands, but also that some droplets are carried upwards to drift downwind. Others may be lifted further by convective air movement. However, charged droplets displaced downwind are likely to be deposited on foliage nearby, especially if an oil-based formulation, which prevents droplets shrinking in flight, is used.

A perceived disadvantage of charged droplets is lack of penetration into the crop canopy. However, for IPM, this may be a real advantage. Much depends on the pest complex and the individual crop involved. On cotton, for example, most of the major pests attack the young buds and tend to be found on the outer section of lateral branches; in cowpeas post-flowering, it is the pods that most need protecting. Electrostatic spraying therefore provides the potential for 'selectivity in space', resulting in greater survival of predators of both the target and other pests. Apart from this potential advantage, it is likely that electrostatic spraying will result in less pesticide reaching the soil in and between the rows. This will reduce the impact of a pesticide on the often very valuable soil-inhabiting beneficials.

More studies are needed to assess the full potential of electrostatically charged sprays. The distribution of charged droplets on crops is sometimes unsatisfactory for the control of certain pests (Cayley et al., 1984). The equipment requires skill to use and limits the range of products that can be applied. Evidence to date suggests that charged, ultra-low volume oil-based sprays are more appropriate for some pesticides, such as the pyrethroids, than others in IPM, where selective deposition of droplets can be exploited. However, development of electrostatic spraying systems has been constrained by patent protection to one multinational company. There have also been difficulties in developing the necessarily oil-based formulations for the system.

There is also the problem of operator contamination to consider. This may arise during preparation of the spray or during actual spraying. The first problem has been tackled by the agrochemical companies with new types of packaging (e.g. water-soluble sachets), recycling of pesticide containers, changes in formulation and a trend towards closed filling systems, especially for large, mechanized equipment. For the small-scale farmer a disposable container, via which the pesticide is automatically diluted, has been developed though not yet commercialized. The development of ultra-low volume (ULV) formulations in closed containers (e.g. the electrodyn 'bozzle') is another attempt to eliminate operator contamination during mixing. Without such closed systems, farmers frequently miscalculate the amount of product to use and may spill pesticide over their hands when measuring out concentrate. Contamination during spraying is a widespread hazard with most types of equipment, and protective clothing and sealed tractor cabs are among the countermeasures which are available to those in temperate climates with adequate finance. In the tropics, protective clothing is seldom worn. Moreover, most small, manually operated sprayers are designed for the hand lance to be carried in front of the operator, who thus walks directly into the sprayed area and is usually heavily contaminated on the front of his or her body. This contamination is often further aggravated by leakages from junctions on the machine and hoses. Holding nozzles downwind, as with hand-held controlled-droplet-application (CDA) sprayers or using a sprayer with a 'tail boom', is a better alternative. However, although such improvements were developed for spraying cotton in Zimbabwe in 1960, they were not widely promoted because of a perception that the operator needed to see where the spray was going.

Few countries have demanded improvements in the cause of operator safety. In the UK, the Health and Safety Executive requested the British Standards Institute to produce a specification for lever-operated knapsack sprayers (BS4711). This is already influencing equipment design; a similar specification for compression sprayers is being prepared. An Asian Development Bank initiative has made a start to improve the specification and safety standards of knapsack sprayers in Asia, but little progress has yet been achieved. Even when better equipment becomes available, it is difficult to persuade farmers to purchase a replacement for still-functioning older models.

The partially selective use of broad-spectrum insecticides has usually been seen as an exercise in formulation, timing or placement. It is clear from the above that application technique can play a major role in respect of placement. More research is surely needed to compare the effects on natural enemies of different application techniques.

Biological control

If the use of insecticides is to be reduced through IPM, then the consequent reduction in pest control clearly has to be replaced in some way. Biological control represents the most widely available source of such additional pest control. Nearly all pests, in all situations, will be subject to some level of this control.

In *Silent Spring*, Rachel Carson saw natural enemies as an ideal alternative to pesticides, since she saw biological control as the main way 'nature' prevents overpopulation by insects. It is true that one rarely sees explosions of plant-feeding insects on noncommercial plants. The often quoted figure is that insects and other herbivores consume less than 1% of the plant biomass in the wild, though other estimates go as high as 7%. Either figure is a stark contrast to the average of 30% for food crops. However, it is important to realize that biological control takes place against a very different background in the two situations. First, our crop plants have, through plant breeding, lost many of their chemical barriers to herbivore attack. Secondly, the plant species mosaic in more natural ecosystems makes the discovery of, and movement between, suitable host plants difficult compared with a monoculture. Thus pest status of a herbivore is usually man-induced; we cannot expect that the reductions in plant genetic and floral diversity wrought by agriculture would favor carnivore populations as they have the populations of herbivores. It is inevitable that evolution should lead to behavioral and other adaptations in both natural enemy and prey to maximize the survival of both trophic levels. Any natural enemy that overexploited its food resource would be destined for the fossil record. Thus we cannot be surprised that such a finely tuned population dynamics relationship between carnivores and their prey is destroyed by the ecological catastrophe of agriculture. The very features of agricultures that encourage pest status discourage natural enemies (pp. 99–123).

Biological control only retains coherent meaning if defined as the use as pest control agents of living organisms that are natural enemies. This is because many other approaches to pest control, such as host-plant resistance and pheromones, are also 'biologically' based.

The agents used for biological control by this strict definition are predators, parasitoids or pathogens.

Predators

Predators are natural enemies that consume several prey during their development. They are usually considerably larger than their prey and are active hunters or ambushers. Many orders of insects include predatory groups and species, especially the Hemiptera, Neuroptera, Diptera and Coleoptera. In the latter three orders it is often the larvae that are the predatory stages (e.g. hover flies). However, it is not uncommon for both larvae and adults to share the predatory habit (e.g. ladybirds). In a few examples (e.g. robber and assassin flies), it is primarily the adults that are predatory. The order Hymenoptera also contains predatory ants and wasps. Other important predators are found in the class Arachnida, the spiders and mites. Polyphagous predators such as spiders and ground beetles can exercise valuable control of pests in many crop situations. That such predators can feed on many different prey species limits their use, in the eyes of many workers, as a single control measure, but they may often be of greater value than more specific predators in the integrated approach.

Parasitoids

Parasitoids complete their development on (exoparasitoids), or more usually within (endoparasitoids), a single host that is eventually destroyed in the process. The larva is thus the stage that feeds on the pest, though in a few examples the adults may also kill prey as predators to obtain food as adults. Most parasitoids are found in just two orders, the Hymenoptera and Diptera (Tachinidae). Hymenopterous parasitoids have received far more attention in biological control than have the parasitoid Diptera.

Pathogens

Diseases of insects are widespread in nature, and break out as epizootics particularly when their insect hosts are stressed, especially by over-crowding. Fungi infect their hosts by penetrating the cuticle; they require both high humidity for the spores to germinate and proximity of infected and uninfected hosts for the disease to spread. The humidity requirement has made soil-inhabiting pests particularly suitable targets. Most successes with fungi have involved Deuteromycetes. All other 'microbial insecticides' rely on ingestion by the host for entry. They have thus evolved an infective stage that can survive outside insects for a little while; they are therefore less humidity-dependent than the fungi. Bacteria have been particularly successful commercially. All are in the genus *Bacillus*, and *B. thuringiensis* is the predominant species. Death of the host results from the liberation of a toxic protein crystal when the wall of the ingested stage of the bacterium (the parasporal body) ruptures in the gut. Among the viruses, most interest has centered on the Baculoviridae because they are totally different in structure and biochemistry to the viral diseases of vertebrates. Most are either granulosis or nuclear polyhedrosis viruses. Protozoa also cause diseases in insects but have two serious drawbacks for insect control. First they are obligate parasites, and cannot be multiplied outside the host; this makes their commercial production uneconomic. Secondly, their effects on their insect host are slow and debilitating, rather than causing the rapid death of the pest that the grower requires. A few genera of insect-parasitic nematodes are also used in biological control. They are included here under 'pathogens', since death of the insect results from a bacterium carried by these nematodes.

Although, like insect natural enemies, pathogens of insects are natur-ally occurring mortality factors, their use in IPM has been limited. Their use is very much more dependent on commercial development than is true for insect natural enemies, yet they are made subject to expensive safety testing and registration procedures. Moreover, a virtue of biological control agents, specificity to target, becomes a market limitation for commercially developed pathogens.

Principles of biological control with insect natural enemies

Biological control is selective, and without the kinds of environmental hazards presented by insecticides. Undesirable side-effects can nevertheless occur, even if rarely. Nontarget insects, for example insects related to the pest but valuable for biological control of weeds, have been adversely affected in at least two instances. Also, a potentially serious disease of sugar cane was introduced into Trinidad on the ovipositor of a parasitoid imported against sugar-cane borer (Bourne, 1953).

Biological control can be self-perpetuating following the original introduction. It may spread over large areas after release and reach targets inaccessible to pesticides (e.g. pests concealed in fruit or stems). The research and establishment costs of biological control are very moderate (less than 10%) compared with the development of a new pesticide, and are often borne by national or international agencies. For the farmer, biological control is often free. It may be the only economic solution for protecting the staple crops of poor farmers in developing countries or in forestry and pasture.

Pests are not without defences against biological control agents.They have immune systems against pathogens, they can encapsulate eggs of parasitoids laid in their bodies, and can sequester toxins from their host plant to render them unpalatable to predators. Falling from the plant or running away are just two of several behavioral defence mechanisms. Pimentel (1964) has argued that natural enemies will select for 'resistant' prey. However, the natural enemies will face similar selection pressure to overcome such resistance, and the natural enemies of today must have coevolved countermeasures to mechanisms for resistance to them. Although resistance to biological control cannot be ruled out, its chances are very small in comparison to the likelihood of resistance to insecticides. We know of no unambiguous evidence of a biological control system with insect natural enemies failing for this reason. Pimentel (1963) cites two examples involving insect natural enemies: strains of prickly pear cactus which were not readily destroyed by the moth *Cactoblastis*, and a parasitoid of larch sawfly in Canada which suffered from increasing encapsulation by its host. Carl (1982) states 'The examples he [Pimentel] quotes are not very convincing,' and 'If his hypothesis were true, biological control would be a pointless endeavour.' However, insect diseases are often used more as pesticides than as self-perpetuating biocontrol agents (p. 96), and examples of pest resistance to diseases have already occurred (p. 96).

Unfortunately, these very potent advantages of biological control are offset by some serious disadvantages. Pre-eminent among these is, excepting most microbial control agents other than some fungal pathogens, an inherent incompatibility with insecticides. Potentially effective biocontrol of one pest may therefore be difficult where a pest complex attacks the crop, and pesticides are still needed against these other problems. In IPM, however, in contrast to single-component biological control, the challenge of reconciling the two 'incompatible' approaches of chemical and biological control can frequently be met (p. 75).

A second disadvantage of biological control is that it is usually slow to act unless natural enemies or pathogens are 'overdosed' as 'biological pesticides'. Otherwise it takes time for them to build up to adequate numbers and spread from the point of release to impact on the pest population. Meanwhile, the pest may increase to intolerable levels, yet the use of pesticides is likely to endanger the biological control system.

Most self-perpetuating biological control systems require the pest to remain present, but at a low level. This level may be too high for the farmer to accept if the pest adversely affects the quality of high-value crop units such as fruits, or if the pest is a problem at low densities as a vector of plant diseases.

Biological control has the further disadvantage of being ever unpredictable. The ability of natural enemies to respond to increases in the density of their prey is often weather dependent; they can easily lose their control impact totally once the population of the pest has reached a critical 'escape' level. Glasshouse growers, using the parasitoid *Encarsia* for control of whitefly, are only too aware of this danger. Under other conditions, it is possible for the natural enemy gradually to reduce its prey population to virtual extermination, and then in turn die out itself. Such an event may not be noticed by the farmer, who has learnt to trust biological control, until new pest individuals arrive on the crop. Such an unexpected pest outbreak occurred in 1953 in Puerto Rico after 15 years of successful control of mulberry scale by ladybird beetle. Of course, this example still compares very favorably with the much shorter effective life of many pesticides before resistance shows in the pest population.

As mentioned earlier, the nature of agriculture makes biological control which is satisfactory to the farmer an unlikely phenomenon; farmers also require far greater pest reductions than are found in natural situations. A biological control intervention has therefore to involve some 'trick' to make the control successful. Biological control in practice therefore is unnatural, and represents an ecological disturbance of the population system just as does the use of pesticides, though admittedly to a lesser degree.

In population dynamics jargon, the key principle of self-perpetuating biological control is the concept of 'density dependence'. Unless the percentage impact of the natural enemy rises with prey density, the pest population cannot be brought under control. Similarly, as pest density then falls again, the impact of a density-dependent mortality factor itself decreases. This results in 'regulation' rather than overexploitation of the pest population. Direct density dependence of a natural enemy thus not only prevents the pest population from becoming too large; it equally prevents it falling to very low levels. The result is oscillation of the prey population around an equilibrium. In agriculture this equilibrium level may still be economically damaging, i.e. biological 'control' is not occurring. Moreover, as pointed out above, any sudden population increase of a pest may change the natural enemy:pest density relationships to an irrevocable extent. A pest epidemic then results from the failure of biological control. Sudden changes in the rate of increase of pest populations may be caused by a sudden change in weather or a sudden change in host-plant quality. Southwood (1977) has likened the regulation of prey by natural enemies to a ridge on which regulation occurs (Figure 5.3). Any escape by the pest on the far side of the 'natural enemy

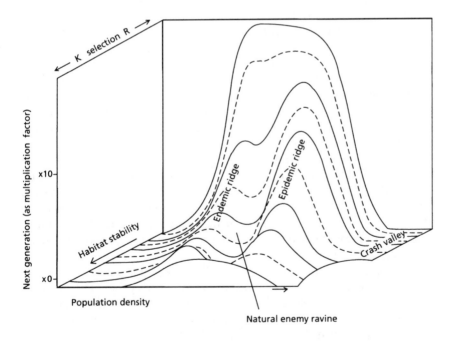

Figure 5.3 *Synoptic model of population growth relationships (modified from Southwood, T.R.E., Proc. 15th Int. Congr. Entomol., Washington, DC., August 1976, p. 45, with permission).*

ravine' inevitably leads to a population explosion up an 'epidemic' ridge. This explosion continues until overpopulation results in competition for resources and the population declines into 'crash valley'.

The density dependence of natural enemies can sometimes be delayed in action, in which case an increase in pest numbers only results in greater natural enemy impact in the next generation or season. This still provides regulation of the pest population, but with greater oscillations in numbers around the long-term equilibrium.

Patchy distribution of pests, as commonly results when the offspring of immigrant pests do not disperse widely, can lead to spatial density dependence. The biological control agent is then concentrated in patches of high host density, where it may have arrived too late to prevent pest populations reaching damaging levels. Meanwhile, new patches of pests develop elsewhere in the crop, i.e. the biological control agent is always 'one step behind' its prey. In the confined space of glasshouses, this limitation for effective biological control can be overcome by artificially introducing the pest evenly before distributing biocontrol agents.

Whether biological control agents show the positive density dependence necessary to achieve regulation of pest populations depends on the nature of the responses shown to pest density. These responses are classified as 'numerical' or 'functional'.

Numerical responses of natural enemies are their reactions to changing prey density in terms of numbers. Such reactions include migration to or aggregation in areas of higher prey abundance, and enhanced production of offspring in such areas. Functional responses, on the other hand, refer to the impact per individual natural enemy. To simplify a little, this response has the components of 'searching time' and 'handling time'. As prey density increases, biocontrol agents may find a higher proportion of their prey since less time has to be spent searching for food or hosts. Handling time refers to the time that a predator or parasitoid spends on each prey individual, both in the time taken for the attack and in actual consumption. This may set an upper limit to the number of prey each individual natural enemy can deal with ('satiation'), but can be to some extent compensated for by less time spent on each individual. For example, some predators only eat parts of their prey when it is abundant instead of extracting all the potential food.

The high incidence of responses that are not positively density dependent provide considerable scope for integrating biological control with other control measures within the context of IPM. This integration introduces the kind of 'trick' referred to earlier that is often required to make biological control effective. Especially, the partially selective use of pesticides, habitat modification (p. 105–113) and some forms of cultural control can shift the numerical response of biocontrol agents in our favor. Partial host-plant resistance, however, although having a similar potential to improve the numerical response, has the unique property of giving us scope for manipulating the functional response of natural enemies as well (p. 151). Such integrations may well convert 'failure' of natural enemies to keep a pest below the economic threshold into success. As pointed out earlier, the massive unnatural intervention in insect population dynamics that agriculture represents usually requires a parallel unnatural manipulation of biological control.

Techniques of biological control

There are several different approaches to biological control; each has its own opportunity for incorporating the principle of 'tricks' to increase the effectiveness of the control achieved.

Inoculation

Often called 'classical biological control', this approach involves the transfer of natural enemies from one environment to another, in the hope that they will establish and that the control will become self-perpetuating. It has been particularly successful in bringing pest species imported from another continent under control, and has worked best with sedentary pests of perennial crops. Although the emphasis has been against pests imported into new agriculture as in California and Australia, it is relevant to any situation where a paucity of natural enemies exists. Such situations include those where intensified agriculture, including widespread pesticide use, has upset a previous pest/natural enemy equilibrium situation. Only a limited number of beneficials are released, with the aim of long-term regulation of the pest population. Inoculation of natural enemies (insects, mites, etc.) has been tried against 292 insect pests (involving 563 beneficial species) and 70 weed species (involving 126 beneficial insect species). This has happened in 168 and 55 countries, respectively. For biological control as a sole control measure, the 'substantial success' rate of establishments has been 40% against insect pests and 31% against weeds (Waage and Greathead, 1988). Classical biological control using pathogens is a very recent development by comparison, but can already claim about four successes against insects and two against weeds.

Classical biological control has had a long history. The first well-documented example of an introduction for biological control purposes dates from 1762, when a myna bird was taken from India to Mauritius to control locusts. Most people, however, regard the control of cottony cushion scale in California (by vedalia beetle introduced from Australia in 1888) as the founding landmark of modern biological control. An initial shipment of 100 ladybirds was all that was needed to give control of the scale insect within as little as 15 months.

Another early example of classical biological control is the importation from Malaysia of the tachinid fly *Bessa remota* to control coconut moth, *Levuana iridescens*, in Fiji in the mid-1920s (Tothill, Taylor and Paine, 1930). This example has the interesting feature that *B. remota* is a parasitoid of a different genus of moth (*Brachartona catoxantha*), also feeding on coconut. In Fiji, 32 750 parasitized *L. iridescens* larvae were released, and complete control over the whole of Fiji was achieved in 2 years.

However, it still took many years before the scientifically based biological control inoculation programs of today were developed. The first step is an evaluation of whether biological control is a viable control strategy. This involves assessing whether the pest is a problem only at high densities, whether it is a pest over large areas, whether it (or a similar species) has been controlled successfully biologically elsewhere, and whether existing natural enemies leave any scope for introducing biological control against a different life-stage of the pest. The question relevant to IPM, whether biological control might succeed in an integrated context, is only just beginning to be asked, 30 years after *Silent Spring* and more than a century after control of cottony cushion scale in California. If biological control appears a viable option, natural enemies of the pest in question are usually sought abroad. The part of the world searched is that assumed to be the evolutionary home of the pest species. However, Hokkanen and Pimentel (1984) argue that 'new associations', where the natural enemy comes from a different area than the pest, or attacks a different but related pest species, are more promising. This is because previous exposure to the natural enemy would be expected to lead to increased resistance to it in the host (p. 87). Hokkanen and Pimentel believe that 'new associations' might increase the success rate of biological control programs from what they estimate (but see Waage and Greathead, 1988, above) to be about one in seven to one in four. The example of coconut moth above is indeed a successful 'new association' accomplished 70 years ago. A well-documented example where the natural enemy came from an area different from the assumed centre of origin of the pest is the control of sugar-cane borer (*Diatraea saccharalis*) in Barbados (Carl, 1982). This pest is assumed to originate from the Amazon Basin. Here attempts at biological control, mainly involving mass releases of the egg parasitoid *Trichogramma*, over 30 years had failed. Success came with the introduction of the alien parasitoid *Apanteles flavipes* from India. Carl points out that this parasitoid fails to suppress *D. saccharalis* in India! Carl concludes: 'success may be achieved where it appears highly improbable.' The debate over 'old' v. 'new' associations has adherents of both viewpoints. Based on an analysis of a database, Greathead (1989) argues that failures of biological control have been more frequent with new than with old associations.

Table 5.1 Criteria for pre-introductory evaluation of natural enemies (from van Lenteren, J. C. (1992) *Biological Control of Pests: Course Manual*, Agricultural University, Wageningen, p. 123, with permission)

Criterion	Type of release program		
	Inoculative	Seasonal inoculative	Inundative
Seasonal synchronization with host	+	−	−
Internal synchronization with host	+	+	−
Climatic adaptation	+	+	+
No negative effects	+	+	+
Good rearing/production method	−	+	+
Host specificity	+	−	−
Great reproductive potential	+	+	−
Good density responsiveness	+	+	±

Most workers in biological control would agree that it is virtually impossible to identify the most suitable agent during a search abroad. This is because the host/prey densities are likely to be much lower and the agent that can respond rapidly to pest populations may be quite rare in the indigenous setting. Thus many potential agents are usually sent back home, to be reared and studied in expensive quarantine facilities to preclude unintentional liberation. It is in these quarantine facilities that attempts are made to increase numbers of the agent and simultaneously evaluate their potential for successful establishment and performance. Table 5.1 shows the criteria that are assessed.

The most important trick introduced during quarantine is the elimination of the beneficial insect's own parasitoids and diseases; this gives the beneficial an unnatural increase rate when released. Climatic adaptation may not be assured by collecting agents at the evolutionary origin of the pest; this is well illustrated by attempts to control walnut aphid in California with the parasitoid *Trioxys pallidus* collected in France in 1961. This succeeded on the coast, but failed in the Central Valley with its greater temperature extremes. A search for agents in an area of greater climatic similarity, the Central Plateau in Iran, revealed an adapted ecotype of the same parasitoid species. This ecotype was successful when released in 1968 into the Central Valley in California (van den Bosch *et al.*, 1970). One way in which beneficials avoid overexploiting their prey is that they often require a larger temperature sum for development, and thus emerge later in the crop season than their prey. It may therefore be worth using another 'trick', and seeking the beneficial in a climate colder than that of the region to which it is to be imported; such beneficials should be adapted to a lower temperature sum for development and thus synchronize better with the appearance of the pest on the crop.

If studies during quarantine identify a good candidate for release, larger populations are bred and releases made at several points, typically of 1000–5000 individuals at each point. Such releases may be repeated at the same sites annually for up to 5 years. In many ways, the successful establishment of biological control agents in a new environment faces the same problems as the establishment of rare species in nature conservation. This latter discipline has shown the importance of having the release sites closer together than the dispersal range of the beneficial. Previously, biological control workers have tried to cover as wide an area as possible with their release points. If these are sufficiently close together, then any sites where some temporary event causes establishments to fail may be recolonized from more successful sites. Release sites must be monitored for about 10 years to check whether the pest is declining. It is also important to ascertain that the biological control agents are not themselves a new resource for indigenous natural enemies, such as hosts for predators and parasitoids (or hyperparasitoids, i.e. parasitoids of released parasitoids).

The control of cottony cushion scale in California by the vedalia ladybird beetle in the late 1880s has already been mentioned as a landmark in early biological control. By contrast, one of the most recent successes is the control of cassava mealybug in Africa (Herren and Neuenschwander, 1991). This shows the contribution to success made by research before release of the chosen agent. The cassava mealybug was accidentally introduced into Africa in the early 1970s. It was first found in Zaire and rapidly spread to 30 African countries. Cassava originated in South America, so it was here that the search for potential biological control agents was concentrated. The mealybug was located as a common insect in northern South America, but its natural enemies there failed to establish in Africa. It was then discovered that the mealybugs in northern South America and Africa were different species. A further search revealed the right species much further south, in southern Brazil and Paraguay. Eventually a small specific parasitoid (*Epidinocarsis lopezi* in the family Encyrtidae) was released. This was not promising in terms of its rate of increase, but the researchers realized that this was outweighed by its ability to locate mealybug-damaged cassava tissue. This enabled it to find the small isolated remnants of the mealybug population bridging the dry season. Here the parasitoid, although unable to keep pace with the pest during the rest of the year, could dramatically reduce the very slow-breeding perennating populations. The parasitoid was first introduced in Nigeria in 1981, was subsequently released at 30 sites, and by 1986 was established and successful in 13 countries.

Inoculation of imported natural enemies is also used to forge 'new associations' (see above) with indigenous pests, although it has often been argued that such pests already have their complex of natural enemies and that there may even be dangers in adding to that complex. An existing enemy might be displaced or interspecies competition between natural enemies increased. Both these outcomes could either improve or lessen control of the pest, but there are two other possibilities. The imported natural enemy could occupy an unfilled niche, when control would probably be improved, or die out, with no change in the level of control. Examples of successful new associations with indigenous pests include the control in Fiji of the indigenous coconut leaf-mining beetle (*Promecotheca reichei*). This was achieved by the parasitoid *Pleurotropis parvulus* from other species of *Promecotheca* in Java (Taylor, 1937). There have been similar successes in the biological control of weeds; for example, of *Leptospermum scoparium* in New Zealand by an introduced scale insect (*Eriococcus orarienses*) (Hoy, 1949). Carl (1982) gives further examples. However, although inoculation seems to have been most successful against imported pests, Waage (1992) believes 'by the next century . . . we will see a substantial increase in classical biological control programmes against indigenous pests'.

Pathogens seem rarely to be used for biological control using the inoculation technique. Soil-inhabiting pathogens, such as some *Metarhizium* species, have been injected to give long-term control of chafer grubs, but generally pathogens are applied to give levels of kill equivalent to insecticides (see below). Perhaps more research is needed to see how pathogens might be used more subtly to introduce disease into pest populations, to flare up as an epizootic when pests get stressed when crowded. This would not be of much value for most crop pests. However, it could be the ideal control for serious problems like locusts and armyworm that spend most of the time in low numbers in wild vegetation and only 'swarm' when food scarcity arises. Swarming might then be preventable by the introduction of pathogens into the wild breeding areas, if it were practical to identify and treat a high proportion of such areas.

Inundation

In this technique, the biological control agent is mass-produced and used against the pest as a single introduction to achieve a high kill (analogous to a pesticide); i.e. an outrageously high inoculum of agent is applied to obtain immediate control without any expectation of a subsequent carry-over to subsequent generations. Indeed, the agent normally dies out soon after application, following the virtual disappearance of the host/prey. It is therefore most often targeted against pests with only one generation on annual crops, as exemplified by the mass release of the egg parasitoid *Trichogramma* against bollworm of cotton in parts of the USA and against the European corn borer in parts of Europe. As the generation time of *Trichogramma* is only about 10 days at 25 °C, and lepidopteran eggs can be produced in vast numbers in a little space, it is possible to have billions of parasitoids ready for release. Many similar releases of *Trichogramma* have also been carried out in developing countries, especially in Latin America (p. 205).

Pathogens are nearly always used as inundative applications. This is mainly because their spread is too slow and limited unless the pest density is very high. Pathogens have the advantage that they can be applied with the equipment the farmer already has for spraying chemical pesticides. They are compatible with most pesticides and are well used in combination with them; it may seem obvious, but there have been instances where it has been overlooked that insect pathogenic fungi may not be compatible with fungicides! Pathogens are particularly valuable for dealing with cases of pesticide resistance or in situations where pesticide use is particularly undesirable. Such situations arise close to harvest and in forestry or pasture where pesticide use is usually uneconomic.

Pathogens are disease organisms, but most can be multiplied outside the insect by fermentation procedures. Unfortunately, their shelf-life is limited and they are subject to rapid destruction in the field at high temperatures, under strong sunlight and from leaf surface exudations from the plant. These are additional features that lead to their use in inundation rather than inoculation.

Table 5.2 shows the features of the various groups of pathogens that need to be considered when planning an inundative application. Commercially, the most successful pathogen has been *Bacillus thuringiensis*, which is used on several million hectares annually world-wide for the control mainly of caterpillars in agriculture and forestry. It is also used extensively in medical entomology for vector control, particularly in Africa. Although a number of strains have been available for some years, more recently the bacterium has become a prime target for genetic modification. Such work has concentrated on the gene responsible for producing the toxic protein. Today, preparations with especial virulence against different target pests are available, but they are increasingly preparations of the toxin without the living bacterium. The increased specificity to target unfortunately adds an additional market limitation. Also, the greater the virulence and specificity of the toxin, the greater is the danger of resistance developing in the target pest; there are already reports of this having occurred (Marrone and MacIntosh, 1993). Resistance of rabbits to myxomatosis happened many years ago. The first documented example in insects was resistance to *Bacillus thuringiensis* in the Indian meal moth, reported in 1979. In 1990, resistance to the same pathogen was reported in the diamond-back moth. Since then, resistance in caterpillars has also been reported elsewhere, including in the USA.

Over 1600 viruses from over 1100 insect and mite species are known, but commercial considerations and registration difficulties have led to very few having been marketed. A virus was used for some time against

Table 5.2 Comparative properties of the principal microbial agents used to control insect and mite pests (from Smits, P. H. and Eilenberg, J. (1992) *Biological Control of Pests: Course Manual*, Agricultural University, Wageningen, p. 144, with permission)

	Bacteria (*Bacillus thuringiensis*)	Nematodes (insect-parasitic rhabditids)	Fungi (Deuteromycetes)	Viruses (Baculoviruses)	Protozoa (Microsporidia)
Host range	Lepidoptera, some Coleoptera and Diptera	Very broad, >1000 insect species	Very broad, but strain specificity	Principally Lepidoptera and Hymenoptera	Very broad
Speed of kill	1 day	2 days	4–10 days	4–10 days	Chronic, >2 weeks
Stability	Spores killed by UV. Crystals persist longer	Unstable to UV and desiccation. Reasonable persistence in soil	Unstable to UV. Good spore survival in soil	Unstable to UV. Prolonged persistence in soil	Unstable to UV
Capacity to spread	Do not spread	Localized spread within soil. Seeks out host	Some spread by air currents, rain and host movement	Rapid local spread and long distance by passive vectors (birds)	Often transmitted vertically through the egg
Production	*In vitro*	*In vitro*	*In vitro*	*In vivo*	*In vivo*
Costs	Cheap	Expensive	Moderate	Moderate	Moderate
Ideal ecosystem	Field and forest: vectors in ponds and rivers	Soils, composts	Soils, protected crops and other humid environments	Forests, field crops, greenhouses	Forests, pastures
Strategy of use	Inundative	Inundative or inoculative	Inundative or inoculative	Inundative or inoculative	Inoculative
Major infection routes:					
mouth	+++	++	+	+++	+++
anus	–	++	–	–	–
spiracle	–	+	+	–	–
integument	–	+	++	–	+
transovarial/ovum	+	–	–	++	+

Heliothis on cotton; today the widespread use is in apple orchards, particularly against codling moth. Viruses are rapidly inactivated by strong radiation, and are best applied at dusk and to the underside of foliage. Although viral diseases multiply and spread through populations better than bacterial diseases, the process still takes 2–3 weeks. This is why viruses are normally used for inundation rather than inoculation. Viruses are attractive vehicles for genetic engineering, including the incorporation of natural pesticides. Recently, snake venom has been transferred to a nuclear polyhedrosis virus to kill forest caterpillars. The aim is to provide a faster kill than achieved by the virus alone. Indeed, the fast kill is necessary to limit the degree to which the virus can replicate in the caterpillars and thus spread snake venom in the environment.

Commercial fungal preparations (*Verticillium lecanii*) have been used against aphids in glasshouses; since aphids feed on the phloem, the bacteriaand viruses (which have to be ingested by pests from the leaf surface) would be totally ineffective. Until recently, high humidity requirements have greatly limited the use of fungi for inundation, but now they may have a future against a much wider range of insect pests.

Despite good results in the laboratory, spray applications of pathogens have often achieved limited success in the field. The causes of failure include: insufficient pathogen reaching the parts of the plant where the pest is to be found, insufficient persistence outside the insect, poor formulation and hostile environmental conditions, particularly high radiation from the sun and low humidity. However, there has recently been a breakthrough in relation to low humidities. Work with the fungus *Metarhizium flavoviridae* in locust control has shown very good results with correct droplet size and an oil-based formulation. The oil prevents the conidial spores of the fungus desiccating before germination (Bateman, 1992).

Concern has been expressed about the safety of releasing insect pathogens, particularly if they are genetically engineered rather than naturally occurring strains. The concern is particularly that they may infect nontarget organisms, including man. Other potential hazards are poisoning from bacterial toxins, allergy and persistence or multiplication in the environment. These risks are taken into account in registration procedures for commercial products. However, it is a little early to be certain that, with this safeguard, the risks are negligible. The matter is discussed more fully in Lundholm and Stackerud (1980).

Seasonal inoculative release

This is used particularly against fast-breeding pests of short-lived crops, and thus has found its major use in greenhouses. A large number of natural enemies, often obtained from specialist suppliers, are liberated to obtain rapid control, but also then to hold pest populations in check for as long as possible. Before release of natural enemies, the pests may even be introduced into the greenhouse to ensure their presence in an even distribution. This is to prevent the biological control showing spatial density dependence (p. 90). The method is frequently combined with chemical control. One pesticide spray may be used towards the end of the season to clean the crop of insects before sale and/or to cope with any gradual loss of biocontrol effect. Natural enemies and protocols for their use have become available during the past 20 years for all the principal pests in European greenhouses, i.e. aphids, whitefly, red spider mite, western flower thrips, mealybugs and leaf miners.

Figure 5.4 illustrates the differences of principle between inoculation, inundation and seasonal inoculative release as biological control techniques.

Conservation

For many decades the inoculation method, very largely as a single control method, predominated. There is now increasing effort towards conserving and enhancing the action of indigenous natural enemies, particularly in relation to their integration with other pest control methods. Conservation has two main aspects:

1. to eliminate or reduce the destruction of natural enemies when pesticides have to be applied; and
2. to manipulate the environment to provide requirements of natural enemies that would otherwise be absent or insufficient.

This latter approach is often called 'habitat modification', and more details are included in the section on 'Biodiversity' which follows. The integration of natural enemies with pesticide use has been discussed earlier (p. 75).

Biodiversity in IPM

The convention on Biological Diversity (UNEP, 1992), recently prepared during the Rio Conference in Brazil, defined biodiversity as 'the variability among living organisms from all sources including *inter alia* terrestrial, marine and other aquatic ecosystems and the ecological complexes of which they are a part; this includes diversity within species, between species and of ecosystems.' Thus the term encompasses different genes, species, ecosystems and their relative abundance.

Habitat destruction and species overexploitation are the main immediate causes of biodiversity loss. The underlying problems are well known:

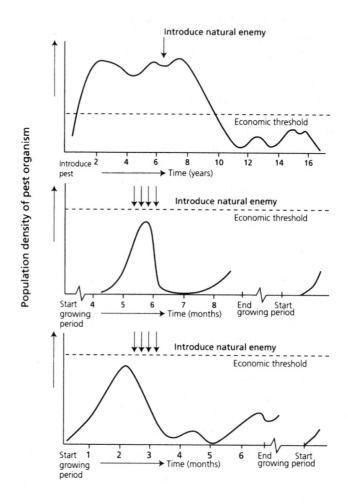

Figure 5.4 *Different biological control strategies. Top, inoculative releases; middle, inundative releases; bottom, seasonal inoculative releases. (From van Lenteren, J.C. and Smits, P.H.,* Biological Control of Pests: Course Manual, *Agricultural University, Wageningen, p. 114, 1992, with permission.)*

1. development pressure, arising from, for example, population growth and economic development;
2. actions economically rational in the short term, or for an individual or country acting alone; however, they are not necessarily sensible for the community or for the world as a whole; also, they m not take account of the needs of future generations;
3. inappropriate policies, like fiscal and other incentives for deforestation, or distorting pricing policies in the agricultural sector.

Definitions of diversity

The simplest concept of diversity is 'richness', the number of species per the number of individual organisms. Biodiversity, however, is a more recent term that relates species number as much to locality as to the composition of a given number of individuals. The concept of 'equitability' is often added to that of 'richness' in assessing diversity. Equitability places more emphasis on the more common organisms and thus the relative frequency of the species in the mixture. Equitability is less important than richness in those aspects of pest management concerned with the preservation of genetic resources. By contrast, it is clearly very important in relation to how far biodiversity impinges on pest problems in farmers' fields.

The fundamental value of biodiversity

A reduction in biodiversity may or may not affect biological productivity or ecological processes. There is no direct and obvious link between the diversity of ecosystems and their importance in maintaining essential global processes. Such processes include biogeochemical recycling, climate regulation and maintenance of soil fertility and water quality. It is not known how far biodiversity can be reduced before crucial ecological processes are affected. Biodiversity is not synonymous with 'biological resources', although it is a characteristic of them – a distinction unfortunately rarely made.

It is a common belief that natural ecosystems are characterized by a great diversity of plant and animal species. The simplification of the ecosytem inherent in agriculture has been coupled with the assumption that pest outbreaks represent ecological 'instability'. This coupling has been used to justify a dogma, deeply rooted in the literature, that diversity begets stability, and strongly expressed in the following quotation: '... the more diverse and complex the ecosystem the more stable it is. And yet we tremble over our wheatfields and cabbage patches with a desperate battery of synthetic chemicals, in an absurd attempt to impede the operation of the immutable law we have just mentioned' (Goldsmith *et al.*, 1972).

Experimental tests of the dogma have shown, however, that the 'law' is not 'immutable', and that the correlation between diversity and stability often fails (van Emden and Williams, 1974). This is because it is the 'quality', and not just the 'quantity', of the diversity which is related to stability. Contrasts between simple and complex ecosystems need to be put into the context of how ecosystems evolve from simple to more diverse up the seral succession. Early in plant succession, floral diversity is low but productivity of biomass is high. Annual crops are a good example; a definition of agriculture might be that it is an attempt to halt the natural process of plant succession. There is a rapid arrival of organisms ('pests') which can advance succession by opening up spatial heterogeneity of the flora to allow the ingress of new producer species. These species, as a consequence of food conversion processes, then channel biomass into respiration. This is the first step in the stabilizing influence that more mature ecosystems have on the less mature. This direction of movement continues with more, but less rapidly reproducing and mobile, herbivores and species from higher trophic levels (i.e. carnivores). It is therefore inevitable that ecosystems evolve, though increasingly slowly, towards greater diversity until a maximum is reached. Adding new species *per se* will, if anything, lead to decreased stability if they merely replace existing biomass through competition (see p. 95 in relation to biological control). Thus simple systems with a limited inflow of energy may be perfectly stable, and equal community organization (expressed as stabilizing the productivity:respiration ratio) will probably be more stable the lower the diversity. The key point is that ecosystems evolve towards both increased stability and increased diversity; the two have a parallel evolution but not a cause and effect relationship. However, because diversity increases the range of outcomes from which evolution can select for greater stability, the chances are greater that a stable system will be more diverse than a simple one. Thus evolved diversity, once destroyed, cannot easily be recreated. However, with due regard to 'quality', man can sometimes add small amounts of the right diversity (habitat management) as a potentially potent force for improved pest management.

The effects of agriculture on diversity

Agriculture requires alteration of the natural habitat to fit the needs of the crops or animals to be produced. Systems of plant cultivation and intensive animal production, both in developed and developing countries, have led to the loss of many plant and animal species. Sometimes they have also done irreparable damage to the land. However, in Africa there are still extensive areas where the natural habitat has been largely preserved in the presence of domestic livestock kept by pastoralists or ranchers.

Some 850 million ha world-wide have been converted to agriculture in the past 100 years. Of these, 90.5 million ha were converted in Africa between 1920 and 1978 (Richards, 1984). The process is ongoing, and cataclysmic forecasts of the rate of loss of modern species regularly appear. One such prediction for the tropics is a 25–33% loss of plant species in the next 30–40 years (Batten, 1983).

The population explosion in Africa demands intensification of production, greater use of land-saving technologies and provision for increasing pressure on the land. Increasing cropping intensity should receive high priority to save land for other uses. This can best be achieved through better management of diversity, including in the temporal and spatial arrangement of cropping patterns.

Today, we face a projected dearth of new groups of synthesized pesticides. The realization that the plant kingdom is a largely untapped source of new molecules with biological activity and short persistence in the environment has led to a great interest in exploiting the plant diversity still remaining, but also rapidly shrinking, in the tropics. A recent review (Simmonds, Evans and Blaney, 1992) lists 58 plant families that between them show about 180 genera containing compounds with insecticidal, antifeedant or insect growth regulator activity. The families Asteraceae, Compositae, Labiatae, Leguminosae, Meliaceae (including the well-known neem tree) and Solanaceae are particularly well represented in the list. Yet this list is clearly only the tip of the iceberg whenne compares the genera that have been investigated with the diversity that is available.

The traditional approach to plant breeding has been to produce cultivars that are genotypically and phenotypically homogeneous. This has resulted in the production of high-yielding cultivars well suited to intensive farming, but with a very narrow genetic base. Thus approximately 70% of US maize is probably based on as few as six inbred lines (Simmonds, 1979).

Hybridization continually produces different combinations of existing genes. As a result, after 50–70 years of breeding, further progress becomes slow and new adaptability (e.g. host-plant resistance) is unlikely unless new genetic variability is introduced into the breeding program. This has to come either from wild relatives of crop species or from the many land races used by subsistence farmers. The genetic variability tied up in these wild relatives and land races is phenomenal; they are naturally outcrossed and adapted to local pests, diseases, climate and soil environment. However, the problem now is the rapid erosion of this great resource of genetic variability. A few crop cultivars are replacing the myriads of land races. The migration of people from the land to urban areas means loss of locally adapted material (Sattaur, 1989). Drought and famine can result in the consumption of seed stocks. Once plant genetic material has been lost, it can never be recovered.

The International Rice Research Institute (IRRI) in the Philippines predicts that the annual world production of rough rice must increase at 1.7% a year by the year 2020 to meet the projected demand. High-yielding varieties already represent 50% of the rice grown in South and South-East Asia. Although the incorporation of resistance to major pests has progressed, the first distributions were bred for high yield under high inputs. Pest and disease problems are often more intense in these cropping systems than in the traditional ones; moreover, the upright crop habit suppresses weeds less effectively than do traditional varieties. Meanwhile, the widespread replacement of traditional varieties has led to a considerable loss of genetic diversity.

Soil cultivation is another change wrought by agriculture, and a particularly devastating one. After all, its aim is temporarily to reduce floral diversity to zero. For one plant species in the UK, *Pulsatilla vulgaris*, ploughing was responsible for its disappearance at 25 out of the 39 sites at which it has become extinct (Hawksworth, 1968).

Pesticides have also been often blamed for loss of diversity, especially for the loss of species of conservation interest. The older organochlorine insecticides (e.g. DDT, lindane, aldrin, dieldrin, endrin) used extensively in the 1950s and 1960s were threats to beneficial insects. Herbicides have dramatically reduced the diversity of annual plants in agricultural areas and therefore also the food of many invertebrates, including beneficial insects. There is, however, no evidence that pesticides have affected biodiversity of arthropods in untreated areas in the way they have reduced the populations of some birds and mammals. Populations of arthropods that are highly dispersive recover rapidly, even on crops if pesticide use is stopped (Burn, 1988).

'Islands' of uncultivated land are important in maintaining biodiversity in otherwise intensively farmed landscapes. In the UK such islands often take the form of hedgerows at the field edges. Great concern has been expressed at the rate of hedgerow removal (7000 km between 1946 and 1970 (Hooper, 1984)) with the enlargement of fields and the replacement of hedges by post-and-wire fencing. Hedges provide a refuge for the biodiversity associated with the woodlands that preceded agricultural development in the UK.

The question that arises in relation to pest management is: 'How much biodiversity needs to be retained in the agricultural landscape as a general principle?' Here the dramatic reduction in the abundance of wild plants in East Anglia in the UK, resulting from hedgerow destruction and use of herbicides, singularly fails to provide an answer. There is no sign that such changes have affected the severity of pest problems or that there has been a need to increase insecticide use. Beneficial insects still occur in the crops. The most likely conclusion is that wild plants are still sufficiently abundant in spite of the declines that have undoubtedly occurred.

Non-crop plants as insect reservoirs

Uncultivated land acts as a reservoir of crop pests in three main ways:

1. Adults overwinter in the leaf litter and other shelter provided in field boundaries and adjacent woodland.
2. Wild plants are sources of alternative food for pests early in the season or between cropping seasons.
3. Wild plants can maintain the pest when different cropping seasons also involve different cropping patterns; for example the summer/

winter alternation of soybeans and wheat across large parts of southern Brazil.

Many pest–weed associations involve weeds related to the crop, for example the ubiquity of grasses and the abundance of cereal crops, wild Malvaceae and cotton, and also wild Solanaceae and potatoes. Destruction of crop-related weeds is a common pest control recommendation, but unfortunately the mobility of insects means that weeds outside a farmer's control may be important. Moreover, many pest insects are sufficiently polyphagous that weeds unrelated to the crop may also be pest reservoirs. For example, *Aphis gossypii* feeds on over 20 unrelated weed species in and around cotton fields.

Associations of sucking insects with weeds related to the crop hold a particular potential for the transmission of crop diseases. Thus in Africa, many examples of disease transmission from wild plants occur over enormous distances. In the USA, beet curly top virus is spread by the leafhopper *Circulifer tenellus* from the western States to Utah and Colorado, where the vector itself is unable to survive the winter (Thresh, 1981).

One positive aspect of the 'pest reservoir' is that wild plants provide an additional arena to the crop for breeding, intraspecific competition and regulation by natural enemies. These will all delay the adaptation of a pest species to host-plant resistance and insecticides.

Of course, non-crop plants also provide a reservoir of beneficial insects. Any pest management program incorporating a component of biological control by indigenous natural enemies depends on the presence of some non-crop plants in the agroecosytem. Alternative prey on weeds may enable natural enemies to colonize the crop earlier and in larger numbers than otherwise (Table 5.3). This is especially important where the crop has been temporarily absent (fallow, crop rotation) or where it has been treated with insecticide.

Sometimes natural enemies require alternate prey (i.e. a second prey species is essential to the life cycle). When this phenomenon is recognized, economic importance can suddenly become attached to an insect previously regarded as 'economically neutral'. An example is the Yponomeutid moth *Swammerdamia lutarea* on hawthorn. This was identified as the overwintering host of *Angitia fenestralis* (an important parasitoid of diamond-back moth on brassicas in the UK) more than 20 years after it was realized that the annual life cycle of the parasitoid could not be completed on diamond-back moth alone (van Emden, 1965).

The practical exploitation of plant diversity

This section will review only those topics where real attempts have been, or are being, made either to utilize existing diversity or to modify habitats

Table 5.3 Examples of increased parasitism due to the presence of alternative hosts (modified from Powell, W. (1986) in *Insect Parasitoids*, (eds J. Waage, and D. Greathead), Academic Press, London, p. 322, with permission)

Parasitoids	Pests	Crops	Alternative hosts
Macrocentrus spp. (Braconidae)	*Cydia molesta*	Peaches	Lepidoptera on weeds
Archytas spp. (Tachinidae)	*Heliothis virescens*	Cotton	Cutworms on flax
Scelionids	*Eurygaster integriceps*	Cereals	Pentatomidae in nearby natural habitats
Lydella grisescens (Tachinidae)	*Ostrinia nubilalis*	Maize	*Papaipema nebris* on giant ragweed
Anagrus epos (Mymaridae)	*Erythroneura elegantula*	Vines	*Dikrella cruentata* on blackberry
Lysiphlebus testaceipes (Aphidiidae)	*Schizaphis graminum*	Sorghum	*Aphis helianthi* on sunflowers
Emersonella niveipes (Eulophidae)	*Chelymorpha cassidea*	Sweet potato	*Stolas* sp. on morning glory
Braconids	*Rhagoletis pomonella*	Apples	Tephritidae on weeds

for pest management purposes. These ideas stem largely from developed agriculture in tempate countries rather than from the tropics. This is no coincidence; it is in the former scenario that plant diversity has disappeared from agricultural landscapes to the greatest extent.

Monitoring

That pest species also feed on wild plants has an obvious use for monitoring the potential for pest status in crops. The classic example must be the monitoring of locust breeding areas to give advance detection of outbreaks. Another example is that, since 1970, winter eggs of *Aphis fabae* have been counted on selected spindle bushes in south-eastern England. Separate forecasts are then provided for each of 16 areas of the likely necessity for chemical treatment of field beans later in the season (p. 74). This is supplemented by a sampling of the aphid populations on spindle in mid to late May.

Trap plants

Trap plants are usually sacrificial crop plants sown at the edges of fields earlier than the main crop area. However, a remarkable example from

Canada concerns the use of sterile brome grass as a trap plant for the wheat stem sawfly, *Cephus cinctus* (cited by van Emden, 1989). The brome grass traps many of the incoming sawflies; they lay their eggs there and the larvae bore into the stems. The elegant feature of this system is that the larvae die in the brome stems before pupating, but after larval parasitoids have emerged. No control of the insect on the trap plants is therefore necessary. Moreover, the grass acts as a filter which converts the pest into beneficial biomass.

Creation of diversity adjacent to the crop

The importance of flowers of wild plants in providing adult food (pollen and nectar) for natural enemies has been recognized since 1938 (Thorpe and Caudle, 1938). Where crops are kept free of weeds and harvested before flowering, flowers outside the crop may be the only available pollen and nectar sources for natural enemies. At least one biological control project has suffered from a lack of suitable flowers for the released beneficials (Wolcott, 1942). The purposeful culture of potted Umbelliferae for placing in cabbage fields has been practised in Russia for cutworm control. Table 5.4 lists some examples where increased parasitization of pests on crops has been attributed to weed sources of nectar. Unfortunately, flower feeding is similarly essential for the adults of several pest

Table 5.4 Examples of increased parasitism due to the presence of adult food sources (modified from Powell, W. (1986) in *Insect Parasitoids*, (eds J. Waage and D. Greathead), Academic Press, London, p. 325, with permission)

Parasitoids	Pests	Crops	Food sources
Tiphia popilliavora (Tiphiidae)	*Phyllophaga* spp. *Lachnosterna* spp.	Various	Nectar from weeds; honeydew from scale insects
Aphelinus mali (Aphelinidae)	Aphids	Apples	Nectar from the honey plants *Phacelia* and *Eryngium*
Apanteles medicaginis (Braconidae)	*Colias philodice*	Alfalfa	Nectar from weeds; honeydew from aphids
Aphytis proclia (Aphelinidae)	*Quadraspidiotus perniciosus*	Orchards	Nectar from the honey plant *Phacelia tanacetifolia*
Various	*Malacosoma americanum* *Cydia pomonella*	Apples	Nectar from weeds
Lixophaga sphenophori (Tachinidae)	*Rhabdoscelus obscurus*	Sugar cane	Nectar from weeds (*Euphorbia* spp.)

species, including some root flies. Eight *Anthriscus sylvestris* plants per meter of field boundary produce sufficient nectar to feed at least 2000 cabbage root flies before oviposition (Coaker and Finch, 1973).

The past 10 years have produced some exciting new research aimed at cereal farmers in the UK. Since it was shown in 1985 that the grey partridge has 2–6 times larger broods in cereal areas where a 6 m strip at the edge of the crop is left unsprayed, a much reduced spraying schedule has been developed for farmers who seek to make profits from partridge shooting. Except when certain highly selective herbicides are sprayed (e.g. for control of grass weeds), the outside boom of the spraying machine is shut off at the edge of the field (Figure 5.5). The loss in cereal yield is not great since crop yield from the headland is always lower than elsewhere. It is also unimportant when set against the profit from shooting (Sotherton, Boatman and Rands, 1989). It is quite clear that the flowering headlands sustain high numbers of predatory arthropods and are heavily visited by Syrphidae, which are attracted to the flowers. Any effect of enhanced biological control further into the crop still has to be shown. There is a danger that the natural enemies could be held in the field margin by the abundance of food, and not disperse into the rest of the crop.

Field boundaries are the preferred overwintering sites for ground beetles, especially those boundaries that combine good drainage and grass

Figure 5.5 *Cereal field with selectively sprayed 6 m margin at left of picture (photograph courtesy of Dr N.W. Sotherton).*

Figure 5.6 *Raised bank sown to* Dactylis glomera *running across a cereal field (photograph courtesy of Dr N.W. Sotherton).*

cover. This has made it possible to design simple 'mini-hedgerows' under post-and-wire fencing. Nothing more complicated is involved than a ridge sown with grasses, *Dactylis glomerata* being particularly suitable. These strips can develop overwintering predator densities approaching 1500 m^{-2}, greater than in the best natural habitats. Since the even dispersal of predators extends for only 100 m from the boundary, experiments have continued with wider ridges at 200 m intervals across cereal fields (Figure 5.6). These within-field ridges are grassy banks 0.4 m high and 1.5 m wide. They do not extend all the way to the field margins, in order to allow machines to move from one side to the other (Wratten, van Emden and Thomas, 1996). An economic evaluation suggests that the costs over 3 years (including yield lost from the area devoted to the strips) would be doubly recouped by 1 year with no spraying against aphids.

Further work is being developed on the drilling of pollen and nectar plants across and around fields (Wratten, van Emden and Thomas, 1996). Here the emphasis has been on the American plant *Phacelia tanacetifolia*, which is very attractive to beneficial insects, and a favorite with bee-keepers. Preliminary results are encouraging in that fields with added *Phacelia* show higher syrphid egg numbers and lower aphid numbers per plant than control fields.

Provision of alternate prey

Where it is known that the annual life cycle of a natural enemy is dependent on a switch of hosts, it can be very simple to make the necessary addition to the local diversity.

The famous practical example here stems from the vineyards of California, and concerns the grape leafhopper, *Erythroneura elegantula* (Doutt and Nakata, 1973). The effective egg parasitoid *Anagrus epos* cannot use eggs of *E. elagantula* for the overwintering generation, because this leafhopper overwinters as an adult. However, the parasitoid will switch to the blackberry leafhopper, *Dikrella cruentata*, which does overwinter in the egg stage. Thus the planting of blackberries was the only specific and purposeful diversification needed to establish biological control of grape leafhopper in the vineyards.

Manipulation of within-crop diversity

Undersowing

Undersowing is practised by farmers for improving soil fertility and to provide winter feed for cattle. Any pest management contribution is therefore an added bonus and would not conflict with farmers' interests as would recommendations to retain weeds in their fields.

In a study of cabbage, with or without clover in the inter-row spaces, cabbage root fly eggs were consistently reduced by 25–64% in the intersown plots (Ryan, Ryan and McNaeidhe, 1980). Laboratory experiments suggested that intersowing had not deterred oviposition; neither did pitfall traps provide evidence of higher numbers of epigeal predators. However, predator exclusion experiments revealed that the presence of clover alone reduced oviposition (by 18%), and that predators further reduced egg numbers to 41% of control values.

Potatoes undersown with perennial rye grass have been found to have aphid populations reduced by up to 66% compared with plants in bare ground. Since the number of colonizing aphids was not affected by undersowing, increased mortality from natural enemies was suspected.

Undersowing has also been used specifically to manipulate natural enemies. Rye grass undersown in wheat was purposefully infested with *Myzus festucae*, followed by a release of the parasitoid *Aphidius rhopalosiphi*. Thus a parasitoid population was established on the rye grass before *Sitobion avenae* invaded the wheat in the spring (Powell, 1983). Populations of the pest aphid on wheat were smallest on those plots which had developed the largest *M. festucae* populations in the spring.

In Switzerland, major improvements in IPM for vineyards with adequate precipitation have been effected by establishing a semi-permanent 'green cover' between the rows. An alternating mowing regime maintains a constant supply of flowering plants on 50% of the ground surface. Floristic (104 species) and faunistic (about 2000 arthropod species) diversity of the vineyard has been greatly increased. Insect pests have remained at a low level, accompanied by a significant increase in beneficial and 'economically neutral' arthropods (Boller, 1992).

Intercropping

Traditional farming has many contrasts with modern farming. The former is characterized by small farms, polycultures, heterogeneous germplasm, little or no artificial fertilizer or other agricultural chemical input, minimum tillage and often varying periods of fallow. The primary concern of the traditional farmer is to produce enough food to feed his family until the next harvest.

Growing two or more useful plants simultaneously in the same area is common practice with small farmers in the tropics. When the crops are intermingled it is called 'mixed intercropping' and where planted in rows, 'row intercropping'. The advantages of polycultures over monocultures are agronomic, socio-economic and nutritional.

Most of the food consumed in Africa (e.g. from 80% of the cultivated area in West Africa), tropical Asia and Latin America is produced in such systems. A mixed or intercropping regime provides a greater total land productivity as well as insurance against the failure or unstable market value of any single crop. In addition, intercropping systems may improve soil fertility, provide alternative sources of nutrition and reduce the incidence of pest attack, reducing pest control costs.

The International Centre of Insect Physiology and Ecology (ICIPE) commenced extensive studies on the effects of intercropping on pest status in 1978. Eight different systems were involved, and the target pests were sorghum shoot fly (*Atherigona soccata*), the stem borer *Chilo partellus* and the cowpea pod borer (*Maruca testulalis*). The experiments showed that, when two crops of similar type and particular insect host range were intercropped, colonization and population increase of pests was favored. However, intercropping with non-host plants caused a considerable reduction in the incidence of most insect pests on the host-plant species.

The ICIPE researchers concluded that scientists (particularly in developed countries) must recognize that there are sound reasons for maintaining the concept of intercropping. After all, it is based on the indigenous knowledge of the farmers in tropical sustainable agriculture. Ignoring this indigenous knowledge and introducing monocrops on large-scale farms in Africa (as happened during the oil boom in Nigeria) quickly led to environmental problems (erosion) and high pest damage, in spite of high inputs of fertilizer and pesticides.

Combinations of tall and short annuals such as sorghum with cowpeas, or maize with beans, are traditional cropping patterns in the tropics. These traditional systems have a low risk of pest outbreak, although the reasons have rarely been elucidated.

The pest control effects of intercropping have been termed 'associational resistance', and two hypotheses have been proposed:

1. *The natural enemy hypothesis.* The diversity of polycultures and micro-environmental effects will result in a greater abundance and diversity of natural enemies than occur in monocultures.
2. *The resource concentration hypothesis.* The total concentration of attractive stimuli for a pest species will be determined by the number of host-plant species present, the relative preference of the pest for each and their density and spatial distribution, and interference effects from non-host plants. Thus recapture over 25 m of mated onion fly females at traps baited with host plant odor was reduced by 80% when the area was grassy as compared with bare soil (Judd and Borden, 1988).

A survey by Risch, Andrew and Altieri (1983) of the effects of intracrop diversity on almost 700 herbivorous species showed a decrease in numbers in 53% of the examples, increased numbers in 18%, no change in 9% and variable responses in 20%. Many of these examples came from the literature on intercropping. In only about 10% of the reports were any ecological mechanisms that might account for the observed differences assessed. In even fewer cases was the impact of natural enemies measured, but in these cases the effect proved to be negligible.

The main effect of intercropping in lowering pest numbers is therefore probably the disruption of visual and odor cues, and the reductions seem to be more effective for host-restricted than for polyphagous pests. However, when intercropping systems have been subject to detailed study by entomologists, other factors have also been identified. The various types of intercropping effects on insects are listed in Table 5.5.

Intercropping maize and beans reduced numbers of leafhoppers (*Empoasca kraemeri*) and leaf beetles (*Diabrotica balteata*) on beans. The reduction was greater when maize was planted before beans than in simultaneous plantings (Altieri, van Schoonhoven and Doll, 1978). It was suggested that the absolute height of the maize plants might be the important determinant, perhaps through shade or effects on air currents.

Several workers have stressed the importance of the diversity *per se.* However, disruption of pest behavior is itself one result of diversity. That highly susceptible cowpea varieties can be protected from pests by sowing them among more resistant cultivars (S.R. Singh, personal communication) may show how sophisticated the cues can be. Researchers using a mosaic of cultivars in screens for host-plant resistance to pests often find it hard to get similar pest pressure to that on crops in the surrounding area.

Nevertheless, in spite of the evidence that even intracultivar diversity

Table 5.5 Overview of intercropping effects on insect pest populations (from van Emden, H. F. and Huis, A. (1992). *Biological Control of Pests: Course Manual*, Agricultural University, Wageningen, p. 144, with permission)

Colonization and establishment
  Visual effects
    crop background
    phenotype
    wider spacing
    camouflage
  Olfactory effects
    masking host plant
    masking host insect
    repellent chemical stimuli

Population development and survival
  Microclimate
    humidity, shade
  Feeding
    quality
    confusing olfactory stimuli
  Natural enemies
    favorable habitat
    alternative host/prey
    supplementary food
  Dispersal
    trapping
    physical interference
    loss of dispersing individuals

can have pest management potential, the choice of partners for intercrops remains important in relation to shared pests. Only when a non-host was included in maize, bean and squash di- and tri-crops, did individual beetle pests show reductions; otherwise increases in pest incidence were observed (Risch, 1980). Cotton and maize together can promote *Heliothis* damage on cotton. However, clever timing can produce the opposite effect. If the emergence of maize tassels coincides with bud formation on cotton, then the attractive tassels cause *Heliothis* to oviposit principally on the maize. On maize, if several larvae enter a cob, they then cannibalize each other, and the insect does not develop as a serious pest (Reed, 1965).

Cultural control

Cultural control is pest control by the manipulation of agricultural practices. It was man's chief weapon against pests before the arrival of modern synthetic pesticides, and thus represents most pest control used in the long 10 000 year history of agriculture. If we condense this history into a time span of just one calendar year, cultural control ruled till 6 a.m. on December 29, and already by 8 p.m. the overuse of the 'new' method (pesticides) had stimulated the publication of *Silent Spring*. The *Review of Applied Entomology*, an abstracting journal published since 1912, shows a sudden switch from experiments on cultural control to insecticide trials in the early 1940s.

The reason for the sudden acceptance of insecticides was, of course, that the new insecticides offered a previously unattainable level of control. Given that cultural measures were also often labor-intensive and thus expensive in the developed world, it is not surprising that they were rapidly abandoned once insecticides became easily available.

Conversely, until recently it has been labor and not pesticides that have been cheap in developing countries in the tropics. Thus cultural measures for pest control are still widely practised there, although continuous cropping limits the opportunities.

Pesticides originated for the developed world in the developed world, much of which is temperate in climate. Thus pests have fewer generations compared with the tropics. Where irrigation can be used in the latter to provide crops in the dry season, the long quiescent season characteristic for pests in temperate climates is missing. Although cultural controls usually give inferior control to pesticides, they do greatly reduce average pest densities in a region. Both these differences result in more rapid breeding of insect pests in tropical than in temperate climates. The pest problems in some tropical areas have already shown that they cannot be restrained where cultural controls have been replaced by total reliance on pesticides. Particularly the larger number of insect generations per year has favored the rapid appearance of resistance to insecticides; the same problem has not been so striking in temperate countries. An important principle of IPM for the tropics is therefore the retention of cultural control measures.

A second principle of IPM relevant to cultural control is not related to maintaining existing agricultural practices, but to the IPM consequences of new ones. The contribution that cultural measures can make to pest control becomes evident, not only in purposeful attempts to reduce pest problems, but equally in changes to the pest spectrum that may occur when agricultural practices change for purely agronomic reasons. Both sides of the cultural measures 'coin' are reviewed in the account that follows. The available strategies on which cultural control can be based are (Coaker, 1987):

1. to make the crop or habitat unacceptable to pests by interfering with their oviposition preferences, host-plant discrimination or location for both adults and larvae;
2. to make the crop unavailable to the pest in space and time by utilizing

knowledge of its life history, especially of migration and overwintering patterns;

3. to reduce survival on the crop by enhancing natural enemies or by altering crop suitability.

Sources of cultural controls

Cultivation

Soil tillage

Many pest insects live or pupate in the soil, selecting a suitable depth in terms of temperature and humidity. Other pests live or shelter in crop debris or weeds before a new crop is planted. When ploughing is carried out, insects in both categories may be buried to a depth from which they cannot emerge. Others die of the temporary drought created in the upper soil layers or are exposed on the surface where they are desiccated or eaten by predators, including birds. There is also some insect mortality from simple abrasion on the soil clumps in motion.

The importance of tillage as a pest control measure has been amply demonstrated in many parts of the world, including the tropics, following the introduction of zero or minimum tillage systems. Here seed is drilled into otherwise undisturbed soil, though often a weedkiller is used as a chemical flame gun to replace the weed control effects of ploughing. Where minimum tillage has been introduced, damage from general soil-living pests such as cutworms, wireworms and slugs has increased dramatically. In Europe, wheat seedlings drilled into herbicide-treated pasture became so seriously attacked by mature frit fly larvae, moving from the dying grasses, that the practice of direct drilling into leys was ended (van Emden, 1989).

Little is known about the effects of ploughing on beneficial insects. However, it is known that adult cereal leaf beetles (*Oulema melanopla*) disperse from the fields to overwinter, whereas their larval and pupal parasites remain in the soil and many are destroyed by ploughing in the spring (Carl, 1979).

Compaction

Running a roller over young plants has been used to stimulate tiller production in cereals to save crops badly 'gapped' by stem borers. Rolling also limits between-plant movement by many soil pests. For example, rolling has been used against shiny cereal leaf beetle (*Nematocerus* spp.).

Mulching

Plant materials (milled bark, crop remnants) or even black polythene have been used to cover the ground between the rows of the crop, mainly to suppress weeds. Mulching reduces drought effects, even in plantation crops, and conditions then become less favorable for the reproduction of aphids and thrips. In coffee, one season without mulch may be enough to elevate thrips to pest status; thrips problems can be very serious in un-mulched cowpea but rarely in mulched crops. The damper conditions also favor natural enemies moving on the soil surface. However, care needs to be taken with the choice of mulch so that it is not a source of plant pathogens for the mulched crop.

Irrigation

In dry seasons or areas, particularly therefore in parts of the tropics, irrigation inevitably means that the crop is the only lush green vegetation in an otherwise barren environment. Pest incidence on cotton sometimes rises dramatically after irrigation, and irrigated crops become a 'magnet' for generalist herbivores. None the less, natural enemies may also benefit from the continuity irrigated crops provide. For example, the irrigated cotton crop in California and Peru enables a bollworm/natural enemy system to persist. This then becomes of equal value to the rain-fed crop in reducing the importance of this pest.

The extreme example of using irrigation to reduce pest numbers was probably the attempt in France many years ago to deal with *Phylloxera* by flooding the vines. Paddy rice, of course, is a crop where water levels can be controlled; raising levels can suffocate Lepidoptera eggs or drive active stages off into the water where they are taken by aquatic predators. In China, ducks may be put into the paddy for this very purpose.

Also, the swelling of soil particles in irrigated crops can kill soil pests with pressure. Additionally, irrigation changes crop physiology to make plants less suitable for thrips and aphids.

Stopping irrigation to give early cotton boll maturity reduces over-wintering numbers of the bollworm *Pectinophora* and restricts the development of late-season populations of whitefly (Matthews, 1989).

Manuring

That well-manured crops are 'resistant' to pests is one central belief of the 'organic farming' movement. This belief is far from erroneous, for good fertilization induces maximum crop tolerance to pests by promoting rapid, healthy plant growth.

First, this shortens any restricted susceptibility window. For example, the cotton stainer, *Dysdercus*, can only breed successfully during a short period in the development of the cotton boll (Pearson, 1958). Most stem-boring larvae, after hatching from the egg, need to enter stems young enough to be penetrated. Yet these stems must also be sufficiently mature to have sufficient suitable internal tissue to sustain the larva for its development.

Secondly, rapid healthy plant growth is essential for plants to achieve maximum compensation for damage that does occur. Here an example is compensation by tea for damage by shot-hole borer, *Xyleborus fornicatus*. Provided nitrogen fertilization is adequate, the weakened branch becomes supported by rapid production of new wound-healing tissue over the gallery entrance (Judenko, 1960).

Thirdly, it is probable that even allelochemical defence mechanisms are maximized. Fourthly, the uniformity and high foliage density of a well-manured crop can reduce colonization by aerially dispersing pests, such as chinch bugs and aphids.

However, just as fertilization promotes plant growth, so it can equally promote the population of small, fast-breeding insects such as aphids, leafhoppers, thrips, whitefly, leaf miners and mites. Nitrogen availability in plants has been recognized as a major limiting factor for many herbivorous arthropods (Southwood, 1973). Cotton agronomists will often tell you that they can identify high nitrogen plots in their experiments by the vastly increased numbers of whitefly. Such disadvantages of nitrogen fertilization, especially in relation to sucking insects (which are limited by levels of soluble nitrogen in the sap stream), can to some extent be minimized by preventing any deficiency of two other nutrients, potassium and phosphate. The former is especially subject to loss from leaching, with the result that nitrogen and potassium are often out of balance unless supplementary potassium fertilization is given.

Sanitation

Destruction of crop residues
This is probably the most important single cultural control measure worth including in IPM programs, if applicable. It has been used particularly extensively against stem-boring insects, in crops including rice, banana, cotton, maize, sorghum and sugar cane.

Destruction of crop residues removes and destroys populations that could otherwise carry over to another crop, and removes shade and shelter which many pests need. It is obviously particularly effective where the pest does not have alternative wild hosts (e.g. the cotton bollworm, *Platyedra*). Destruction of the old cotton crop and a synchronized gap before cotton is replanted are mandatory in many countries, including many in Africa. However, the regulation is often neither policed nor enforced.

Stalk destruction is commonly practised in maize against the stem borers *Busseola* and *Heliothis*; the latter, however, pupates in the soil and stalk destruction should therefore be combined with ploughing. Two other crops where crop sanitation is used are cocoa and banana. In these crops, respectively, stems are peeled to control rhinoceros beetle (*Oryctes*) or banana weevil (*Cosmopolites sordidus*); the peelings dry out fast and, once the pests are dead, the peelings can be used as a mulch.

However biologically feasible it might be to break the life cycle of a pest by crop residue destruction, the measure is unlikely to be practical where those same residues have an important place in the local economy. Thus cereal straw (e.g. sorghum) is often a valuable building material, especially in parts of Africa where it is used for fencing, roofing and even walling. A good example is the yellow-headed borer of deepwater rice in Bangladesh, the caterpillars of which tunnel in the long stems below floodwater level. This pest could probably be controlled easily by burning the straw once the floods have receded and the crop has been harvested. However, with a virtual absence of trees on the flood plains, the straw is an essential resource for the local population, and becomes the fuel for cooking the very rice grains the straw supported.

Destruction of standing crop material

Destroying infested plants or parts of plants within the crop ('roguing') is an obvious way of limiting the spread of an infestation, and is particularly suitable for small crop areas. In sisal in East Africa, one contribution to control of sisal weevil (*Scyphophorus*) has been the roguing of attacked dead or dying plants. Similarly, the yellow-headed Cerambycid borer (*Dirphya nigricornis*) of coffee has been controlled by cutting out branches on which the larval frass is visible (Le Pelley, 1959).

With several pests feeding in the fruit of orchard and plantation crops, the plant abscises the infested fruits long before the uninfested ones are picked at harvest. These fallen fruits are often an important source of the pests concerned, and gathering the fruits on the ground and destroying them can reduce pest populations significantly. This can have particular benefit in citrus and coffee against fruit flies and coffee berry borer respectively. The cost of such a laborious exercise has become prohibitive in large citrus and coffee plantations, only to be replaced by the cost of a heavy pesticide program. Other pests in banana and cocoa breed in fallen, rotting leaf and stem material, removal of which is obviously recommended. In Europe, before the breeding of resistant blackcurrant varieties, a major control measure for blackcurrant gall mite was the grubbing and burning of whole bushes. The mite carries the yield-reducing reversion virus, and it was known that a bush with more than 12% buds showing symptoms ('big-buds') could not be saved by spraying pesticides. Control with pesticides is difficult anyhow, for the peak mite presence outside the buds and on the leaves coincides with flowering and pollination. This is because the mites swarm on to the leaves to disperse by phoresy on the pollinating bees.

Destruction of weed hosts

That weeds are important alternative (optional) or alternate (obligatory) hosts for many crop pests has already been discussed under 'biodiversity'. Sometimes pests that are a local problem for a particular farmer can be traced to specific weeds on the farm; such problems can be dealt with by dealing with those weeds. Cotton farmers are often advised to destroy volunteer cotton and Malvaceous weeds around their fields, especially as a control for the cotton stainer bug, *Dysdercus*. However, weed destruction is often impractical and anyway not too effective. This is because many pests are sufficiently mobile that weeds outside a farmer's control are probably as important a source of pests as his own.

Pruning

Many pests of plantation crops are reduced if the density of the foliage is limited by pruning. Particularly citrus and coffee pest management benefits from pruning, and pest problems alleviated include antestia bug of coffee. Where biological control of scale insects is practised (e.g. of helmet scale of coffee), it is advisable not to remove the prunings from the plantation floor until the parasitoids have emerged from parasitized pest individuals.

Mixed cultivation

Intercropping

This has already been discussed under biodiversity (see earlier). It only needs to be added here that shade trees, often planted for agronomic reasons in low plantation crops such as coffee, generally encourage biological control of, for example, antestia bug. Where coconut palms are used as shade for cashew or citrus, the presence of the lower crops encourages the ant *Oecophylla*, a predator of the coconut bug (*Pseudotheraptus wayi*) (Le Pelley, 1959).

Strip farming

The farming of different crops in small strips is still common in developing countries. The practice arises partly from cultivation purely to meet family demand and partly from the break up of larger crop areas by inheritance traditions. Such strip farming systems therefore tend to persist, even without any understanding of their valuable pest-reducing characteristics:

1. A strip of a different crop can act as a barrier to the spread of pests, though of course the choice of crops is important in respect of shared pests.
2. The crops are likely to share unspecialized natural enemies (e.g.

predatory bugs, ladybirds, ground beetles and spiders) which can move if pests begin to build up on an adjacent crop strip.

The abandonment of strip farming in Peru over 40 years ago has been blamed for the bollworm outbreak that followed on the cotton mono-cultures that replaced the older systems, and re-diversifying the cotton agroecosystem there greatly reduced bollworm incidence.

However, strip farming may also suffer from greater pest problems than other cropping systems. This can happen where adjacent strips share important pests (e.g. wheat and maize) or where the pests breed and feed mainly at crop edges (e.g. grasshoppers). In a strip, the edge is much of the crop area.

Trap crops

The aim of trap crops is to concentrate the pest in a smaller area. Here it can be destroyed with insecticide or preferably with a measure to which the pest cannot develop resistance, such as a flame gun, ploughing in or feeding the trap crop to animals. The use as traps of plants other than the crop species has been discussed earlier as an aspect of 'biodiversity'. However, sacrificial plants of the same crop are often used.

Earlier sowing, as of maize for corn earworm, is one technique of making the sacrificial trap plants more attractive than the main crop. The trapping ability of earlier sown plants is often mainly due to the height differential between the trap and main crop plants. The higher early sown plants filter out aerially borne arriving pests. Also, that trap crops tend to occupy the field edges holds pests moving on to the field in the trap rows.

Spraying the sacrificial plants with attractants for the pest is another technique currently gaining interest; this is discussed under semiochemicals (p. 129).

Trap crops need not be specially sown plants. Heaps of cotton seed have been tried, although with very limited success, to trap cotton stainer bugs, and cut banana stems have been laid on the ground to trap banana weevil.

Crop rotation and isolation

Attempts to separate the crop from its pests and diseases represent one of the oldest forms of cultural pest control. The main plantation crops of the world are now grown in continental isolation from their origin, to escape the pest, and particularly the disease, problems of the indigenous situation. By contrast, less dramatic isolations within the distribution range of pests have tended to fail. This is because many pests also utilize wild hosts, and such hosts have been present in the new areas in which the crops were planted.

For similar reasons, and because many pests are highly mobile, crop rotation on a farm often does little more than reduce or delay attack. However, crop rotation can be highly effective in the control of nonmobile pests, especially those in the soil. Crop rotation is still the most effective control for some eelworm problems. The common temperate rotation of cereals, legumes and root crops does not mean that any of these crops are missing on a farm during the season. None the less, crop rotation can be effective against the soil-inhabiting wireworms, chafer grubs and leatherjackets. Also, the various crop midges (e.g. of sorghum and wheat) are sufficiently weak fliers to be controlled by crop rotation.

Rotation relies for success on there being only a few shared pests across the rotation. Thus, of the 50 serious pests of the maize, wheat and red clover rotation, only three are important pests of all three crops.

There is at least one example, however, where a pest benefits from crop rotation, presumably because a rotation-based strain has been selected over centuries. This is the case of the wheat bulb fly (*Leptohylemyia coarctata*). Strangely, this fly does not lay eggs in the soil of its host crop (wheat), but only in fallow soil or in the soil of other crops in the rotation, such as brassicas or root crops. Thus the pest is not a problem where wheat follows wheat, as has become possible with modern herbicides and direct drilling techniques.

Sowing and harvesting practices

Variation in sowing and planting practice
Where conditions allow, particularly in relation to availability of water, a change in the sowing/planting date may have pest control value. Either the crop escapes pest attack (e.g. by avoiding the main oviposition period of the pest) or it may be in a resistant growth stage when the pest appears. Thus delays in sowing wheat have provided very effective control of the hessian fly (*Mayetiola destructor*), which has a very short peak of flight activity for oviposition (Metcalf, Flint and Metcalf, 1951). Similarly, in Indonesia, delayed sowing of rice until after the first flight peak of the white borer (*Scirpophaga*) has been practised. By contrast, early planting of sugar cane, provided this is carried out for a whole region, enables the crop to be beyond its susceptible stage when borers attack.

However, tampering with the normal phenology of the crop to control one problem may make the crop more vulnerable to attack by another. For example, early planted maize to give tolerance to maize streak virus vectored by leafhoppers may lead to increased damage by borers; also bollworm (*Heliothis*) is a greater problem in cotton planted early to lessen damage from *Lygus* bugs.

Thus it is not surprising that changes in sowing date introduced for agronomic reasons can be accompanied by changes in pest problems. A classic example of this has been a general switch from spring-sown cereals in Europe to cereals sown ever earlier in the autumn or even late summer ('winter' cereals). The major crop protection problem this has created is that of the aphid-transmitted barley yellow dwarf virus (BYDV). This is transmitted to spring-sown cereals by aphids arriving in the summer, and the virus does not have sufficient replication time to produce severe symptoms. However, in winter cereals, the plants have emerged in the autumn when a different aphid (*Rhopalosiphum padi*) migrates from wild grasses and introduces BYDV. This early virus infection can cause significant yield losses. A second, less important, consequence of the switch from spring to winter cereals was the appearance, as a new pest, of the stem-boring grass and cereal fly *Opomyza florum*. This is a denizen of wild grasses, but now finds cereal stems at a suitable growth stage when it seeks to oviposit for its overwintering generation in late summer.

Seed and planting rate
In general, close plant cover reduces colonization by passively dispersing insects such as aphids and thrips. There is clearly a 'bare ground' effect, though whether this affects insect orientation or microclimatic variables, including thermal air currents, is unknown. Virus incidence has often been reported as reduced in dense plant stands. A good example here is rosette virus of groundnuts transmitted by *Aphis craccivora* (A'Brook, 1968).

Many crops are sown in excess of optimum density; interplant competition then results in a plateau in yield ha^{-1} over quite a wide range of plant densities. Cereals are a good example where, provided pest attack is on individual plants rather than patches of plants, many plants can be lost before yield is affected. The introduction of precision drilling in crops such as cereals, therefore, has increased the importance of seedling pests such as slugs and frit fly.

Sugar beet used to be a crop requiring extensive laborious thinning to stand, partly because each polygerm seed produced a variable number of seedlings. Together, precision drilling and the breeding of monogerm cultivars have removed the need for thinning. As a consequence, seedling pests (including Symphyla and pigmy beetle) have become of new major importance.

Variation in harvesting practice
Early harvesting can enable crop residue destruction to take place while pests are still in or on the old crop plants, and before they can emerge. Damage to wheat caused by wheat stem sawfly (in terms of lodging of the weakened stems) can be minimized by early harvesting.

In the USA, the early harvesting of cotton is a major pest control measure to reduce bollworm problems. This requires not only early maturing varieties, but also synchrony of opening of the bolls. Cultural measures include uniform planting, early termination of irrigation and the application of leaf desiccants and defoliants.

Conclusions on cultural control

It may be possible to devise cultural control measures with pest control potential, but they may not then be acceptable to the farmer. They may be too labor-intensive, or they may not fit with his yield aspirations or his other agronomic practices. They may also affect other uses he has for the crop, the site, his machinery or his labor.

It is agronomists, soil scientists and cropping systems scientists who experiment with new cultural systems for farmers. Perhaps the crop protection scientist should gather data on these experiments, rather than setting up his own experiments on cultural controls.

Not only may a new and potentially acceptable cultural form of pest control be identified by this approach, but at the very least any escalation of old problems or development of new ones inherent in the new cultural technique would be noticed before widespread use by farmers.

Semiochemicals

Semiochemicals are chemical signals that are released externally by organisms, with information content for other organisms; they are sometimes also called 'behavior modifying' or 'behavior controlling' chemicals. Semiochemicals are called 'pheromones' if they effect communication between members of the same species, and 'allelochemicals' if they effect interaction between members of different species. Table 5.6 gives additional divisions for these two categories.

Pheromones

Pheromones have the particular advantages in IPM that they are highly specific to species, leave no undesirable residues in the environment and are effective in minute quantities. Sources of pheromone used for manipulating insect behavior have been caged insects, extracts from insects or, increasingly, synthetic production of the pheromone itself or a chemical mimic of it. The high specificity of pheromones arises not only from the production by different insects of different compounds (often long-chain unsaturated aldehydes, alcohols or acetates), but also from more subtle variations of the same compound (especially isometry and in rate of release) and precise ratios released of blends of several compounds. In these blends, components that elicit a response on their own are known

Table 5.6 Classification of semiochemicals (Modified from Griffiths, D. C. (1990) *Proc. Br. Crop Prot. Conf., Pests and Diseases, Brighton, 1990*, p. 487, with permission)

Semiochemicals (signal chemicals)
  Pheromones (act on same species)
    sex pheromones
    alarm pheromones
    epideictic pheromones
  Allelochemicals (act on different species)
    allomones (favor emitter)
    kairomones (favor receiver)
    synomones (favor both)
    apneumones (from nonliving sources)

as 'primary components'. 'Secondary components' are ones that elicit no response on their own but greatly enhance the response to a mixture. It is likely that the ease with which primary components can be identified has led to many single-component pheromones being synthesized. These can often be used even if not identical to the natural pheromone, which is perhaps only rarely a single component.

Sex pheromones

The sex pheromones produced by female Lepidoptera were the first insect-produced semiochemicals to be identified (Butenandt *et al.*, 1959). Their potential in IPM is great, though as yet largely unrealized in practice. Naturalists had for long recognized that male moths fly upwind towards females over large distances. However, it was the arrival of the technique of gas–liquid chromatography that first made it possible to prove the involvement of chemicals in the air at very low concentrations. The word 'pheromone' was not coined until 1959, and applied research on pheromones and other semiochemicals had not even begun at the time *Silent Spring* was published. Semiochemicals thus represent an addition to the IPM armory unknown to Rachel Carson.

Since the initial identification of female-produced sex pheromones in the Lepidoptera, such pheromones have been identified in an increasing number of insect orders. It is likely they are used by nearly all insects. After an exploration of the insect orders, it has been the turn of the sexes. There are now an increasing number of reports of male-produced sex pheromones, though these seem to have a much shorter effective range than those produced by females.

So far only the female-produced sex pheromones have been tested in the field for IPM purposes. Most use of synthesized pheromones has been for pest monitoring. This aspect has been described earlier (p. 74), with attention drawn to the limitations imposed by the attraction only of males. However, there are also three possibilities for using sex pheromones directly for pest control; these are as oviposition deterrents, for mass trapping of individuals and for interfering with mate-finding (the 'confusion' technique).

Oviposition deterrents

Relatively high concentrations of female pheromone, perhaps by simulating the competition of high female pest densities, appear to have the possibility of suppressing, or at least greatly reducing, oviposition. However, this approach is still very much at the research level.

Mass trapping

This most frequently involves the use of a high density of pheromone traps, often treated with insecticide. The aim is to remove sufficient males from a pest population in order to reduce greatly the fertility of the females. Unfortunately, the likelihood of success is remote. Mathematical models suggest that a 90% capture of males is required to hold a population at its existing level, let alone reduce it.

This is illustrated by a Norwegian exercise, when pheromone traps for bark beetles were set out in nearly 4000 ha of spruce forest (Lie and Bakke, 1981). The exercise involved several thousand people, and 36 million beetles were killed; however, no reduction in the pest population was effected.

Perhaps the most successful example of mass trapping of males is that of bollworm in the USA. This is because spraying is only needed if more than 10% of the bolls are attacked. Pheromone traps were set out at 12 ha^{-1} in the spring, increasing to 50 ha^{-1} as the season progressed. In one major trial, whereas pesticides were needed on 45% of the fields without traps, they were only needed on 9% of fields with traps.

By contrast, grape berry moth in the USA remains a pest even at quite low densities. When pheromone traps were set out on a 14 m grid and no pesticide was used, grape infestation was reduced to 6.4% compared to 15.5% damage in an area without traps; however, 6.4% infestation is still too high to satisfy the growers.

Confusion

In this technique, mate-finding by the males is disrupted by creating artificial pheromone trails or by saturating the whole crop environment with sex pheromone. Large numbers of fixed sources of pheromone (akin to traps) have been used in the past. Increasingly, however, pheromone is now distributed by spraying encapsulated droplets (which tend to release pheromone for a few days only) or broadcasting hollow polymer fibers sealed at one end and a few centimeters in length.

Cotton, both in the USA and in the tropics, has again been a major target crop for the confusion technique (Henneberry et al., 1981). Between 1976 and 1978, the use of 1.75 cm fibers applied at 7 g ha^{-1} every 7 days was evaluated in California. This evaluation clearly showed an inherent problem in the technique; it becomes less effective as the number of natural sources of female pheromone increases. In 1976, 1200 ha were treated with pheromone and needed no insecticide, whereas other cotton areas did. In 1977, the pheromone treated area was expanded to 9000 ha. As in the previous year, the application of pheromone proved very successful. However, a few fields next to untreated fields, where bollworm densities were high, did need insecticide. In 1978, the area treated with pheromone was further increased to 20 000 ha. This proved a very favorable year for the pest, and sufficiently high populations to force insecticide treatment developed in July or August even on fields treated with pheromone. Nevertheless, pheromone provided a valuable delay in the need for the first insecticide treatment.

Most people probably now accept that the confusion technique is unlikely to be totally reliable as a replacement for insecticides, but that it is very likely to enable a considerable reduction in pesticide use to be achieved. Unfortunately pheromones, partly because of their limited market, are expensive and become uneconomic if pesticide is also needed.

Alarm pheromones

The aphid alarm pheromone (β-farnesene) has been researched extensively. The principal application of applying this compound is to increase the restlessness of aphids, and thus increase their contact with residual insecticides or fungal entomopathogens.

There has also been some interest in exploiting the production of β-farnesene by some plants in plant resistance breeding.

Epideictic pheromones

Such pheromones affect the spatial distribution of members of the same species; aggregation pheromones are the best known in this group. Often both sexes are affected, giving epideictic pheromones a considerable advantage over sex pheromones in pest control.

A practical example comes from cotton. An aggregation pheromone produced by the male boll weevil, but attractive to both sexes, is used for trapping out (Ridgway, Inscoe and Dickerson, 1990). This example of trapping out has been very much more successful than similar attempts for other pests with sex pheromones (see above).

Allelochemicals

Allelochemicals are important factors in host-plant resistance (see earlier) but both volatile and nonvolatile allelochemicals also have considerable potential for manipulating the behavior of pests and beneficial insects. Kairomones, allomones and synomones are terms given to allelochemicals that respectively give some advantage to the receiving species, the emitting species or both.

Volatile allelochemicals

Often, volatile kairomones are produced by herbivorous insects, especially in their oral secretions, their excreta and their constructions (e.g. cocoons) or from damaged plant tissue. These kairomones aid the location of the herbivores by their natural enemies; there is therefore considerable interest in utilizing such volatiles to manipulate natural enemies and increase their effectiveness.

Sex pheromones of herbivores are not infrequently used as kairomones by natural enemies (e.g. by beetle predators of bark beetles). Recent work in the UK has shown that at least one polyphagous aphid parasitoid (*Praon volucre*) can be attracted in large numbers to traps baited with nepetalactone and nepetalactol, major components of the male sex pheromone of many aphid species. Also, the alarm pheromone for aphids is thought to attract hover flies.

Volatile allelochemicals released by plants are widely used as cues by both pest and beneficial insects. The detection by herbivores of their host plants and avoidance of non-hosts is to a large extent governed by plant volatiles. However, it is difficult to see a practical use of these semio-chemicals in pest control, since the very volatiles that would deter one pest species from colonizing a plant may well, and often do, attract another.

More practical potential attaches to the use made by biological control agents of these same volatiles. The general importance of this interaction between plants and particularly parasitoids has been the subject of extensive research in the past decade. For many parasitoids, location from a distance of odors emanating from the host plant of their prey has often been shown to be more important than odors from the prey itself. It is even possible experimentally to trick a parasitoid into attacking the wrong prey species. This can be done by offering the prey together with the plant odor to which the parasitoid is attracted. The plant odor preferences of a parasitoid seem to relate to the plant on which that parasitoid individual itself developed. Experiments involving the transfer of parasitized hosts at various times after parasitization suggest that the preference is fixed during the parasitoid's larval development. The plant preference of emerging parasitoids can therefore be manipulated, even to the extent of 'training' a parasitoid population to favor a particular crop cultivar (Wickremasinghe and van Emden, 1992). Such between-cultivar discrimination, however, only influences choice; in the absence of the preferred cultivar, other cultivars of the crop are usually totally acceptable, unless their volatile spectrum is greatly different. The practical implications are not, therefore, that we can train biological control agents to spread from a point of release to seek a particular cultivar, but that we need to test, while breeding for plant resistance to pests, that we have not altered the volatile spectrum to the point that our new crop cultivar is no longer recognized by indigenous natural enemies. There is increasing interest in 'banker' plants outside the crop or interspersed with the crop in glasshouses to provide reservoirs of beneficials on alternative prey. For these, we need to check that the natural enemies concerned will, after all, readily disperse into an adjacent crop producing different volatiles.

Nonvolatile allelochemicals

Such allelochemicals in plants tend to act as 'antifeedants', i.e. they do not kill the pest directly but inhibit its feeding behavior. This is probably the result of the allelochemical inhibiting the response of the pest's gustatory sensors to phagostimulants in the plant. One useful attribute of many antifeedants is that they are not also repellents – the pests therefore remain on the plant till they starve. The pests are then available as food for natural enemies for a time. Disadvantages are that antifeedants are not usually persistent or systemic when sprayed. They would need frequent application to give an adequate length of time of protection, and areas missed by the spray or new growth appearing after spraying will be attacked.

Partly because of these disadvantages, synthetic chemicals with antifeedant properties have not met with commercial success, and more potential attaches to the nonvolatile allelochemicals produced by plants. These are widely distributed through the plant tissues, a property that can now, theoretically, be transferred to crops lacking allelochemicals by gene transfer techniques.

There is already one widely known and marketed plant antifeedant. This is azidarachtin from the seeds of the neem tree in the family Meliaceae (Rembold and Schmutterer, 1981), a family that also provides other antifeedants. Neem is toxic to insects if they ingest sufficient quantities, but its strong antifeedant effect means that such quantities are rarely ingested. Other families with promising antifeedants include the Labiatae, Polygonaceae and the Piperaceae. Ajuganin, from the Labiate genus *Ajuga* is currently being researched, as is polygodial from the African tree *Warburgia*. The latter is active against aphids, which is a useful property; most antifeedants affect chewing rather than sucking insects.

Antifeedants, even if produced within the plant, do not have a long-lasting effect. Insects will eventually eat the foliage through some form of 'habituation'. Antifeedants cannot therefore simply replace traditional insecticides, and require strategies for their use to take account of their transient effects. Thus, on a laboratory scale (Griffiths *et al.*, 1991), it has proved possible to apply antifeedants to the tops of plants to drive chewing insects down to the lower leaves. Here, damage is less detrimental to yield than damage to the young leaves. On the older leaves the pests could then be killed with a localized application of a selective pesticide.

This experimental approach may be transferable to the field in a modified form known as the 'push-pull' strategy, more arcanely designated the 'stimulo-deterrent diversionary strategy'. Here the modification involves adding a 'pull' towards trap plants to the 'push' away from the crop provided by the antifeedant. This involves applying plant-derived attractive volatiles to the trap crops together with nonvolatile phagostimulant allelochemicals. On the trap crops, pesticides, or preferably an insect pathogen, would be used.

The nonvolatile semiochemicals that could be used to manipulate beneficial insects are usually contact stimuli which arrest the natural enemy. They are usually kairomones produced by the insects themselves, often in their excreta.

Probably the most explored of the contact kairomones of insect origin is the honeydew of aphids. Honeydew releases a volatile breakdown product of tryptophan attractive to natural enemies, but the contact stimulus to arrest natural enemies of aphids is also important. Even in the absence of aphids, natural enemies can be held in an area where honeydew has been applied. This stimulus appears to be rather nonspecific; the honeydew of most aphid species will arrest most natural enemy species. It remains to be seen how far a practical application can be found for this phenomenon. Unfortunately, large amounts of honeydew can interfere with the searching behavior of parasitoids of aphids. The parasitoid becomes contaminated with honeydew and has to spend a considerable proportion (this can be nearly half) of its time in cleaning behavior, seriously reducing the time spent searching.

Genetic controls

Several pest control methods seek to interfere with the reproduction of pests by affecting the sex cells directly and thus inducing sterility. The idea stems from the mid-1950s, and had already been used successfully in one instance by the time *Silent Spring* appeared. This one success is given a high profile by Rachel Carson. It had resulted in the elimination of the species in question, and there were then high hopes for the future of the approach. Indeed, proponents of such techniques have erected the term 'total population management' (Knipling, 1966), with its acronym TPM, to compete with IPM. Not many people know this, which carries its own message. One of the main virtues of genetic controls is that they operate on the 'one to many principle'. This (we think) simply means that treating one individual can transmit lethality to many individuals in terms of future reproductive potential. Why a predator eating an aphid, already carrying three telescoped future generations within it, is not an example of the same principle is not entirely clear. Most genetic control methods are also extremely specific to the genetics of the target pest population. The other characteristic that applies generally to genetic methods is that the treatment can only be applied to a limited number of individuals. Thus, in relation to most natural pest populations, the methods are best used when pest populations are naturally low or after they have been reduced by insecticide application.

γ-Radiation sterilization

This was the method that had been successful before *Silent Spring* and which Rachel Carson mentions with such enthusiasm.

The early successful use of γ-radiation sterilization of males (the 'sterile insect release method' or SIRM) was against *Cochliomyia hominovorax*, the screw-worm fly of cattle. The adult fly lays its eggs in wounds, and the resultant maggots feed on the flesh of the animal (a condition known as 'myiasis', causing severe weakening and loss in weight). This can lead to death of the cattle if they are not treated. Before the SIRM program, the estimated annual loss to farmers from screw-worm was estimated at about US$200 million. The first trials took place on the island of Curacao in the early 1950s, since here there was a finite wild fly population without any problems of immigration. The release of large numbers of artificially cultured males, sterilized by exposure to a 5000 Röntgen unit cobalt bomb, resulted in eradication of the pest across the island in just 8 weeks (Kirchberg, 1955). After the success of this pilot trial, the technique was transferred to the real target on the US mainland. Here the screw-worm population in the southern United States and Mexico was first reduced with insecticide. At the height of the SIRM campaign, more than 50 million flies were reared, sterilized and released every week from a fly 'factory' on a disused airfield.

SIRM is based on the release of artificially reared sterile males to mate with the females in the wild population, matings that result in zero egg production. Knipling (1955) illustrated the theory quantitatively by assuming a wild population had been reduced to 1 million females. This was followed by a release of 2 million males at each generation. In the first generation, sterile:fertile matings would be in the ratio 2:1, producing only 333 333 females in the next generation. Another release of 2 million sterile males would lead to sterile:fertile matings in the ratio 6:1, and to ratios of 42:1 and 1087:1 in F_3 and F_4, respectively. Numbers of wild females would therefore decrease from 2 million and 333 333 to 47 619, 1107 and <1 (i.e. extermination) in the successive generations after commencement of the program. Thus the technique does rely on swamping the wild population with enormous numbers of sterile males. Knipling further listed the conditions he considered were necessary for a successful SIRM program as:

1. a method for mass rearing males;
2. the released males must disperse rapidly through the wild population;
3. sterilization must not affect sexual competitiveness;
4. preferably the females mate only once.

It is perhaps the fact that *Cochliomyia* is an unusual insect in complying with conditions 3 and 4 that resulted in the campaign against it being so successful.

The fly was eradicated successfully from the southern USA, and efforts have continued to prevent reinvasion by the pest from Mexico and to extend the eradication to all areas north of the Panama Canal. Flies are released from the 'factory' to 'intercept' invading females along the USA–Mexico border in a strip 400 km wide and more than 3000 km long (Snow and Whitten, 1979). Sterile males are also provided for the extension of the program. At peak production, the factory can produce 150 million sterilized pupae a week from an input of 50 tons of meat and blood.

However, 90 000 cases of screw-worm were reported from the cleared area in 1972 (Snow and Whitten, 1979). It became clear that continual laboratory inbreeding had weakened the released stock with the result that condition 3 was failing, i.e. the females would now mate less readily with sterile males. Because of this experience, the rearing stock is now periodically revitalized by addition of new wild-collected flies.

The success of the technique was repeated in Africa, when the fly escaped from the western hemisphere into Libya, probably through importation of infected animals. As screw-worm attacks both animals and humans, the rapid build-up of the pest in Libya (to a peak of nearly 3000 new cases per month in September 1991) posed a very serious threat indeed. It was difficult, for political reasons, to involve in Libya the only real expertise in SIRM, which resided in the US. Instead, an international multi-donor program led by FAO was organized to eradicate screw-worm from Libya using SIRM (FAO, 1991). Releases of sterile males started in December 1990, and the last case of screw-worm myiasis was reported in April 1991. Sterile male releases ceased in October 1991 when the program was declared a success.

It was inevitable that the original much-publicized success of SIRM against screw-worm in southern America would lead to trials against other pests in many parts of the world. These programs have nearly always ended in inadequate success, but this research is still being actively pursued, particularly through and at the International Atomic Agency in Vienna.

Various fruit flies have been eradicated from several small Pacific islands, and in 1962 success was reported against sugar-cane borer (Lepidoptera) in Louisiana, USA. Partial successes have been reported against several other insects, including species of mosquitoes, stable fly, cockroaches and tsetse flies.

Next to screw-worm, the Mediterranean fruit fly (medfly, *Ceratitis capitata*) has been the principal target for SIRM. Programs have been attempted in Central and South America, the USA and Egypt, in order of amount of effort. Again, these programs have usually begun with extensive insecticide use to reduce wild populations. However, SIRM has not achieved eradication. SIRM projects against tsetse fly have been funded in Africa since the mid-1970s. In every case, local eradication has been achieved. However, the lack of natural ecological barriers to the movement of the fly would necessitate maintaining a high-cost continuous sterile male release program, and reinvasion always occurred. A recently funded attempt on the island of Zanzibar has a greater potential for success.

When one adds to Knipling's (1955) criteria for success of SIRM (see above) that the male insect should not be economically harmful and can survive the levels of irradiation necessary to achieve sterilization, it becomes obvious that the chances of the spectacular success achieved against screw-worm being repeated with most other insects must be slight.

Any feasibility of SIRM is limited to extremely damaging pests that have the biological characteristics required for success, and are located in an ecologically or geographically delimited area. The high initial cost of the method can then be recouped over years following elimination without additional recurrent cost. As part of several controls in the IPM context, SIRM is probably uneconomic because of the high recurrent cost, especially of the facilities needed to produce sterile males.

The practical difficulties of separating the sexes of sterilized insects mean that SIRM mostly involves the release of both males and females. Efforts are underway to find ways of limiting production to sterile males. This would greatly improve the economic efficiency of any sterile insect facility. It would also avoid problems of sterile females mating with sterile males and reduce any crop damage (e.g. attempted oviposition in fruit by fruit fly) caused by sterile females. It would be valuable to find a female-specific deleterious trait, such as high temperature sensitivity or susceptibility to a chemical. Then the insects destined for sterilization could, at some stage, be exposed to the lethal factor to kill all females.

Radiation-induced translocation

This has three advantages over sterilization. Both males and females can be used, only one release is needed and this needs to be at only half the wild population. So far research on the method has been limited to medical entomology, but there seems no reason why it should not be applied to pests of agriculture.

γ-Radiation is used to create translocations (an exchange of chromatin between chromosomes). The translocated individuals are then mated to produce homozygotes for the translocation. Large numbers are then bred and both sexes are released to mate with the wild population. Half the potential offspring of such matings fail to be produced, as they receive the translocation from only one parent and are nonviable. Half the remaining viable insects carry the translocation in the homozygous form to hand the lethality on to subsequent generations (Curtis, 1968).

Genetic engineering

An analogous approach is to seed the natural pest population with laboratory-reared individuals that carry a deleterious trait (e.g. male sterility) transferred by gene transfer methods. Such traits would, of course, be expected to be eliminated rapidly from the wild population by natural selection. However, geneticists have discovered methods for increasing the chances that such deleterious traits will be maintained in the population. One mechanism involved is meiotic drive. One chromosome, or part of a chromosome, is transmitted to the next generation in a higher proportion than Mendelian genetics would predict, thus assuring its spread. The deleterious gene is inserted in the genetic material involved in meiotic drive (Cockburn, Howells and Whitten, 1989).

Chemical sterilization

Sterilizing chemicals are cheaper to use than radiation. They can be applied to insects in the field without the need to release large numbers of insects reared in the laboratory. Although atomic energy institutes are interested in funding the use of radiation, chemosterilants have the advantage that they can be manufactured and sold by the agrochemical industry. This industry devoted considerable resources to an extensive search for effective chemosterilants in the 1960s and 1970s. Many compounds with sterilant activity were discovered, particularly among alkylating agents and, to a lesser extent, antimetabolites. The former group alkylate hydrogen groups in nucleic acid synthesis, producing lethal factors in the sperm. The typical result is seen, for example, when glasshouse red spider mites are treated with apholate, one of the alkylating agents that was marketed. Matings by the sterile males result in male offspring and nonviable eggs representing the female fraction of the population. Among the antimetabolites, which also interfere with nucleic acid synthesis by competing as substrates for enzymes, flouracil was marketed. However, these early chemosterilants have now been withdrawn.

The alkylating agents and antimetabolites, particularly the former, are unfortunately also mutagenic in man. The concept of treating the field population directly by spraying has therefore been replaced by localizing the chemosterilant. A trap, baited with a specific attractant for the pest in question, is commonly used. This attractant is often a sex pheromone, which is appropriate since it attracts males, the target sex for sterilization. However, other attractants have also been used. For example, attempts to sterilize cabbage root flies involved using traps with both visual (yellow) and olfactory (mustard oil volatiles) attraction, though the technique was never used commercially.

Although the search for chemosterilants has mostly resulted in male sterilants, there are examples, for example amethopterine against Diptera, where the female is the more susceptible sex.

It is clear, therefore, that the development of chemosterilants by industry has not proved a successful commercial venture.

Host-plant resistance

As a principle of IPM, host-plant resistance has particularly valuable properties. It is, to a large extent, compatible with other control measures, and may often show useful synergism when used in combination with other methods (p. 150). Host-plant resistance also requires no special skills or equipment; it often consists in merely the replacement by the farmer of one seed stock by another. It is also a very environmentally friendly form of pest control; many mechanisms of host-plant resistance will have no effect on the environment or human health. Clearly one has to guard against the potential danger that exists in this respect (p. 145).

Although the above properties of host-plant resistance make it valuable in agriculture world-wide, the method has received an especial impetus in the tropics. This is not only because it is cheap and simple for the farmer to use; it is also because it can produce dramatic yield increases in crops suffering high pest damage without necessarily attaining the full physiological yield potential the crop possesses. Such maximum yields are more easily obtained with pesticides. This makes host-plant resistance less attractive to farmers in developed countries, who have the money and technology to use pesticide if they so wish. Host-plant resistance was therefore seen as the key (and originally even as the only) pest control thrust for the international research institutes. These were established in developing countries by the Consultative Group on International Agricultural Research (CGIAR), with plant breeding as the principal method of crop improvement. It is only recently that crop protection scientists at these institutes have been encouraged to research more broadly in IPM.

There can be considerable variation in the degree of susceptibility to pests shown by different varieties of a crop. This has been known to farmers for a very long time. The many local land races of crops grown by farmers in the tropics are selections their ancestors made, over past centuries, as able to yield in spite of intense pest pressure. They are important sources of pest resistance, and it is important that such genetic diversity is not lost as they are replaced with potentially higher yielding varieties.

Host-plant resistance has long been a subject of research and practical exploitation by plant pathologists and nematologists. This is because the early chemical control of these crop problems was rarely as complete or reliable as that of insect pests. Although some pest-resistant varieties have been developed and grown for many years, the real impetus for host-plant resistance to pests has come with the general acceptance of the need considerably to reduce our use of agrochemicals. Host-plant resistance does not appear to be mentioned in *Silent Spring* (p. 41).

The 1951 publication of *Insect Resistance in Crop Plants* by R.H. Painter was the first definitive treatise on the subject. Painter presented a classification of insect host-plant resistance that is still accepted and has practical use. The three types of resistance are:

1. *Nonpreference.* A later alternative name, *antixenosis* (resistance to colonization), is now more commonly used. The rationale of the newer term is that (like the other types of resistance) it is a property of the plant, rather than of the insect. Antixenosis (Kogan and Ortman, 1978) refers to plant properties that reduce colonization by pests arriving in search of food or oviposition sites. 'Nonpreference' and 'antixenosis' are not strictly synonymous. Nonpreference refers to the insect's behavior in a choice situation, whereas antixenosis includes van Marrewijk and de Ponti's (1980) concept of 'nonacceptance' (in a no-choice situation).
2. *Antibiosis* (resistance to biological processes). Antibioisis reduces the survival or 'performance' of the pest on the plant. 'Performance' includes such pest characteristics as generation time, growth, fecundity and longevity. Several of these indicators can be combined with mortality into a single statistic, the 'potential rate of natural increase'.
3. *Tolerance.* This is the ability of a plant to show a reduced response in terms of damage (often measured as crop yield), compared with another, given the same pest burden both in numbers and time.

At first sight, tolerance would seem an ideal form of plant resistance, in that there is pressure on neither the plant nor the pest to counter-adapt.

However, tolerant varieties have the danger that a farmer would become a source to other farmers of the pests and diseases he is allowing to multiply and has no incentive to control. It is sometimes the only practical solution to certain plant diseases, including viruses, but is better replaced by antibiosis if this is available, even though the latter is more likely than tolerance to lead to biotype problems (p. 146). Nor do the phenomena necessarily occur in isolation. Plants with antibiosis often also show antixenosis, and this combination is usually regarded as the form of plant resistance most worth developing once it is identified.

As is discussed later, there are other classifications of plant resistance more used by plant pathologists and related to the genes possessed by both insects and plants. Such classifications are theoretically equally applicable to host-plant resistance to insects as they are to resistance to diseases. However, two points need to be made. First, breeding for plant resistance rarely involves the introduction of a new characteristic; it more usually involves the enhancement of a characteristic already present. Even so-called susceptible varieties present the insect with considerable problems (Southwood, 1973) of tissue hardness, nutrient limitations or unpalatable or toxic secondary compounds. Passively dispersing pest species (which land on most plants and show their acceptance by whether or not they remain) often abandon varieties rated as 'susceptible' at rates of 50–95%. Secondly, pests respond to plant characteristics more tangible than genes. As pointed out earlier, susceptibility in the same genotype can change with time and environment; there are often identifiable anatomical, physiological or biochemical mechanisms involved. One mechanism may require one or several genes in the plant; the same number of genes is not necessarily required in a pest for it to overcome that one mechanism.

The development of plant resistance requires sources of resistance, a technique for identifying resistance and methods for transferring it to varieties adapted for commercial production.

Sources of plant resistance

For most world crops, particularly important sources of genetic variation are the germplasm banks. These are collections of either regional or world-wide variation, kept as small stocks of seeds, budwood, etc., stored or regenerated as necessary to maintain viability on a continuing and permanent basis. In 1973 CGIAR, conscious of the danger of losing genetic diversity as crop land races were replaced by newer varieties, established the International Board for Plant Genetic Resources (IBPGR). This has the aim of collecting together existing genetic diversity and maintaining it in germplasm banks distributed between the international research institutes in the tropics. IBPGR produces a list of world germplasm banks (IBPGR, 1984) which lists over 100 centers covering between them some 60 crops. Another useful list of germplasm banks for 22 crops has been produced by Harlan and Starks (1980). Additionally, the Information Science/Genetic Resources Program at the University of Colorado, USA, maintains a database on the world germplasm collections. Germplasm banks accumulate collections of seeds from farmers' fields, commercial varieties before they become obsolete and cease to be available from seed merchants, and new hybridizations produced by the plant breeders but rejected for commercial development. Increasingly, the 'wider gene pool' is also assembled; modern methods of gene transfer mean that crossing barriers between crops and genetically very different related genera and species have become less serious.

Apart from germplasm banks, commercial seed merchants market a range of varieties bred for characteristics other than pest and disease resistance. Many of these have never been tested for the latter property, and may well show a range of susceptibility. They are well worth testing, especially since a partial rather than a high level of resistance may be what is needed in IPM programs (p. 150).

Another way of obtaining genetic diversity is 'random outcrossing'. Here individual plants from the genetic diversity available are planted with, at intervals across the field, rows of a single agronomically adapted variety unable to produce pollen. For many crops, the plant breeders have produced such 'male-sterile lines'; if not, then manual emasculation will be necessary. Any seed produced on the male-sterile line is a cross between that line and an unknown pollen donor from the other plants on the field, i.e. it is potentially a new variety.

Other techniques for producing variation involve inducing mutations in seeds by irradiation with X-rays or by chemical treatment of the seeds, especially with colchicine or ethyl dimethyl sulfate. Many such mutations are, of course, not viable.

Finally, plants derived from callus tissue produced in cell and tissue culture show considerable genetic variability through the phenomenon of somaclonal variation. It has been possible to find resistance to sugar-cane borer (*Diatraea saccharalis*) among 2000 somaclonal variants from a susceptible cultivar (White and Irvine, 1987).

It has been mentioned that the wider gene pool, especially wild ancestors of present-day crop plants, are a rich source of genes for host-plant resistance. Some of these genes have been retained in peasant land races, but many have been lost in the course of crop improvement over centuries. At first sight it is surprising that man did not retain valuable resistance characteristics as he selected for more productive types. There are, however, several good reasons why this loss of resistance occurred:

Table 5.7 Contrast between some wild and cultivated plants in levels of secondary compounds

Chemical (units)	Plant	Quantity	Reference
2-Tridecanone (μg cm^{-2} leaf surface)	*Lycopersicon esculentum* (tomato)	0.1	Kennedy (1986)
	L. hirsutum	44.6	
Allylisothiocyanate (μg g^{-1} dry weight)	*Sisymbrium officinale*	3660	van Emden (1990b)
	Brussels sprout	167	
	January King cabbage	847	

1. Until this century, by which time much of the selection had occurred, breeding was done in ignorance of the basic principles of genetics. It was not realized that unproductive types might well contain some valuable characters that it would be possible to retain (see 'Yield penalty', p. 145). Loss of resistance was particularly rapid in the early stages of plant improvement, when man faced the initial challenge of converting the small seeds of grasses into 'grains', or tiny fruits and pods into much larger structures.
2. During 'domestication', unpalatable or toxic chemicals had to be greatly reduced during the plant breeding process. These same chemicals deter colonization by insects on the wild ancestors of our crop plants. An example is the taste contrast between wild and cultivated lettuce. Table 5.7 gives some examples of the contrast in quantity of plant secondary compounds in wild and cultivated species.
3. Any resistance in crops that was selected early in domestication may since have been nullified by adaptation in the pest or through the importation of new races of pest species from abroad. Crop plants have been subject to considerable transport and exchange of breeding material across the world.
4. The important pests of a crop today are not necessarily the same as those that were present in the early stages of crop breeding. Importation, changes in agricultural technologies (such as abandonment of rotations) and the side-effects of pesticides on natural enemies of previously non-pest species have all changed the pest spectrum of major crops. It would be possible to list many such changes, even since *Silent Spring*. Major pests today that were unimportant then include *Heliothis*, brown planthopper of rice and whitefly on cotton.

Detection of resistance characters

This usually involves a process known as 'screening', where different lines of a crop are exposed to pest pressure. Preferably, this is done in the field, though sometimes such work is carried out in the glasshouse to achieve greater pest densities. Glasshouse data, however, can often be misleading.

Because many lines may be available for testing (perhaps hundreds), and because germplasm collections can often supply only a few seeds of each accession, the first stage is usually unreplicated 'negative' screening. Here, short single rows of a few plants of each accession are planted among similar rows of a known susceptible variety planted at intervals. Because of the large number of accessions usually involved, the results of the screen tend to be scored on a visual scale (often 1–5), comparing each small row with a nearby susceptible check row. This type of screen is called 'negative' because it can only identify susceptible lines. Lines with low damage scores may merely have escaped attack by chance in the trial. However, a badly damaged line must be susceptible in spite of the small sample and lack of replication. Such lines can then be omitted, and the remaining ones screened for a second, and after further eliminations, a third time by the same technique. For a line to escape by chance three times is an unlikely event, and at this point the most promising lines are selected and multiplied for replicated plot field trials. Chiarappa (1971) has compiled protocols for evaluating such trials for a variety of crop pests, but a further useful criterion is the yield of each line with and without insecticide protection.

From such trials, the most promising few lines should then be tested at a number of sites spanning as far as possible the climatic and soil variation of the areas in which the crop is principally grown.

Transfer of genetic resistance

Once plant resistance has been identified, it may be immediately useful. Partial resistance found in an existing cultivar may be sufficient to introduce into an IPM program. More usually, the resistance is detected in a wild relative of the crop or in an agronomically poorly adapted variety. This then becomes an input into a program to transfer the gene into adapted varieties. The following are common strategies for the transfer of genetic resistance:

Grafting

In perennials, resistance may often be transferred by grafting a susceptible commercial variety on to a rootstock of the resistant line. The classic example here is *Phylloxera* resistance in vines based on American rootstocks.

Plant breeding

Pedigree breeding

This is most appropriate where a single dominant gene is involved. The hybrids of resistant × adapted crosses are self-fertilized for five or six generations, with selection to combine adaptation and resistance.

Bulk breeding

Where resistance is based on several genes, many progenies of hybrids between resistant and adapted parents are selected if they show resistance, and are self-fertilized. Then a number (often 10) of the most resistant lines considered the most acceptable for other characters are selected, and re-hybridized in all combinations. Selection and re-hybridization processes are then continued until both adaptability and resistance are considered satisfactory.

Convergent crossing

This aims progressively to increase the percentage of the adapted parent's genotype. This is accomplished through repeated back-crossing of resistant selections to the adapted parent. Finally, paired lines with different sources of resistance are crossed, then the offspring of these paired crosses, and so on until one final line is produced.

Transgenic plants

Biotechnological methods of gene transfer enable specific genes to be introduced into commercially adapted crop varieties without the lengthy process of traditional plant breeding. They may even enable resistance to be transferred from nonrelated plants (Levin, 1979; Gatehouse, Boulter and Hilder, 1992). In the past decade, these techniques have become routine and reliable. This has stimulated a new interest in producing pest-resistant crop varieties, particularly since the multinational agrochemical companies have identified genetic engineering as a poten-tially profitable diversification of their crop protection interests. Some concerns this raises in relation to host-plant resistance to pests and diseases will be raised later (p. 152). There are also valid ecological concerns that the transgenes might 'escape' by invasion or pollination into plants in semi-natural and natural habitats. Because of these, as well as other, concerns, the release of transgenic crop varieties has become subject to rigorous governmental risk assessment not applied to traditional plant breeding.

There are many approaches to the production of transgenic plants; the methods that have most been applied to transferring desirable characteristics into crop species are described briefly below:

Indirect gene transfer

The most common method used to produce transgenic plants is that of transfer mediated by *Agrobacterium tumefasciens*. This is the soil-borne bacterium that causes crown-gall disease. For genetic modification proce-dures, the T-DNA of the bacterial genes, which codes for enzymes involved in the induction of the galls, is removed to produce 'disarmed' plasmids.

The removed genes are replaced by plant-resistance genes. Material of the adapted variety (e.g. leaf segments, leaf discs, even isolated protoplasts) is then incubated with the disarmed bacteria for 1–2 days. Gene transfer using *Agrobacterium* has not been as successful with monocotyledons (including the world's cereals) or with grain legumes. It appears that these plants do not respond to *Agrobacterium* infection in the same way as most dicotyledons (roughly representing 'broad-leaved plants' as opposed to plants such as grasses).

Direct gene transfer

'Electroporation' is a technique exploiting the fact that many plant species will regenerate from protoplasts, and that protoplasts can be induced to integrate foreign DNA when exposed to a suitable external electric field. A similar effect can be produced with exposure to certain chemicals, including polyethylene glycol ('chemically-mediated DNA uptake'). Another important direct method of plant transformation is 'biolistics'. Here fine 4 μm spherical tungsten or gold particles coated with the appropriate DNA are fired into plant tissues, by an instrument akin to a minute shotgun, at velocities at which the particles can penetrate plant cell walls.

Mechanisms of plant resistance to pests

Often the anatomical, biochemical or physiological mechanism of resistance is unknown or poorly understood. A knowledge of the mechanism can, however, be useful in two ways. First, monitoring the inheritance or segregation of the resistance by assaying for the mechanism is often much quicker than measuring pest populations. Secondly, identifying the mechanism makes possible predictions about the likelihood of problems arising from the use of that resistance character (see later).

Figure 5.7 lists the major mechanisms, linking them to the three resistance categories of antixenosis, antibiosis and tolerance. This book is not the place for an extensive review of the mechanisms of host-plant resistance; however, since not all the titles on Figure 5.7 are self-explanatory, a very brief account (with characteristic examples) is presented below. This is based on van Emden (1987), and also provides appropriate literature references:

1. *Color.* Here resistance has been based on foliage color (e.g. red cabbages) and the UV reflectance of white flowers, both of which are important in determining the attractiveness of crops to some pests.
2. *Palatability at host selection.* The secondary substances that attract some insects to a crop presumably deter acceptance of the plant by the vast majority of other insects. A classic example is gossypol in

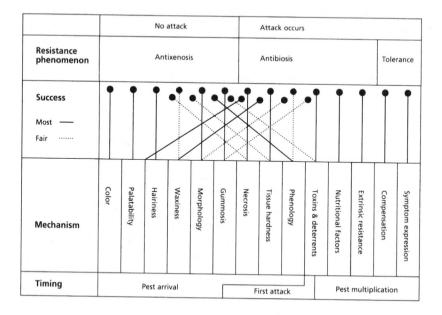

Figure 5.7 *Classification of mechanisms of plant resistance to pests (from van Emden, H.F., in* Integrated Pest Management *(eds A.J. Burn, T.H. Coaker and P.C. Jepson), Academic Press, London, p. 39, 1987, with permission).*

cotton, which attracts boll weevil, *Anthonomus grandis.* Cottons bred for reduced levels of gossypol are unfortunately then highly susceptible to *Heliothis* spp. and blister beetles (*Epicauta* spp.). Levels of the quinone DIMBOA (2,4-hydroxy-7-methoxy-1,4-benzoxazin-3-one) in maize varieties seem to confer some resistance to oviposition by *Ostrinia nubilalis.*

3. *Hairiness.* In the early 1920s, a close relationship was established for cotton varieties between hairiness of leaves and reduced oviposition by leafhoppers. This led to the breeding of resistant cottons based on this character, though doubt was later cast on any causative relationship. Hooked hairs on certain bean varieties trap landing aphids and leafhoppers. Although glandular hairs in the Solanaceae have a similar entrapping effect, they also produce a liquid that hardens on the tarsi and mouthparts of aphids and leafhoppers.

4. *Waxiness.* Several workers have found that glossy-leaved brassica and cereal varieties are resistant to insects compared with the normal waxy types. This effect may not be due to the presence of less wax, but to higher levels of diketones and hydroxydiketones in the cuticular wax of the glossy-leaved types.

5. *Major morphological characters.* The best example here is the extensive cotton breeding program, largely with *Heliothis* as the target. Some resistance has been obtained by breeding in (a) the character of narrow, twisted bracts from the Frego mutant and (b) the nectariless character (referring to the absence of leaf nectaries) (Matthews, 1989). Both characters reduce the attractiveness of cotton to *Heliothis*. Another example is that cowpea varieties with long peduncles and erect pods show some resistance to the pyralid pod borer *Maruca testulalis*. The larvae of this moth penetrate the pods most successfully wherever these are in contact with each other or the foliage.

6. *Gummosis.* The pods of some legume varieties produce gum when damaged; this seems to drown young bruchid larvae attempting to penetrate. In other crops, various exudations (e.g. latex) can also cause insect mortality.

7. *Necrosis.* Hypersensitivity is a form of plant resistance well established in relation to both plant pathogens and pests. Local damage, as caused for example by stylet insertion by a sucking insect, liberates both substrates and enzymes for a polyphenolase reaction. The reaction causes rapid local death (necrosis) of cells that either inhibits or deters further stylet penetration. Such local hypersensitivity is involved in resistance of certain tomato varieties to the nematode *Meloidogyne incognita* and of apple varieties resistant to woolly and rosy apple aphid. The whole-plant necrosis of, for example, some cereal variety seedlings when attacked by stem borers, may be a resistance mechanism against pests. If these arrive randomly in crops with high plant populations, the spread of infestation through the stand is limited, and the loss of isolated plants may not affect yield ha^{-1} (p. 73).

8. *Tissue hardness.* Even minor differences in stem solidity greatly affect susceptibility of cereals to stem borers. Often, for example in resistance of rice to *Chilo*, the silica content of the tissues seems to be an important component of 'hardness'.

9. *Phenological resistance.* Such resistance can result from the rate of development of a particular cultivar or the timing of phenological events such as flowering. Varieties may appear resistant in the field simply because they may be at a less attractive stage for ovipositing adults or less susceptible when pest attack occurs. Such 'pseudo-resistant' varieties of pigeon pea, faba bean and pea have been found for *Heliothis*, bruchids and pea moth, respectively.

10. *Toxins and feeding deterrents.* Resistance can involve inorganics (e.g. selenium), primary and intermediate metabolites (e.g. cysteine and certain aromatic amino acids) and secondary substances (particularly isoprenoids, alkaloids, protease inhibitors, glycosides, flavonoids, condensed tannins and stilbenes). The many examples in this category include DIMBOA in maize resistance to *Ostrinia nubilalis* and the steroidal alkaloid demissine against *Leptinotarsa decemlineata.*

11. *Nutritional factors.* Sucking insects can be adversely affected by nutritional changes in the sap without necessarily also affecting the value of the plant as food for man or his domestic animals. For example, resistance to pea aphid (*Acyrthosiphon pisum*) in three pea varieties has been linked to lower concentrations of free amino acids, and lentil varieties with low nitrogen:reducing sugar ratios are resistant to bean weevil, a chewing insect.

12. *Extrinsic resistance.* This is the interaction of plant resistance with a second factor causing pest mortality, the most common example being biological control. This is an important phenomenon in relation to the role of plant resistance in IPM, and is discussed more fully on p. 151. Two highly crop-specific examples of extrinsic resistance can be cited. A ladybird (*Cryptolaemus*), which normally feeds at cotton leaf nectaries, becomes a predator of *Heliothis* on nectariless varieties. The second example concerns the partially aphid-resistant 'leafless' pea varieties. Ladybirds can maintain a surer footing than on normal pea leaves, from which a high percentage fall (Kareiva and Sahakian, 1990), and thus achieve greater predation of aphids.

13. *Compensation.* Tolerance to pest attack is often based on some form of compensation. For example, when plants or organs compete with one another, the crop elements surviving attack can grow larger. Full compensation for damage can also occur when the attacked source of assimilates is larger than the physiological sink for yield. Thus plants with a high leaf area index such as turnips can compensate in root yield for damage by *Plutella maculipennis.* The reverse situation, when the sink is larger than the source, pertains in many crops with indeterminate growth which produce flowers and even young fruiting structures to excess. Here insect attack reduces the amount of natural shedding without necessarily affecting yield, though perhaps delaying harvest (e.g. cotton).

14. *Symptom expression.* Tolerance may take the form of a suppression of the severity of the plant reaction to, for example, toxic insect saliva. Such tolerance is seen in cowpea varieties tolerant to leafhopper

attack, which normally results in debilitating 'hopperburn'. It has been suggested that tolerance to aphids of certain wheat cultivars, on which infested leaves do not curl, is the result of lower levels of plant growth substances in the leaves.

Problems of plant resistance

Yield penalty

Many mechanisms of plant resistance (including tolerance) appear to involve some diversion of resources to extra structures or production of chemicals. 'Energy or other resources which the plant diverts for defence cannot be used for growth and reproduction' (Hodkinson and Hughes, 1982). It has been argued that any cost of resistance is not significant in the total yield economy of the plant, yet ICRISAT (1980) reported 'we already have some evidence to show that selection of pigeon peas for yield in pesticide-protected trials may produce materials with greater pest susceptibility.' In the same report, data were given on 31 pigeon pea varieties. These data show a highly significant negative relationship between potential yield (i.e. under insecticide protection) and resistance to insect damage to the pods. For 90% resistance, a potential 31% yield loss can be predicted. Partial plant resistance (of the kind compatible with IPM rather than a single control measure) therefore appears to give greater compatibility with yield aspirations.

Health hazard

This is clearly a potential problem, perhaps especially when plant chemicals are transferred transgenically. New or higher levels of existing toxins may be incorporated into varieties destined for consumption by man or his animals (p. 144). There could also be detrimental changes of levels of vitamins and essential nutrients. In recent years, a new cultivar of potato with a high alkaloid content in the tubers was withdrawn in America following illness of consumers. It seems strange that toxicity testing, a legal requirement on the producers of synthetic and bio pesticides, is not necessary for the plant breeder. The USA has, however, nutrient criteria that must be met by any new variety of a crop which contributes 5% or more to the average US diet of any nutrient. These criteria state that the quantity of any nutrient must not have been reduced by 20% or more, and that no toxicant is increased by more than 10% (Spiher, 1975).

Disadvantageous effects on natural enemies

Some secondary plant compounds involved in host-plant resistance can be toxic to predators or to parasitoids within hosts (Herzog and Funderburk, 1985). If not lethal, sterility of the beneficial insect may be induced. This sounds a warning for the transgenic production of resistant varieties; the transfer of such secondary plant compounds has, up to now, been the principal approach. Also, it can be shown that rearing in the smaller hosts on resistant varieties results in smaller and less fecund parasitoids, and that the effect increases in subsequent generations. Such problems again suggest merit in not seeking higher levels of plant resistance than are necessary in an IPM context.

Problem trading

Any anatomical or physiological change in the plant to impart increased resistance to a pest may simultaneously increase its susceptibility to another pest or a disease. Well-known examples are the hairy cotton varieties resistant to leafhoppers but extra-susceptible to aphids (Dunnam and Clark, 1938), and Frego bract cotton resistant to boll weevil but highly susceptible to *Lygus* plant bugs (Tingey, Leigh and Hyer, 1975). Many wilt-resistant lucerne varieties have proved especially susceptible to spotted alfalfa aphid (*Therioaphis trifolii*) (van den Bosch and Messenger, 1973).

The need to 'pyramid' (i.e. combine) resistance to different organisms complicates and slows the breeding of a new variety. Pyramiding may anyway be impossible if the mechanisms are mutually exclusive (e.g. hairiness and smoothness of foliage).

Biotypes

Biotypes capable of exploiting a resistant variety are analogous to many examples of strains resistant to pesticides. A high selection pressure causes a race of the pest previously at low gene frequencies in the population to increase in abundance.

The appearance of biotypes adapted to resistant varieties has been a severe problem with plant pathogens for many years; several hundred such biotypes are known for wheat rust, *Puccinia graminis*.

Tolerant biotypes are always quoted as a major limitation to plant resistance breeding for insect control. The most frequently quoted example is the release of the rice variety IR26, which showed a high level of resistance to brown planthopper and green leafhopper. This new variety

Table 5.8 Insect biotypes and plant resistance

Insect	Crop	No. of biotypes	Reference
Acyrthosiphon pisum	Lucerne	2	Cartier *et al.* (1965)
(pea aphid)	Pea	5	Frazer (1972)
	Pea	3	Markkula and Roukka (1971)
Amphorophora rubi	Raspberry	4	Keep and Knight (1967)
(rubus aphid)			
Aphis craccivora	Cowpea	2	Ansari (1984)
(groundnut aphid)			
Brevicoryne brassicae	Sprouts	2	Lammerink (1968)
(cabbage aphid)	Sprouts	7	Dunn and Kempton (1972)
Dysaphis devecta	Apple	3	Briggs and Alston (1969)
(rosy leaf curling aphid)			
Eriosoma lanigerum	Apple	3	Briggs and Alston (1969)
(woolly aphid)	Apple	3	Sen Gupta (1969)
	Apple	?	Knight *et al.* (1962)
Mayetiola destructor	Wheat	9	Everson and Gallun (1980)
(hessian fly)			
Nephotettix cinctipes	Rice	2	Sato and Sogawa (1981)
(green rice leafhopper)			
Nephotettix virescens	Rice	3	Heinrichs and Rapusas
(green leafhopper)			(1985); Takita and Hashim (1985)
Nilapavarta lugens	Rice	4	Pathak and Saxena (1980)
(brown planthopper)			
Pachydiplosis oryzae	Rice	4?	Pathak and Saxena (1980)
(rice gall midge)			
Rhopalosiphum maidis	Sorghum	5	Painter and Pathak (1962);
(corn leaf aphid)	and maize		Wilde and Feese (1973)
Schizaphis graminum	Wheat	8	Wood, Chada and Saxena
(greenbug)			(1969); Puterka *et al.* (1988)
Sitobion avenae	Wheat	3	Lowe (1981)
Therioaphis trifolii	Lucerne	9	Nielson and Don (1974);
(spotted alfalfa aphid)			Manglitz *et al.* (1966)
Viteus vitifolii	Grapes (USA)	2	Stevenson (1970)
(phylloxera)	Grapes (Europe)	?	Anon. (1994)

was planted over large areas in the Philippines, Vietnam and Indonesia. The resistance was monogenic, and was only effective for 2–3 years. Yet there are many other pest-resistant varieties which lasted for considerable periods of time. Since the late nineteenth century, European grape varieties grafted onto phylloxera-resistant rootstocks remained free from the pest until only very recently (Anon., 1994), when a resistance-breaking biotype appeared on a few hundred hectares of vineyards in the Rhein and Main region of Germany. In other examples, the resistance is still effective after many years of use. Biotype problems have really arisen remarkably infrequently with insect pests (Table 5.8). Even in Table 5.8, many of the examples derive from purposefully testing other strains of a pest rather than from the breakdown of a resistant variety in the field. The list (compiled in 1988, with one later addition) is certain to be incomplete, but, equally, reports cannot all be guaranteed free from identification errors.

Reasons why the biotype problem with insects is relatively rare compared with pathogens probably include:

1. The frequent association of antixenosis with antibiosis. Some of the mobile insects responding to the antixenosis can find a host plant elsewhere.
2. Although a single mechanism for some plant resistances has been supposed, other unsuspected mechanisms have often later been shown to operate simultaneously. It is probably fairly rare for a resistant variety to show only a hypersensitive reaction, only to contain a specific toxin, etc. Biotypes in relation to any one mechanism may therefore exist without these becoming apparent problems.
3. In relation to the gene-for-gene hypothesis, much resistance to insects requires quantification to assess it, and thus would appear as 'horizontal' (see below).
4. Insect generation times are slow compared with pathogens. It is noticeable that the biotype problem has arisen most frequently with aphids, among the fastest-breeding pests.

The concept of resistance-breaking biotypes of plant diseases is related to the gene-for-gene hypothesis (Flor, 1942). This hypothesis states that major genes for resistance in a plant are matched by corresponding genes in the disease for overcoming that resistance. Thus the plant carrying a resistance gene only appears resistant to those biotypes of the disease that have an 'avirulent' allele at the corresponding gene locus; if the allele is 'virulent', the plant appears susceptible. Where resistance to a pest or disease is mediated by a single gene or a few genes (monogenic or oligogenic resistance), it becomes clear-cut and race-specific (for example Table 5.9). Van der Plank (1963) termed this 'vertical' resistance. Polygenic or 'horizontal' resistance is race nonspecific. The interaction

Table 5.9 Response of three biotypes of *Amphorophora rubi* to host genes and gene combinations in the raspberry (from Keep, E. and Knight, R. L. (1967) *Euphytica*, **19**, 209, with permission)

Genes	Strain of *A. rubi*		
	1	2	3
A_1	R	S	R
A_2	S	R	S
$A_3 + A_4$	S	R	S
A_5	R	S	S
A_6	R	S	S
A_7	R	S	S
$A_1 + A_2$	R	R	R
$A_1 + A_3$	R	R	R
$A_1 + A_4$	R	S	R

Plant response: R, resistant; S, susceptible.

between the several plant and pest/disease alleles shows as a quantitative range of resistance to the various biotypes (Table 5.10). Unfortunately, most plant breeding for yield and quality is carried out under chemical protection. As selections are then made without reference to the effects of pests and diseases, horizontal resistance becomes eroded in most plant breeding programs.

Clearly vertical resistance is much more likely to suffer from biotype problems than horizontal resistance. However, monogenic or oligogenic resistance is very much easier to transfer into adapted varieties, especially by transgenic methods, and is likely to provide higher levels of resistance.

Table 5.10 The horizontal resistance model (modified from Robinson, 1980). Figures in body of table represent level of plant susceptibility of each combination as a percentage of maximum (from van Emden, H. F. (1987) in *Integrated Pest Management*, (eds A. J. Burn, T. H. Coaker and P. C. Jepson), Academic Press, London, p. 52, with permission)

	Host: percentage 'resistance' alleles					
	0	20	40	60	80	100
Pest: percentage 'virulence' alleles						
100	100	80	60	40	20	0
80	80	64	48	32	16	0
60	60	48	36	24	12	0
40	40	32	24	16	8	0
20	20	16	12	8	4	0
0	0	0	0	0	0	0

Plant resistance in relation to insect vectors of plant diseases

It is often suggested in the literature (e.g. van Marrewijk and de Ponti, 1980) that insect-borne plant diseases, especially if of nonpersistent transmission, will spread more rapidly in varieties resistant to the vector. The argument is that the pest may be more restless and increase probing. A solution might be to breed high acceptability into varieties antibiotic to the vector.

Atiri, Ekpo and Thottappilly (1984) showed experimentally (in cages) that cowpea mosaic virus was spread more rapidly in two cowpea varieties resistant to *Aphis craccivora* than in a susceptible variety, although the resistant varieties carried only about one-tenth the number of aphids found on the susceptible. However, no field example can be cited where the introduction of a variety resistant to a vector has increased the spread of a plant disease. On the contrary, there are several examples from the field where disease incidence has clearly been reduced (Knight, Keep and Briggs, 1959; Hagel, Silbernagel and Burke, 1972; Lowe, 1975).

Interactions of plant resistance with other IPM methods

'Overdosing' is usually regarded as the prerogative of chemical pest control. However, as pointed out earlier, strong plant resistance (particularly if allelochemically based and dependent on one mechanism) may incur the maximum yield penalty, damage biological control and lead to the most rapid appearance of adapted biotypes. Further (see below), such resistance may also induce tolerance to insecticides.

In the IPM context, therefore, there seems little merit in seeking greater levels of host-plant resistance than are needed in combination with other methods; partial plant resistance may therefore be a suitable goal. Insecticides and biological control have both been identified as having the potential for potentiation with such partial plant resistance (van Emden, 1990a).

Insecticides and plant resistance

Insects on resistant varieties are usually smaller than their counterparts on susceptible ones. As insecticide toxicity depends on the body weight of the target, it would be expected that pests show enhanced susceptibility on resistant plant varieties. That this does occur, has been established for both sucking and chewing pests. Even small levels of plant resistance seem to allow the same kill as on a susceptible variety to be achieved with dose reductions of about one-third. Just as one example of a field result, the leafhopper *Empoasca dolichi* on partially resistant and susceptible cowpea varieties was sprayed with the same concentration of dimethoate. Only

75% mortality was recorded on the susceptible variety, while mortality was 94% on the resistant one (Raman, 1977).

The reverse phenomenon has also been shown. However, decreased rather than increased insecticide susceptibility on resistant varieties appears to be associated with allelochemically based host-plant resistance. This suggests that induction of resistance in insects by exposure to a naturally occurring toxin occurs in the same way as it does by exposure to a synthetic one (insecticide).

Biological control and plant resistance

Reference was made earlier to deleterious effects on natural enemies of feeding on prey on resistant plants. Such effects result from toxic allelochemicals, or (with respect to parasitoids) from developing in the smaller prey found on resistant varieties.

Reference was also made under the plant resistance mechanism 'extrinsic resistance' to crop- and situation-specific beneficial interactions between plant resistance and biological control; however, there are also more general potentiations between partial plant resistance and natural enemies that will apply in many situations. Such potentiations can be defined as pest population reductions on partially resistant plant varieties greater than expected from the equivalent reduction on a susceptible variety.

Van Emden (1990a) summarizes the literature reporting such potentiation of biological control on partially resistant varieties. Sometimes varieties intrinsically less than 20% resistant show population reductions of 60% compared with susceptible varieties, where biological control is acting on both. Explanations for this unexpected result include:

1. *Numerical responses of natural enemies.* There is increasing evidence that parasitoids, in particular, respond to plant odors, and use these as cues for locating their prey from a distance rather than cues from the prey itself. Recent work (p. 127) has even shown that parasitoids will select between cultivars of the same crop, responding to volatiles from the cultivar on which they emerged. This will result in parasitoids arriving at partially resistant varieties in spite of their lower pest densities. However, the phenomenon also carries the danger that plant resistance based on allelochemicals may well deter parasitoids.
2. *Functional responses of natural enemies.* Not surprisingly, a predator will require more of the smaller prey on resistant varieties before satiation than of the larger prey on susceptible varieties. With smaller prey, the impact of predation also rises with density to higher prey densities than with larger prey. In addition, it has been found, both with aphids and caterpillars, that the prey are more easily dislodged when disturbed by natural enemies on partially resistant varieties. The result

is mortality additional to direct mortality from consumption of the prey.

Gould, Kennedy and Johnson (1991) have pointed out that any such potentiation of biological control by plant resistance will affect unadapted genotypes of the pest more than adapted ones. Thus, the reduced fitness of such strains due to the plant resistance, compared with adapted genotypes, will be further reduced by the biological control. This is bound to lead to an acceleration in the spread of genes adapted to the plant resistance. It can equally be argued that positive synergism between partial plant resistance and insecticides will accelerate tolerance developing to both insecticide and plant resistance.

Such arguments should not discourage us from using plant resistance in an IPM context. First, adaptation to plant resistance will be faster if we replace partial plant resistance, plus biological control or insecticides, with stronger plant resistance alone to obtain the same control level. Secondly, partial plant resistance with a multigenic basis would greatly reduce the rate of pest adaptation.

The three-way interaction

Natural enemies may or may not be killed by lower concentrations of a toxin than those which kill their prey (p. 78). More important in the plant resistance context is the steeper slope of their mortality response to increasing pesticide concentration (again, see p. 78). This is especially so in relation to achieving the same kill of the pest by a lower pesticide dose on resistant than on susceptible varieties (see above). These considerations combine to afford a remarkable opportunity for increasing the natural enemy:pest ratio on resistant varieties whenever insecticides are used (compare Figures 5.2 and 5.8). The curve for pest mortality shifts to lower concentrations on resistant varieties (unless the resistance is based on allelochemicals, see earlier). However, a similar shift for the natural enemy only occurs to the extent of any weight reduction resulting from feeding on smaller prey.

Conclusions on plant resistance in IPM

The interactions plant resistance shows with other control methods in IPM may be advantageous or disadvantageous. Yet varieties are rarely tested for extrinsic resistance in breeding programs. Moreover, such programs tend to seek (especially if transgenic methods are used) levels of plant resistance higher than are necessary for use in IPM (van Emden, 1991).

We could seek resistant varieties on which pests are more easily killed with insecticide, on which we can make insecticide applications more

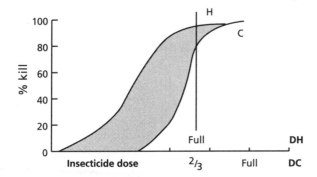

Figure 5.8 *Theoretical effect of host-plant resistance on the selectivity of pesticide applications (compare Figure 5.2). H, dose mortality curve of a herbivore; C, dose mortality curve of predator or parasitoid; DC, dose scale for carnivore; DH, dose scale for herbivore; selectivity 'window' is stippled. (From van Emden, H.F., Proc. Br. Crop Prot. Conf., Pests and Diseases, Brighton, 1990, p. 945, with permission.)*

selective in relation to natural enemies and on which biological control would be enhanced.

More commonly, we develop resistant varieties on which pests are tolerant to insecticide and which are unfavorable environments for biological control.

There are, alas, few commercial reasons for choosing the alternative that is better IPM. Integrating methods has little appeal for the end-user, the farmer. It has no appeal for commercial organizations seeking to market pest control in the seed by transgenic methods. It would only interest plant breeders as a bonus property of varieties being developed for quite other desirable characters.

Insect growth regulators (IGRs)

These are chemicals that can be applied to pest populations in the same way as insecticides. However, instead of being traditional toxins, IGRs interfere with the processes of growth and development of insects, usually by their close relationship to an insect's own internal hormones or by acting as antagonists of the latter (Staal, 1975).

The late Professor Carroll Williams, an eminent insect hormone specialist in the USA, proposed that insect hormones or their analogs would make excellent control agents. He argued that they should be specific, at least to insects, and therefore safe to man; also resistance to such materials would be unlikely to develop. The story, which is probably apocryphal, is told that Williams, while visiting the River Plata in South America, was impressed by the lack of aquatic insects in the river. He analyzed a sample of the water, and detected high levels of insect molting hormone that he thought had come from aquatic vegetation. The paucity of insects was attributed to these hormone levels. The story goes on that molting hormone is not in fact found in the river, but that the persistent material got into the analyzed sample from contamination on Williams' own hands!

Whatever the truth of this unlikely tale, the idea that internal hormones that regulate growth and development of insects could be used against them was eagerly taken up by others. These included researchers in insect endocrinology as well as in industry, who saw a new opportunity for marketing 'chemicals'. Hormones and hormone mimics were synthesized in some number, and several were marketed as the 'third generation insecticides'.

Ecdysone

Ecdysone (molting hormone) regulates the destruction of the old insect cuticle and the development and hardening of the new. Although ecdysones interfere lethally with these processes in experiments, there proved to be commercial problems in exploiting them, particularly the extremely high costs of synthesis (Staal, 1987).

Juvenile hormone (JH)

JH is again involved in metamorphosis by inhibiting maturity to the adult while it remains present in the body of the insect. By contrast with ecdysone, JHs are simpler to synthesize, as are analogs that possess the advantage over the natural materials of being stable in sunlight. They are of low toxicity to vertebrates and break down into harmless substances in the environment. Typical effects of these compounds as pest control agents are usually seen at larval to pupal metamorphosis, where lethal deformations and even intergrades between larva and pupa appear. Indeed, the compounds need to be applied to the late larval stage of the insect. Other uses of JH analogs are to disrupt embryogenesis in insects; thus they may also be effective as chemosterilants. Two JH analogs that have been marketed and used extensively are methoprene and kinoprene. Both have been used particularly in public health and stored food products (Menn, Raina and Edwards, 1989), two applications where the low toxicity to humans of JH analogs is a major advantage.

Unfortunately, Williams' prediction that resistance to such compounds would not appear has been shown to be wrong. Resistance to many marketed JH analogs appeared far faster than would be normal for a traditional pesticide.

JH antagonists

A problem of JH analogs is that they may prolong larval life and thus feeding, and the larva is so often the damaging stage of the pest. This problem, as well as the resistance danger, does not apply to JH antagonists (Staal, 1987). Younger larvae are more affected than with JH, and metamorphose into miniature pupae, with larger larvae metamorphosing into sterile adults. This precocious development is the origin of the name 'precocenes' for the JH antagonists. With the potent advantages of antagonists, much research has been devoted to them over what is now many years. However, no commercially viable 'precocene' has been developed, mainly because activity of the compounds tested so far has been too low to be practical.

Cuticle inhibitors

These represent the most important IGRs to have been developed commercially, and for most practical crop protectionists cuticle inhibitors are the only IGRs. These compounds, the benzoylphenyl ureas, and their action, were discovered as long ago as about 1970. Although the formation of the new skin proceeds normally, cuticle inhibitors disrupt the actual molting process itself. Affected insects either die totally within the old cuticle or only partly emerge from it.

For many years one compound, diflubenzuron (dimilin), was the only cuticle inhibitor on the market. It has been widely used, particularly against caterpillars, and has recently been joined by several newer similar products. Several agrochemical companies are now involved in R & D and marketing of cuticle inhibitors. They are still more expensive than traditional insecticides, but have sufficient selectivity, including in favor of parasitoids, that they are being used more widely. They are particularly valuable as a part of IPM menus, or where the pest has become resistant to insecticides.

Putting it all together

This chapter has explored the individual control techniques that can contribute to IPM. It has, where appropriate, emphasized that the combination of two or more measures may lead to more than additive results, i.e. the effect of the combination may be greater than the sum of the parts.

In the 30 years since *Silent Spring*, in which Rachel Carson discussed the alternatives to chemical pest control also as alternatives to each other, more control possibilities than were known in the late 1950s have emerged. The well-worn phrase today used in other contexts, 'We have the technology', clearly also applies to IPM, but far less clear is what we do with the technology we have. Many textbooks on IPM do not tackle the integration of the various methods very seriously. It is in this integration, rather than in technical advances in methods, that IPM scientists have failed to have their potential impact.

Menu systems

The most highly developed pest management packages are probably to be found in the USA. Cotton (Bottrell and Adkisson, 1977), the crop which led the way in pesticide-induced disasters and thereby subsequently in IPM, is probably the most comprehensive example. It is a 'menu' system, inviting farmers, or the crop protection consultants they employ, to choose elements as appropriate for their conditions.

Any use of insecticides is firmly related to scouting and thus to economic thresholds, which are known for all the regular pests. When thresholds are exceeded, farmers have advice available on what are at least partially selective insecticides so that natural enemies are conserved as far as possible. An obvious choice, before its use was discouraged because it is an organochlorine compound, was endosulfan which, as mentioned in Chapter 3, is selective in favor of parasitoids in the order Hymenoptera.

Although any item in the menu that contributes to fewer pesticide applications and the use of better compounds will contribute to increased biological control, the menu also includes biological control specifically. In the case of cotton, this is periodic releases of the parasitoid *Trichogramma* against eggs of Lepidoptera, particularly of bollworm.

Many plant characteristics have been identified as leading to pest resistance in cotton, and a number have been incorporated into commercially acceptable varieties. Among the best known are early maturing cottons to enable the crop to escape late season bollworm attack, and 'Frego bract' and 'nectariless varieties', both resistant again to bollworm.

Several menu items are ones of cultural techniques. Most are designed to capitalize on the possibilities early harvest gives for minimizing bollworm damage. Such earliness of maturity may be inducible by early termination of irrigation and the spraying of leaf desiccants to synchronize early boll opening. Early harvesting needs to be coupled with early destruction of the stalks remaining in the field after harvest. Other cultural measures available are insecticide-treated trap crops against boll weevil, and the possibility of holding *Lygus* bugs on any adjacent lucerne (and thus away from cotton) by strip harvesting rather than total combining of the former.

The sterile-male release has been developed for control of boll weevil, as has trapping out of boll weevils using the male-produced aggregation pheromone. The confusion technique with female-produced sex pheromone against bollworm can often make spraying unnecessary, or at least delay the need to spray when bollworm population pressure is high.

Such a menu system can be offered to farmers because of the resources that have been invested in research on cotton insect pests and their control at many research stations and universities over many years, including before *Silent Spring*. It is not the kind of system that could be developed *de novo* in response to a need for IPM within any reasonable time-scale, even in the USA. The second feature of such a system is that it represents a synthesis of target-specific controls; it is perhaps characteristic of a menu system that exploitation of useful synergism between components has to be foregone. Thirdly, the system is very crop-specific; the same menu cannot be applied in any crop other than cotton. Fourthly, a feature of the USA cotton menu is that several options, singly or in combination, are specified for single pest species. Rather than 'killing two birds with one stone' it is more a question of 'hurling a basketful of stones at a single bird'.

Another well-developed menu-based system is available, in various forms, in many developed countries for orchards (Hardman, 1992). Again, it is crop-specific and the outcome of putting together the results of many years of research. It is interesting that the growers themselves have often requested this menu from the scientists. Such requests stem from a fear of insecticide resistance in the pests as the number of insecticides registered for use in orchards has shrunk dramatically.

The orchard IPM menus are built around the use of insecticide-resistant predatory mites for the control of red spider mite, and use of pheromone traps for monitoring moth pests. Other components developed around this central core are more selective materials, such as cuticle inhibitors and *Baccillus thuringiensis* (often in combination with a low dose of pyrethroid insecticides), both for caterpillar control. Additionally there is emphasis on careful timing of sprays when pests do occur, to avoid times when biological control of other pests is most intense.

Computer-generated IPM

Developed countries such as the USA have not only synthesized such menu-based IPM systems, but have also investigated how far IPM solutions might be synthesized by computers. The computer programs are based on systems analysis, life-table studies and mathematical modeling. Theoretically, if the causes of population change can be understood adequately, the dynamics of the populations can be modeled with the influence of any abiotic driving variable, such as the weather, included. As the effect of any pest control measure or measures could then also be modeled, the whole system could then be 'managed' to best advantage.

The immediate reaction of many scientists is that meaningful modeling of a single pest population, let alone all the important pests of the crop as well as their natural enemies, would require the collection and interpretation of an impossibly large volume of life-table and other data. However, in practice it turns out that predictable interrelationships between the various elements in such models enable satisfactory simulation to be achieved at a far lower level of cybernetic complexity than would at first sight appear to be essential. It has sometimes been possible to use just a few key parameters to predict pesticide outbreaks or the need for pesticide treatment. Some very usable computer models of crop growth and the effect thereof on pest populations have already been achieved for several crops. These models have already proved themselves in their ability to predict whether and when economic thresholds will be reached in the field.

One of the fullest such IPM models has been developed for apples at Michigan State University in the USA (Croft, Howes and Welch, 1976). The model simulates the assimilated flux of the whole tree, and also the phenology of growth of the leaves, shoots, roots and fruit. This tree model is driven by input of weather data, and is coupled with developmental models for some eight pest species under the name of the Predictive Extension Timing Estimator system (an unusual title, until one realizes the acronym will be 'PETE'!). As many US orchards use predators to control their mite pests, four prey–predator models can be linked into the system. When felt appropriate, synchronization of the model with the field situation occurs, using data from monitoring of orchards. Models such as PETE can include estimates of control costs and benefits, allowing decisions to be made between control alternatives with regard to both long- and short-term consequences.

The experimental approach

An experimental field approach to IPM packages has been developed at the University of Reading, UK, with the aim of producing an IPM protocol for a given crop. The protocol is based on the results of field trials rather than on foreknowledge and collection of basic biological and ecological data (van Emden, 1989). Of course, to carry out a field trial with even just six crop varieties, and four insecticides each at three doses, already involves 72 treatments. If each treatment is then tested with and without release of a biological control agent and with and without a change in cultural measures (e.g. added floral diversity), even two replicates would require nearly 600 plots. The approach adopted has sought to deal with this multiplicity of possibilities in three ways:

1.   The experimental variables are based on a 'pest management triad' (Figure 5.9) of beneficial interactions between just three IPM

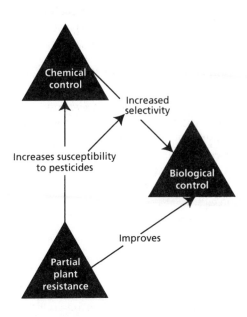

Figure 5.9 *The pest management triad: the interaction of chemical control, biological control and partial host-plant resistance (from van Emden, H.F.,* Pest Control, *Cambridge University Press, Cambridge, p. 100, 1989, with permission).*

    components – plant resistance, biological control and pesticides (but including, for example, pathogens).

2. In the first instance, at least, biological control is by existing indigenous natural enemies.
3. Experiments with a limited number of variables are performed in sequence. These first tackle cultural control, then selection of variety. Then follows a sequence of insecticide trials, each examining just one defined aspect (choice of material, dose rate, application method, timing, placement, etc.).

The experimental approach seeks to engender variation for screening for where IPM efficiency results, largely because of interaction between methods. The criterion for evaluation is the interval between the needs for pesticides, just as much as the immediate effect. It is also recognized that many IPM combinations are never tested, and that the sequence of experiments plays a large part in making this selection. Other pest control methods, including pheromones and release of biological control agents are not excluded. However, they are regarded as too crop-specific to form part of the generalized sequence of trials and remain available for inclusion after the basic IPM protocol has been developed.

In many ways, as will be evident in the next chapter, this latter approach is the nearest of those mentioned above to the way in which IPM has been implemented in the developing world. Some technology from developing countries has been transferable in shared crops, e.g. cotton, and biological control agents found effective elsewhere have been multiplied and released. However, much of the development of IPM has been based on conserving biological control through cultural measures and a modification of insecticide use, with the introduction of host-plant resistance against major pests. Yet there is an important difference between what has happened in the developing countries and the experimental approach mentioned above. The question asked in the developing countries has been 'Does this work?', based on the preconception that it might. The question asked in the experimental approach is 'What does work?' among a wide range of trials without too many preconceptions.

References

A'Brook, J. (1968) The effect of plant spacing on the numbers of aphids trapped over the groundnut crop. *Ann. Appl. Biol.*, **61**, 289–94.
Altieri, M.A., van Schoonhoven, A. and Doll, J.D. (1978) A review of insect prevalence in maize (*Zea mays* L.) and bean (*Phaseolus vulgaris* L.) in polycultural systems. *Field Crops Res.*, **1**, 33–49.
Anon. (1994) Die Rückkehr der Reblaus. *Profil*, **October,** 11.
Ansari, A.F. (1984) Biology of *Aphis craccivora* (Koch) and varietal resistance of cowpeas. Ph.D. Thesis, University of Reading.
Atiri, G.I., Ekpo, E.J.A. and Thottappilly, G. (1984) The effect of aphid-resistance in cowpea on infestation and development of *Aphis craccivora* and the transmission of cowpea aphid-borne mosaic virus. *Ann. Appl. Biol.*, **104**, 339–46.
Bateman, R.P. (1992) Controlled droplet application of mycopesticides to locusts, in *Biological Control of Locusts and Grasshoppers*, (eds C.J. Lomer and C. Prior), CABI, Wallingford, pp. 249–54.
Batten, M. (1983) The rush is on to study jungles. *Int. Wildlife*, **12** (3), 16–19.
Boller, E.F. (1992) The role of integrated pest management in integrated production of viticulture in Europe. *Proc. Br. Crop Prot. Conf., Pests and Diseases, Brighton, 1992*, pp. 499–506.
Bottrell, D.G. and Adkisson, P.L. (1977) Cotton insect pest management. *Ann. Rev. Entomol.*, **22**, 451–81.
Bourne, B.A. (1953) Studies on the dissemination of sugar cane diseases. *Sugar J.*, **16** (3), 19, 22.
Briggs, J.B. and Alston, F.H. (1969) Sources of pest resistance in apple cultivars, *Rep. E. Malling Res. Stn, 1966*, 170–1.
Burn, A.J. (1988) Effects of scale on the measurement of pesticide effects on invertebrate predators and parasites, in *Field Methods for the Study of Environmental Effects of Pesticides*, (eds M.P. Greaves, P.W. Greig-Smith and B.D. Smith), BCPC, Thornton Heath, pp. 109–17.
Butenandt, A., Beckmann, R., Stamm, D. and Hecker, E. (1959) Über den Sexual-Lockstoff des Seidenspinners *Bombyx mori*. Reinanstellung und Konstitution. *Z. Naturforsch., B: Anorg. Chem., Biochem., Biophys.,Biol.*, **14**, 283–4.

Carl, K.P. (1979) The importance of cultural measures for the biological control of the cereal leaf beetle *Oulema melanoplus* (Col. Chrysomelidae). *Mitt. Schweiz. Entomol. Ges.*, **52**, 443.

Carl, K.P. (1982) Biological control of native pests by introduced natural enemies. *Biocontrol News and Information*, **3**, 192–200.

Cartier, J.J., Isaak, A., Painter, R.H. and Sorensen, E.L. (1965) Biotypes of pea aphid *Acyrthosiphon pisum* (Harris) in relation to alfalfa clones. *Can. Entomol.*, **97**, 754–60.

Cayley, G.R., Etheridge, P., Griffiths, D.C. *et al.* (1984) A review of the performance of electrostatically charged rotary atomisers on different crops. *Ann. Appl. Biol.*, **105**, 279–86.

Chiarappa, L. (1971) *Crop Loss Assessment Methods*, CAB, Farnham Royal (and later supplements).

Coaker, T.H. (1987) Cultural methods: the crop, in *Integrated Pest Management*, (eds A.J. Burn, T.H. Coaker and P.C. Jepson), Academic Press, London, pp. 69–88.

Coaker, T.H. and Finch, S. (1973) The association of the cabbage rootfly with its food and host plants, in *Insect/Plant Relationships*, (ed. H.F. van Emden), Blackwell Scientific Publications, Oxford, pp. 119–28.

Cockburn, A.F., Howells, A.J. and Whitten, M.J. (1989) Recombinant DNA technology and genetic control of pest insects, in *Genetical Biochemical Aspects of Pest Management*, (ed. G. Russell), Intercept, Andover, pp. 211–41.

Croft, B.A. and Brown, A.W A. (1975) Responses of arthropod natural enemies to insecticides. *Ann. Rev. Entomol.*, **20**, 285–335.

Croft, B.A., Howes, J.L. and Welch, S.M. (1976) A computer-based extension pest management delivery system. *Environ. Entomol.*, **5**, 20–34.

Curtis, C.F. (1968) A possible genetic method for the control of insect pests, with special reference to tsetse flies (*Glossina* spp.). *Bull. Entomol. Res.*, **57**, 509–23.

Doble, S.J., Matthews, G.A., Rutherford, I. and Southcombe, E.S.E. (1985) A system for classifying hydraulic nozzles and other atomizers into categories of spray quality. *Proc. Br. Crop Prot. Conf., Weeds, Brighton*, 1985, pp. 1125–33.

Doutt, R.L. and Nakata, J. (1973) The *Rubus* leafhopper and its egg parasitoid: an endemic biotic system useful in grape-pest management. *Environ. Entomol.*, **2**, 381–6.

Dunn, J.A. and Kempton, D.P.H. (1972) Resistance to attack by *Brevicoryne brassicae* (L.) on Brussels sprouts. *Ann. Appl. Biol.*, **72**, 1–11.

Dunnam, E.W. and Clark, J.C. (1938) The cotton aphid in relation to the pilosity of cotton leaves. *J. Econ. Entomol.*, **31**, 663–6.

Emmett, B.J. (1984) Pheromones in UK farm pest control, in *Statistical and Mathematical Methods in Population Dynamics and Pest Control*, (ed. R. Cavalloro), Balkema, Rotterdam, pp. 47–57.

Everson, E.H. and Gallun, R.L. (1980) Breeding approaches in wheat, in *Breeding Plants Resistant to Insects and Mites*, (eds F.G. Maxwell and P.R. Jennings), Wiley, New York, pp. 513–33.

FAO (1991) *Programme for the Eradication of the New World Screwworm* Cochliomyia hominivorax *from North Africa, Report for the Period 1 June to 31 August*, FAO, Rome.

Flor, H.H. (1942) Inheritance of pathogenicity in *Malampsora lini. Phytopathology*, **32**, 653–9.

Frazer, B.D. (1972) Population dynamics and recognition of biotypes in the pea aphid (Homoptera: Aphididae). *Can. Entomol.*, **104**, 1729–33.

Gatehouse, A.M.R., Boulter, D. and Hilder, V.A. (1992) Potential of plant-derived genes in the genetic manipulation of crops for insect resistance, in *Plant Genetic Manipulation for Crop Protection*, (eds A.M.R. Gatehouse, V.A. Hilder and D. Boulter), CABI, Wallingford, pp. 155–81.

Goldsmith, E., Allen, R., Allaby, M. *et al.* (1972) A blueprint for survival. Introduction: The need for change. *Ecologist*, **2**, 2–7.

Gould, F., Kennedy, G.G. and Johnson, M.T. (1991) Effects of natural enemies on the rate of herbivore adaptation to resistant host plants. *Entomol. Exp. Appl.*, **58**, 1–14.

Greathead, D.J. (1989) Biological control as an introduction phenomenon: a preliminary examination of programmes against Homoptera. *Entomologist*, **108**, 28–37.

Griffiths, D.C., Maniar, S.P., Merritt, L.A. *et al.* (1991) Laboratory evaluation of pest management strategies combining antifeedants with insect growth regulator insecticides. *Crop Prot.*, **10**, 145–51.

Hagel, G.T., Silbernagel, M.J. and Burke, D.W. (1972) Resistance to aphids, mites, and thrips in field beans to infection by aphid-borne viruses. *Bull. USDA Agric. Res. Serv. No. 33–139*, 1–4.

Hardman, J.M. (1992) Apple pest management in North America: challenge and response. *Proc. Br. Crop. Prot. Conf., Pests and Diseases, Brighton, 1992*, pp. 507–16.

Harlan, J.R. and Starks, K.J. (1980) Germplasm resources and needs, in *Breeding Plants for Resistance to Insects and Mites*, (eds F.G. Maxwell and P.R. Jennings), Wiley, New York, pp. 253–73.

Hawksworth, D.L. (1968) Man's impact on the British flora and fauna. *Outlook on Agric.*, **8**, 23–8.

Headley, J.C. and Hoy, M.A. (1987) Benefit/cost analysis of an integrated mite management program for almonds. *J. Econ. Entomol.*, **80**, 555–9.

Heinrichs, E.A. and Rapusas, H.R. (1985) Cross-virulence of *Nephotettix virescens* (Homoptera: Cicadellidae) biotypes among some rice cultivars with the same major-resistance gene. *Environ. Entomol.*, **14**, 696–700.

Henneberry, T.J., Bariola, L.A., Flint, H.M. *et al.* (1981) Pink bollworm and tobacco budworm mating disruption studies on cotton, in *Management of Insect Pests with Semiochemicals. Concepts and Practice*, (ed. E.R. Mitchell), Plenum, New York, pp. 267–83.

Herren, H.R. and Neuenschwander, P. (1991) Biological control of cassava pests in Africa. *Ann. Rev. Entomol.*, **36**, 257–83.

Herzog, D.C. and Funderburk, J.E. (1985) Plant resistance and cultural practice interactions with biological control, in *Biological Control in Agricultural IPM Systems*, (eds M.A. Hoy and D.C. Herzog), Academic Press, Orlando, pp. 67–88.

Hodkinson, I.D. and Hughes, M.K. (1982) *Insect Herbivory*, Chapman & Hall, London, p. 23.

Hokkanen, H. and Pimentel, D. (1984) New approach to selecting biological control agents. *Can. Entomol.*, **116**, 1109–21.

Hooper, M.D. (1984) What are the main recent impacts of agriculture on wildlife? Could they have been predicted for the future?, in *Agriculture and the Environment*, (ed. D. Jenkins), Institute of Terrestrial Ecology, Cambridge, pp. 33–6.

Hoy, J.M. (1949) Control of manuka by blight. *N. Z. J. Agric.*, **79**, 303–13.

IBPGR (1984) *Institutes Conserving Crop Germplasm: The IBPGR Global Network of Genebanks*. IBPGR, Rome.

ICRISAT (1980) *Annual Report, 1978–79*. ICRISAT, Patancheru.

Judd, G.J.R. and Borden, J.H. (1988) Long-range host-finding behaviour of the onion fly *Delia antiqua* (Diptera: Anthomyiidae): ecological and physiological constraints. *J. Appl. Ecol.*, **25**, 829–45.

Judenko, E. (1960) Shot-hole borer (*Xyleborus fornicatus* Eich.) and clones. *Tea Quart.*, **31**, 72–5.

Kareiva, P. and Sahakian, R. (1990) Tritrophic effects of a simple architectural mutation in pea plants. *Nature, Lond.*, **345**, 433–4.

Keep, E. and Knight, R.L. (1967) A new gene from *Rubus occidentalis* L. for resistance to strains 1, 2 and 3 of the Rubus aphid, *Amphorophora rubi* Kalt. *Euphytica*, **16**, 209–14.

Kennedy, G.G. (1986) Consequences of modifying biochemically mediated insect resistance in *Lycopersicon* species, in *Natural Resistance of Plants to Pests*, (eds M.B. Green and P.A. Hedin), American Chemical Society, Washington, DC, pp. 130–41.

Kirchberg, E. (1955) Neuartige Fliegenbekämpfung auf Curacao mit sterilisierten Männchen. *Orion*, **19/20**, 767–9.

Knight, R.L., Briggs, J.B., Massee, A.M. and Tydeman, H.M. (1962) The inheritance of resistance to woolly aphid, *Eriosoma lanigerum* (Hsmnn.), in the apple. *J. Hort. Sci.*, **37**, 207–18.

Knight, R.L., Keep, E. and Briggs, J.B. (1959) Genetics of resistance to *Amphorophora rubi* (Kalt.) in the raspberry. *J. Genet.*, **56**, 261–80.

Knipling, E.F. (1955) Possibilities of insect control or eradication through the use of sexually sterile males. *J. Econ. Entomol.*, **48**, 459–62.

Knipling, E. F. (1966) The entomologist's arsenal. *Bull. Entomol. Soc. Am.*, **12**, 45–51.

Kogan, M. and Ortman, E.F. (1978) Antixenosis – a new term proposed to define Painter's 'non-preference' modality of resistance. *Bull. Entomol. Soc. Am.*, **24**, 175–6.

Lammerink, J. (1968) A new biotype of cabbage aphid (*Brevicoryne brassicae* (L.)) on aphid resistant rape (*Brassica napus* L.). *N.Z. J. Agric. Res.*, **11**, 341–4.

Le Pelley, R.H. (1959) *Agricultural Insects of East Africa*, East Africa High Commission, Nairobi.

Levin, B.R. (1979) Problems and promise in genetic engineering in its potential applications to insect management, in *Genetics in Relation to Insect Management*, (eds M.A. Hoy and J. McKelvey Jr), Rockefeller Foundation, New York, pp. 170–5.

Lewis, F.H. and Hickey, K.D. (1964) Pesticide application from one side on deciduous fruit trees. *Penn. Fruit News*, **43**, 13–24.

Lie, R. and Bakke, A. (1981) Practical results from the mass trapping of *Ips typographus* in Scandinavia, in *Management of Insect Pests with Semiochemicals. Concepts and Practice*, (ed. E.R. Mitchell), Plenum, New York, pp. 175–81.

Lowe, H.J.B. (1975) Infestation of aphid-resistant and susceptible sugar beet by *Myzus persicae* in the field. *Z. Angew. Entomol.*, **79**, 376–83.

Lowe, H.J.B. (1981) Resistance and susceptibility to colour forms of the aphis *Sitobion avenae* in spring and winter wheats (*Triticum aestivum*). *Ann. Appl. Biol.*, **99**, 87–98.

Lundholm, B. and Stackerud, M. (eds) (1980) Environmental protection and biological forms of control of pest organisms. *Ecol. Bull.*, *No. 31*.

Manglitz, G.R., Calkins, C.O., Walstrom, R.J. *et al.* (1966) Holocyclic strains of the spotted alfalfa aphid in Nebraska and adjacent States. *J. Econ. Entomol.*, **59**, 636–9.

Markkula, M. and Roukka, K. (1971) Resistance of plants to the pea aphid *Acyrthosiphon pisum* Harris (Hom., Aphididae). III. Fecundity on different pea varieties. *Ann. Agric. Fenn.*, **10**, 33–7.

Marrone, P.G. and MacIntosh, S.C. (1993) Resistance to *Bacillus thuringiensis* and resistance management, in Bacillus thuringiensis, *an Environmental Pesticide*, (eds P.F. Entwistle, J.S. Cory, M.J. Bailey and S. Higgs), Wiley, New York, 221–35.

Matthews, G.A. (1989) *Cotton Insect Pests and their Management*, Longmans, Harlow.

Menn, J.J., Raina, A.K. and Edwards, J.P. (1989) Juvenoids and neuropeptides as insect control agents, in *Progress and Prospects in Insect Control*, (ed. M.R. McFarlane), British Crop Protection Council, Farnham, pp. 87–106.

Metcalf, C.L., Flint, W.P. and Metcalf, R.L. (1951) *Destructive and Useful Insects: Their Habits and Control*, McGraw-Hill, New York.

Morse, S. (1989) The integration of partial plant resistance with biological control by an indigenous natural enemy complex in affecting populations of cowpea aphid (*Aphis craccivora* Koch). Ph.D. Thesis, University of Reading.

Nielson, M.W. and Don, H. (1974) A new virulent biotype of the spotted alfalfa aphid in Arizona. *J. Econ. Entomol.*, **67**, 64–6.

Painter, R.H. *(1951) Insect Resistance in Crop Plants*, Macmillan, New York.

Painter, R.H. and Pathak, M.D. (1962) The Distinguishing Features And Significance Of The Four Biotypes Of The Corn Leaf Aphid, *Rhopalosiphum maidis* (Fitch). *Proc. 11th Int. Congr. Entomol., Vienna, 1960, vol.* 2, pp. 110–15.

Pathak, M.D. and Saxena, R.C. (1980) Breeding approaches in rice, in *Breeding Plants Resistant to Insects and Mites*, (eds F.G. Maxwell and P.R. Jennings), Wiley, New York, pp. 421–55.

Pearson, E.O. (1958) *The Insect Pests of Cotton in Tropical Africa*, Empire Cotton Growing Corporation and Commonwealth Institute of Entomology, London.

Pimentel, D. (1963) Introducing parasites and predators to control native pests. *Can. Entomol.*, **95**, 785–92.

Pimentel, D. (1964) Population ecology and the genetic feed-back mechanism, in *Genetics Today*, (ed. S.J. Geerts), Pergamon, New York, vol. 2, pp. 483–8.

Powell, W. (1983) The role of parasitoids in limiting cereal aphid populations, in *Aphid Antagonists*, (ed. R. Cavalloro), Balkema, Rotterdam, pp. 50–6.

Puterka, G.J., Peters, D.C., Kerns, D.L. *et al.* (1988) Designation of two new greenbug (Homoptera: Aphididae) biotypes G and H. *J. Econ. Entomol.*, **81**, 1754–9.

Raman, K.V. (1977) Studies on host plant resistance of cowpeas to leafhoppers. Ph.D. thesis, University of Reading.

Reed, W. (1965) *Heliothis armigera* (Hb.) (Noctuidae) in western Tanganyika II. Ecology and natural and chemical control. *Bull. Entomol. Res.*, **56**, 117–50.

Rembold, H. and Schmutterer, H. (1981) Disruption of insect growth by neem seed components, in *Regulation of Insect Growth and Behaviour*, (ed. F. Sehnal), Technical University Press, Wroclaw.

Richards, J.F. (1984) Global patterns of land conversions. *Environment*, **26** (9), 6–13, 34–8.

Ridgway, R.L., Inscoe, M.N. and Dickerson, W.A. (1990) Role of the boll weevil pheromone in pest management, in *Behaviour-Modifying Chemicals for Insect Management*, (eds R.L. Ridgway, R.M. Silverstein and M.N. Inscoe), Dekker, New York, pp. 437–71.

Risch, S.J. (1980) The population dynamics of several herbivorous beetles in a tropical agroecosystem: the effect of intercropping corn, beans and squash in Costa Rica. *J. Appl. Ecol.*, **17**, 593–612.

Risch, S.J., Andrew, D. and Altieri, M.A. (1983) Agroecosystem diversity and pest control: data, tentative conclusions and new research directions. *Environ. Entomol.*, **12**, 625–9.

Robinson, R.A. (1980) The pathosystem concept, in *Breeding Plants Resistant to Insects and Mites*, (eds F.G. Maxwell and P.R. Jennings), Wiley, New York, pp. 157–81.

Ryan, J., Ryan, M.F. and McNaeidhe, F. (1980) The effect of interrow plant cover on populations of cabbage root fly, *Delia brassicae* (Wied.). *J. Appl. Ecol.*, **17**, 31–40.

Sato, A. and Sogawa, K. (1981) Biotypic variations in the green rice leafhopper, *Nephotettix virescens* (Uhler) (Homoptera: Deltocephalidae). *Appl. Entomol. Zool.*, **16**, 55–7.

Sattaur, O. (1989) Genes on deposit: saving for the future. *New Scientist*, **123**, 37–41.

Sen Gupta, G.C. *(1969)* The recognition of biotypes of the woolly aphid, *Eriosoma lanigerum* (Hausmann), in South Australia by their differential ability to colonise varieties of apple rootstock, and an investigation of some possible factors in the susceptibility of varieties to these insects. Ph.D. Thesis, University of Adelaide.

Simmonds, M.S.J., Evans, H.C. and Blaney, W.M. (1992) Pesticides for the year 2000: mycochemicals and botanicals, in *Pest Management and the Environment in the Year 2000*, (eds A.S.A. Kadir and H.S. Barlow), CABI, Wallingford, pp. 127–64.

Simmonds, N.W. (1979) *Principles of Crop Improvement*, Longman, London.

Snow, J.W. and Whitten, C.J. (1979) Status of the screwworm (Diptera: Calliphoridae) control program in the southwestern United States during 1977. *J. Med. Entomol.*, **15**, 518–20.

Sotherton, N.W., Boatman, N.D. and Rands, M.R.W. (1989) The 'conservation headland' experiment in cereal ecosystems. *Entomologist*, **108**, 135–43.

Southwood, T.R.E. (1973) The insect/plant relationship – an evolutionary perspective, in *Insect/Plant Relationships*, (ed. H.F. van Emden), Blackwell Scientific Publications, Oxford, pp. 3–30.

Southwood, T.R.E. (1977) Entomology and mankind. *Proc. 15th Int. Congr. Entomol., Washington, DC, August, 1976*, pp. 36–51.

Spiher, A.T., Jr (1975) The growing of GRAS. *HortSci.*, **10**, 241–2.

Staal, G.B. (1975) Anti-juvenile hormone agents. *Ann. Rev. Entomol.*, **31**, 391–429.

Staal, G.B. (1987) Juveniles and anti-juvenile hormone agents as IGRs, in *Integrated Pest Management*: Quo Vadis? *An International Perspective*, PARASITIS, Geneva, pp. 227–92.

Stern, V.M., Smith, R.F., van den Bosch, R. and Hagen, K.S. (1959) The integrated control concept, *Hilgardia*, **29**, 81–101.

Stevenson, A.B. (1970) Strains of the grape phylloxera in Ontario with different effects on the foliage of certain grape cultivars. *J. Econ. Entomol.*, **63**, 135–8.

Takita, T. and Hashim, H. (1985) Relationship between laboratory-developed biotypes of green leafhopper and resistant varieties of rice in Malaysia. *Jap. Agric. Res. Quart.*, **19**, 219–23.

Taylor, T.C.H. (1937) *The Biological Control of an Insect in Fiji. An Account of the Coconut Leaf-Mining Beetle and its Parasite Complex*, Imperial Institute of Entomology, London.

Thorpe, W.H. and Caudle, H.B. (1938) A study of the olfactory responses of insect parasites to the food plant of their host. *Parasitology*, **30**, 523–8.

Thresh, J.M. (1981) The role of weeds and wild plants in the epidemiology of plant virus diseases, in *Pests, Pathogens and Vegetation*, (ed. M.J. Thresh), Pitman, London, pp. 53–70.

Tingey, W.M., Leigh, T.F. and Hyer, A.H. (1975) *Lygus hesperus*: growth, survival and egg laying resistance of cotton genotypes. *J. Econ. Entomol.*, **68**, 28–30.

Tothill, J.D., Taylor, T.C.H. and Paine, W. (1930) *The Coconut Moth in Fiji. A History of its Control by Means of Parasites*, Imperial Bureau of Entomology, London.

UNEP (1992) Convention on Biological Diversity, 5th June 1992, Mimeograph No. 92-7807, UNEP, Geneva.

van den Bosch, R., Frazer, B.D., Davis, C.S. *et al.* (1970) *Trioxys pallidus* – An effective new walnut aphid parasite from Iran. *Calif. Agric.*, **28**, 8–10.

van den Bosch, R. and Messenger, P.S. (1973) *Biological Control*, Intext, New York.

van der Plank, J.E. (1963) *Plant Diseases: Epidemics and Control*, Academic Press, New York.

van Emden, H.F. (1965) The role of uncultivated land in the biology of crop pests and beneficial insects. *Sci. Hort.*, **17**, 121–36.

van Emden, H.F. (1978) Ecological information necessary for pest management in grain legumes, in *Pests of Grain Legumes: Ecology and Control*, (eds S.R. Singh, T.A. Taylor and H.F. van Emden), Academic Press, London, pp. 297–307.

van Emden, H.F. (1987) Cultural methods: the plant, in *Integrated Pest Management*, (eds A.J. Burn, T.H. Coaker and P.C. Jepson), Academic Press, London, 27–68.

van Emden, H.F. (1989) *Pest Control*, Cambridge University Press, Cambridge.

van Emden, H.F. (1990a) The interaction of host plant resistance to insects with other control measures. *Proc. Br. Crop Prot. Conf., Pest and Diseases, Brighton, 1990*, pp. 939–48.

van Emden, H.F. (1990b) Limitations on insect herbivore abundance in natural vegetation, in *Pests, Pathogens and Plant Communities,* (eds J.J. Burdon and S.R. Leather), Blackwell Scientific Publications, London, pp. 15–30.

van Emden, H.F. (1991) The role of host plant resistance in insect pest mis-management. *Bull. Entomol. Res.,* **81,** 123–6.

van Emden, H.F. and Williams, G. (1974) Insect stability and diversity in agro-ecosystems. *Ann. Rev. Entomol.,* **19,** 455–75.

van Marrewijk, G.A.M. and de Ponti, O.M.B. (1980) Possibilities and limitations of breeding for pest resistance. *Med. Fac. Landbouw. Rijksuniv. Gent,* **40,** 229–47.

Waage, J.K. (1992) Biological control in the year 2000, in *Pest Management and the Environment in the Year 2000,* (eds A.S.A. Kadir and H.S. Barlow), CABI, Wallingford, pp. 329–40.

Waage, J.K. and Greathead, D.J. (1988) Biological control. *Phil. Trans. R. Soc. Lond.* (B), **318,** 111–26.

Way, M.J., Cammell, M.E., Alford, D.V. *et al.* (1981) Use of forecasting in chemical control of black bean aphid, *Aphis fabae* Scop. on spring-sown field beans, *Vicia faba* L. *Pl. Pathol.,* **26,** 1–7.

Wheatley, P.E. (1963) The giant coffee looper *Ascotis selenaria reciprocaria. E. Afr. Agric. For. J.,* **29,** 143–6.

White, W.H. and Irvine, J.E. (1987) Evaluation of variation in resistance to sugarcane borer (Lepidoptera: Pyralidae) in a population of sugarcane derived from tissue culture. *J. Econ. Entomol.,* **80,** 182–4.

Wickremasinghe, M.G.V. and van Emden, H.F. (1992) Reactions of adult female parasitoids, particularly *Aphidius rhopalosiphi,* to volatile chemical cues from the host plants of their aphid prey. *Physiol. Entomol.,* **17,** 297–304.

Wilde, G. and Feese, H. (1973) A new corn leaf aphid biotype and its effect on some cereal and small grains. *J. Econ. Entomol.,* **66,** 570–1.

Wolcott, G.N. (1942) The requirements of parasites for more than hosts. *Science, N.Y.,* **96,** 317–18.

Wood, E.A. Jr, Chada, H.L. and Saxena, P.N. (1969) Reactions of small grains and grain sorghum to three greenbug biotypes. *Rep. Okla. Agric. Exp. Stn, no. 618.*

Wratten, S.D., van Emden, H.F. and Thomas, M.B. (1996) Within-field and border refugia for the enhancement of natural enemies, in *Enhancing Biological Control of Arthropod Pests through Habitat Management,* (eds C. Pickett and R.L. Bugg), AG Access, California (in press).

CHAPTER 6

The practice of pest management in developing countries

Changes in crop protection in South and South-East Asia, Africa and South America are reviewed with particular reference to pesticide use and the emergence of IPM. The latter is discussed for each region by reference to case studies of particular crops and livestock systems. Reasons for the differing evolution of IPM in developing and developed countries are mentioned.

South and South-East Asia

In recent years, several workshops on IPM in the region have been held, displaying the concern about the continued use of pesticides in Asia. In 1991 alone there were three international meetings, one in Japan and two in Malaysia.

However, crop protection in Asia is still dominated by an increasing use of insecticides, and the issues raised by Rachel Carson in *Silent Spring* 30 years ago are equally relevant today. Twenty years of the 'green revolution' and unecologically sound control methods have 'hooked' farmers on to agrochemicals. Pesticides were big business in Asia at the beginning of the 1960s and are even bigger business now. The principal objective of crop protection policy for agriculture in Asian countries is to minimize crop losses due to pest attack. Thus government recommendations are still limited very largely to pesticide treatment and not IPM methods. Farmers are encouraged to use technology packages that involve fixed chemical inputs at fixed times as part of the credit package. The result is that prophylactic calendar-based spraying is still common. So is application from the air, in spite of all that is known about the ecological damage this practice can cause. Pesticide use in areas other than agriculture, for example public health, is also increasing.

Promotion of pesticides emphasizes the extra yields obtainable and the convenience. Moreover, farmers and the public are given the impression that pesticide use is under strict government control. Often governments spread the view that pesticides, if used properly, pose no risk to the environment.

Not only do credit packages for farmers perpetuate pesticide use, but government policies on subsidies also encourage farmers to use chemicals excessively. The matter is compounded by crop insurance programs that compensate for pest damage only if the prescribed prophylactic pesticide applications have been made. By contrast, where subsidies for pesticide use have been removed, as in Indonesia, the economic advantage of IPM has tended to lead to its adoption. However, most governments in the region try to ensure the desired levels of crop production by encouraging pesticide use while, simultaneously, paying lip service to the desirability of IPM.

Growth in pesticide use in Asia appears to have been faster in the past 30 years than in most other parts of the world. The Asian Development Bank (1987) reported that the average market growth rate for pesticides in the Asian and Pacific region was 5–7% per annum between 1980 and 1985. This compares with the growth by only 4.5% per annum of the world pesticide market (Johnson, 1991). Yet in this period markets in Indonesia and Pakistan grew by as much as 20–30% each year. This trend

Table 6.1 The pesticide market in Asia in comparison with the world market, 1990 (data of the GIFAP Asia Working Group)

Region	Market (US$ $\times 10^9$)	% Share
North America	5.4	21.9
Western Europe	6.6	26.7
Eastern Europe	1.9	7.7
South/South-East Asia	**6.6**	**26.7**
Africa and West Asia	1.4	5.7
Latin America	2.8	11.3
Total world market	24.7	

has continued since 1985 in most Asian countries except in Indonesia, where pesticide use has actually decreased.

In 1990, the Asian share of the world pesticide market was estimated at 26.7% (Table 6.1), split as 45.8% insecticides, 21.9% fungicides, 29.2% herbicides and 3.1% others (Table 6.2). Half this pesticide market is in Japan, especially for fungicides and herbicides.

For pesticides, South and South-East Asia is predominantly a market for insecticides (Figure 6.1), especially for use on rice, cotton and vegetables. An exception is Malaysia, where insecticides only account for 13% of the pesticide market, due to heavy use of herbicides on plantation crops (rubber, oil palm and cocoa). In Asia over 40% of pesticide is used on rice, with fruits and vegetables accounting for 33% and cotton 8.5%. It must be remembered, however, that there is considerable regional variation; thus in India and Pakistan, cotton represents 50 and 70%, respectively, of pesticide use.

The global value of the pesticide market for rice was estimated at US$2400 million in 1988, with Asia taking 90% of this. However, 60% of Asia's share is for Japan, which produces only 3% of the world's rice. On this small amount of rice, nearly half the world's rice insecticides and about two-thirds of rice fungicides and herbicides are used. In 1988, the annual expenditure per hectare for rice agrochemicals was US$680 in Japan. This compared with US$200 in South Korea, only US$14 in the

Table 6.2 The pesticide market in Asia by product group, 1990 (data of the GIFAP Asia Working Group)

Pesticide group	Market (US$ $\times 10^9$)	% Share
Insecticide	3.0	45.8
Fungicide	1.4	21.9
Herbicide	1.9	29.2
Others	0.3	3.1
Total Asian market	6.6	

(a)

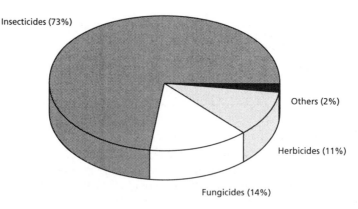

Insecticides (73%)

Others (2%)

Herbicides (11%)

Fungicides (14%)

Total = 113 640 mt kl⁻¹

(b)

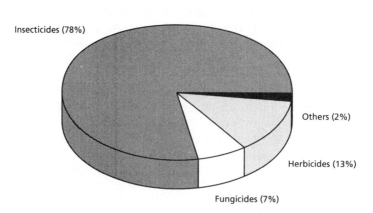

Insecticides (78%)

Others (2%)

Herbicides (13%)

Fungicides (7%)

Total = 750 394 mt kl⁻¹

Figure 6.1 *Combined production and import of pesticides for seven countries in Asia (India, Indonesia, Pakistan, Philippines, Sri Lanka, Republic of Korea and Thailand) in 1988. (a) Technical material; (b) formulated product (data from Anon.,* Renpap Gazette, **1**(3), 1991).

Philippines, US$5 in China, Thailand and Indonesia, and US$2–3 in the Indian subcontinent and Vietnam.

Overall, however, India and China are the largest consumers of pesticides in Asia. India is the second largest pesticide manufacturer in the region after Japan, with China next. In India, eight key formulations account for more than 85% of the total pesticide production. Three alone, HCH, DDT and malathion account for 55% of production. In China, locally produced pesticides form about 90% of the materials used. These 650 000 tons of pesticide are produced in over 200 factories and involve more than 400 formulations. DDT has been banned in agriculture and the use of HCH has been restricted. Yet DDT is still produced for use in public health and so any disappearance in agriculture is only theoretical. Korea is another major producer of pesticide, meeting about two-thirds of its own demand. In Pakistan, only DDT and HCH of the 200-plus products on the market are manufactured locally. Similarly, in Thailand nearly all technical material other than paraquat is imported. Indonesia produces about 13 300 tons of technical grade pesticides, and all other countries in the region are completely dependent on imports to meet their pesticide requirements.

Typical of Asia apart from Indonesia, where pesticide consumption decreased by 50% between 1986 and 1988 (Anon., 1991a), is an increase in pesticide consumption. This rose from 15.4 g ha^{-1} in 1960–61 to 440 g ha^{-1} in 1989–90. Although such consumption is low, compared with the 1.5–3.0 kg ha^{-1} of the USA and western Europe or the 10–12 kg ha^{-1} of Japan, it is worrying that pesticide residues in food and milk in India are the highest in the world. Huge quantities of HCH, DDT and malathion are used in pubic health, yet efficacy must be dubious. Mosquitoes have high insecticide resistance levels, and the applications have human and environmental hazards. The principal user agencies, the National Malaria Eradication Programme and the Food Corporation of India, are not short of sound scientific arguments for replacing these insecticides, yet they do not appear to be able to overcome political pressures.

Thus all kinds of pesticides, including DDT, continue to be easily available in the Asian markets. These markets include countries such as India and Sri Lanka which have tried to ban the use of DDT in agriculture. DDT has, of course, been singled out as a particular hazard to the environment. However, it (like HCH and methyl parathion) is cheap for subsistence farmers. DDT has the additional benefit of being particularly safe for the small farmer, whose only application method is the pesticide dust in a nylon stocking. However, farmers growing cash crops are generally provided with newer pesticides.

Thus pesticide use has been seen both by farmers and governments as an effective method with clear decision criteria and under the farmer's control. By contrast, IPM appears complicated, requiring far more knowledge and sophistication than is required for pesticide application.

An interesting study by Waibel (1990) has used rice in Thailand as an example for analyzing the preconditions for a 'take-off' of IPM technology. This scenario is typical for most countries in the region; in that policy has been in favor of continued increase in pesticide use. The four requirements for successful implementation of IPM were identified (Wardhani, 1991) as:

1. political will to develop, issue and maintain strong policy support for IPM;
2. an adequate base of field research;
3. emphasis on human resource development;
4. sufficient resources to sustain implementation and further development.

To this list we would like to add a fifth requirement: removal of subsidies for pesticides.

The most primitive level of IPM is that economic thresholds are developed as the basis for decision making. Most of the countries in the region have national plant protection services and extension, usually under their ministries of agriculture. Pest survey programs have been established in most countries (Pfuhl, 1988; Sri-Aruntoi, 1988; Saha, 1991). In India, survey data are gathered at 32 central stations for pest monitoring and forecasting of the need to take control measures. Similarly, there are 156 well-equipped forecasting units in Korea for rice pests. Similar warning systems are operated for rice pests in Indonesia, the Philippines, Malaysia and Thailand. China has monitoring programs for cotton, rice and vegetables.

In spite of all this, there is no convincing evidence that farmers in Asia have relinquished prophylactic spraying and embraced economic thresholds. On average, pesticides are applied at only 30% of the recommended threshold (Waibel and Meenakanit, 1988); much lower thresholds are, of course, part of this average. The use of trained personnel to monitor for the farmer has been tried in some IPM programs, but, in the end, economic thresholds really have to be farmer-operated.

Almost all the countries of the region have, as pointed out earlier, declared IPM as part of their agricultural development. They have accepted (in theory anyway) that IPM is the only rational approach to correct injudicious use of pesticides and to prevent future environmental catastrophe. In spite of all the negative pressures in the region discussed above, the process of turning IPM from a concept into practice has been started and many IPM programs have been developed to an advanced stage in the past 20 years, or are actually being implemented. FAO has helped by preparing guidelines for IPM in rice, cotton, maize, sorghum and vegetables and has also helped many countries in following such guidelines.

The economic viability of IPM has also been shown, yet large-scale adoption by the farmers remains limited.

The responsibility for IPM does not only rest with international agencies and governments. It also rests with researchers, the agriculture industry as a whole, farmers, environmentalists, health pressure groups and even consumers. Currently, however, in the Asian scenario, it remains up to government organizations to play the pivotal role in ensuring that IPM is developed, by providing financial support and appropriate incentives.

Case studies

Rice

Rice is probably the world's most diverse crop, with about 120 000 varieties. About half the world's rice-land is irrigated and this produces more than 70% of the harvest. More than 90% of the world's rice is grown on 145 million hectares (11% of the world agricultural hectareage) in Asia and provides 35-60% of the calories consumed by 2.7 billion Asian people.

Crop losses of rice to pests in Asia vary from 20 to 50% of production (Cramer, 1967; Litsinger *et al.*, 1987). The major pests are the brown planthopper (BPH) (*Nilapavarta lugens*), the rice hispa beetle (*Dicladispa armigera*) and several stem borers, particularly species of *Sesamia*, *Chilo* and *Scirpophaga*. Rodents are also important pests. Rice blast (*Pyricularia oryzae*), sheath blight (*Corticium sasakii*), bacterial blight (*Xanthomonas oryzae*) and virus diseases (particularly tungro) are the most important diseases. In addition, yield is limited by weeds.

The example of the Philippines is also typical of other regions of Asia (Shepard, 1990). More and more insecticide has been used by rice farmers over the past 30 years. Insecticide use was very low in the early 1950s, but had built up by 1965 to the point where 60% of irrigated rice-land was being treated; by the mid-1980s this had increased to over 95%. Some of these insecticides devastated the natural enemies of BPH, which also thrived on the continuous cropping of rice and the heavily fertilized susceptible rice varieties being grown. Thus this previously non-pest insect was promoted into the most serious pest of the region (Norton and Way, 1990).

Integrated pest control programs for rice are now being operated in many Asian countries. These are based on resistant rice varieties with some chemical control. Cultural controls have been recommended, but play a minor role in the programs, and the involvement of biological control has hardly been researched at all.

Host-plant resistance has dominated as the key tactic of IPM in rice. 'Rice is without equal in the extent to which insect resistance has been incorporated into high yielding varieties and the extent to which these varieties are being commercially grown by farmers over millions of hectares in Southeast Asia' (Heinrichs, 1988). However, it must be remembered that this major development in rice showed severe problems at first. The excessively extensive cultivation of a single resistant rice genotype led to the rapid appearance of an adapted strain of BPH with devastating results. Now the extensive planting of resistant varieties referred to in the above quotation involves several varieties with different sources of resistance to BPH.

In India too, the keystone of IPM packages in rice has been the use of resistant varieties. Other components have been cultural measures such as community rice nurseries and proper planting and spacing, conservation of natural enemies, and use of economic thresholds with selective pesticides. These technology packages have been made effective by training extension workers and farmers, and by field demonstrations emphasizing that IPM can improve economic returns as well as benefit the ecosystem.

The importance of conserving natural enemies in rice fields is now recognized. It has been claimed (Anon., 1991b) that a typical rice field in the tropics supports 800 species of beneficials, including spiders, wasps, ants and pathogens. These would control 95% of the insect pests, if conserved. In studies on the biodiversity of the fauna in irrigated rice fields, it was found that diversity of insects increased with crop age till maximum tillering. However, it was significantly lower in fields sprayed with insecticides. When populations of beneficials such as spiders and predatory bugs were high, hopper densities were low. With fewer predators, herbivore pests, such as BPH re-establishing after insecticide sprays, can increase rapidly.

The effectiveness of natural biocontrol in the absence of spraying is probably at least part of the explanation of effects in Bangladesh when the 100% subsidy on pesticides was reduced to 50% in 1974 and completely abolished 5 years later. The area treated with pesticide decreased from 5.12 million ha in 1972–73 to only 263 000 ha in 1986–87, but without any appreciable reduction in rice yields (Figure 6.2). Indeed, in recent years there has been a steady increase in rice yields in Bangladesh, irrespective of the quantity of pesticide used. Better new varieties, better fertilizer management and increased irrigation have all been involved in these yield increases.

In India, rice is now cropped continuously in the southern, central and coastal areas. These changes, coupled with higher-density planting and increased application of inorganic fertilizers accentuated crop protection problems. In particular, stem borers, gall midge, BPH, hispa beetle and weeds such as *Echinochloa* emerged as major problems, and epidemics of bacterial blight, tungro virus and blast occurred. It was inevitable that farmers turned to pesticides, to the extent that today 17% of India's use

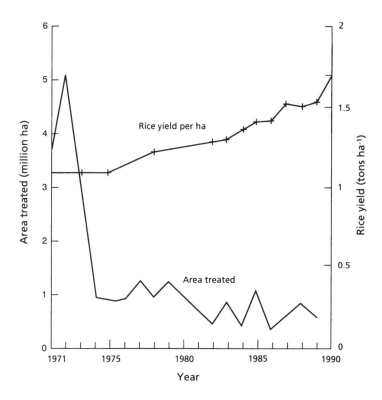

Figure 6.2 *Area of rice treated with pesticides and the rice yield per hectare, Bangladesh, 1971–90.*

of pesticide is on the rice crop. This intensive pesticide use has had serious ecological consequences. Minor pests have become major ones (e.g. leaf folder, white-backed planthopper and green leafhopper) and BPH has been resurgent through resistance to insecticides and loss of natural enemies. It is a hopeful sign that pesticide usage per hectare has recently declined from 263 g in 1985 to 190 in 1989–90 (Anon., 1992).

Mention has already been made of the Indonesian success in introducing IPM and reducing insecticide usage on rice. Although Indonesia used to be the world's largest importer of rice, the large-scale adoption of the 'green revolution' technology led to self-sufficiency by 1984. Rice now occupies about 60% of the total area devoted to food crops. The early years of this intensification were based on the use of pesticides, 3–4 times during each season, for pest control. Pest problems increased immensely and the previously minor pest, BPH, emerged as a major one, causing the loss of 350 000 tons of milled rice valued at more than US$100 million in the

1976–77 season. Additionally, tungro virus transmitted by the green leafhopper (*Nephotettix virescens*) became a major threat. In West Java, where aerial spraying had been used against stem borer, the white stem borer, *Scirpophaga innotata*, affected 75 000 ha, causing more than 40% yield loss.

When the devastating outbreak of BHP as a new major pest occurred, the problem was tackled with a blanket of aerial spraying in Java and northern Sumatra. Similarly, during the outbreak of rice tungro virus in Bali in 1981–82, 12 000 ha were immediately sprayed by air to control the green leafhopper vector.

After such outbreaks of BPH, pesticides became an insurance against attack, and farmers were supplied at heavily subsidized rates as well as given training in the use of application equipment. By 1984, the resulting overuse of pesticides was beginning to show the inevitable results and not long afterwards, in 1986, there was an explosion of BPH over most rice-growing areas of Indonesia. Insecticide subsidies of US$100 million per year were spent, but outbreaks of BPH nevertheless continued. It was subsequently shown that BPH outbreaks were actually engendered by such massive use of pesticides (Anon., 1991a).

Although Indonesia adopted IPM as official policy as early as 1979, subsidies for pesticides, and therefore pesticide use, remained heavy until 1986. The situation changed from that year, following a Presidential Decree on pesticide use (Chapter 8). Initially many chemicals were banned, and this was followed by a reduction and eventual abolition of subsidies for the remainder.

Rice production is now the highest in Indonesia's history and is 15% higher than 1986 levels. Pesticides are now no longer a major part of production technology packages, and their use has fallen by 60% since 1986 (Figure 6.3).

Besides the example from Indonesia, several agencies and institutions are now involved in different aspects of IPM for rice in the region. The FAO Intercountry Programme for Integrated Pest Control in Rice in South and Southeast Asia was started in 1980 and has trained over 400 000 farmers in IPM. This program has succeeded in reducing the use of broad-spectrum insecticides on rice in participating countries. Governments in these countries have also begun reducing subsidies for pesticides.

Another initiative, the Integrated Pest Management for Rice Network (IPM-R), coordinated by IRRI, was started in 1990. Each participating country has organized a team of research scientists and extension specialists involving plant protection science, social sciences and the communication skills (IRRI, 1991).

Because of all these coordinating actions, IPM is currently being practised on about 4.9 million ha of rice (Norton and Way, 1990). The positive experiences of IPM gathered in the past 10 years are being

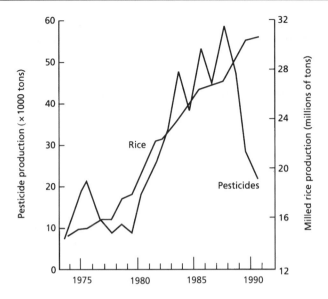

Figure 6.3 *Pesticide usage and rice production, Indonesia, 1973–91.*

extended to larger areas in several countries. Since rice is usually only part of the cropping system, it is desirable to extend the IPM concept across the other crops in the system also. Particularly in relation to the conservation of natural enemies, such an extension of pesticide reduction is essential.

Rice is a very 'political' crop in Asia. Most of it is grown for national consumption; only 2.4% is traded on the world market. Any crop failure therefore has repercussions on the stability of food availability in the region. IRRI (1989) has estimated that an increase in production of more than 20% over the 1987 level of 460 million tons is required by the year 2000. An increase of over 65% will be required by 2020. The implications for pest management are twofold (Norton and Way, 1990). First, some of the increased production required could be obtained by reducing the present level of losses from pests, which is perhaps as much as 50% of potential yield. Secondly, changes in crop production technology and more intensive rice cultivation will become necessary; the effects of these measures on future pest problems as well as the farmers' ability to deal with them need consideration ahead of the event.

Vegetables

The Asian-Pacific region produces more than 200 million tons of vegetables, which is about 47% of world production. China and India together account for about three-quarters of the region's vegetable production. As well as providing important supplementary nutrition for farmers, vegetables also represent an important source of income. Many are produced in the region and include:

1. crucifers such as cabbage, Chinese cabbage, cauliflower and radish;
2. Solanaceae such as tomatoes, potatoes, aubergines and peppers;
3. cucurbits, mainly cucumber, watermelon and pumpkin;
4. legumes such as peas and beans.

Onion, garlic and okra are also important.

The dominant vegetable pest of the region is diamond-back moth (DBM), *Plutella xylostella*, a pest of most crucifers that has become resistant to most pesticides. Pests of vegetables tend to be host-specific, and other serious problems are the melon fly (*Bactrocera cucurbitae*) and *Thrips palmi* on cucurbits, the bean fly (*Ophiomyia phaseoli*) on beans and the Pyralid *Leucinoda orbonalis* on aubergine. Among polyphagous pests, *Heliothis armigera* is a major problem in tomatoes and the armyworm (*Spodoptera exigua*) and cutworms (*Agrotis* spp.) are important in many vegetable crops.

Excessive use of pesticides on vegetables has caused great problems and concern throughout Asia. The FAO report of the Expert Consultation on IPM in Vegetable Crops (FAO, 1989) concluded that crop protection in vegetables in Asia had entered the 'crisis phase'. A wide range of insecticides and fungicides, often applied at far higher dosages than recommended rates and not infrequently as cocktails, are sprayed with unfortunate frequency. Often, because of farmers' ignorance or in the quest for cosmetic quality, they are sprayed close to harvest. The resulting problems have been rising production costs, and development by pests of resistance to pesticides. Large-scale environmental contamination has occurred and there has been an increased human health hazard, including poisoning of farmers and high toxic residues in market produce (Guan-Soon, 1990). In India, the pesticides in use include HCH, methyl parathion, nicotine sulfate, elemental sulfur and several organomercury compounds. Some of the highest residue levels ever recorded have been found in vegetables from urban markets in India. As much as 35 p.p.m. of DDT has been found in spinach. Farmers have even been known to apply pesticide (methyl parathion) after harvest to give cauliflower curds an extra white appearance.

The FAO report stated:

> ... progressive resistance development to pesticides in major vegetable pests is a very serious problem. It constitutes an important underlying cause of the pesticide dependency in farmers. An insidious danger is that it subtly drives farmers deeper and deeper into a state of desperation to constantly seek for stronger pesticides, while at the same time trapping them with the vicious cycle of pesticide dependency.

IPM for vegetables has reached different stages of development in different countries. The most developed programs are on crucifers, because of the ubiquitous difficulty in controlling DBM with chemicals.

The DBM threat to vegetable production is approaching the pest threat in cotton in the 1950s in Central America, which destroyed the cotton industry at that time (p. 38). For 20 years, farmers had successfully controlled DBM with pesticides. However, this has led to the development of resistance to nearly all groups of insecticides in recent years. With the absence of new alternatives, farmers then increased the insecticide pressure on the pest by mixing chemicals and/or applying higher doses. Embarking on the pesticide treadmill has led inevitably to resurgence of DBM through destruction of natural enemies and progressively increased levels of pesticide resistance. Loss of natural enemies has, of course, also favored other previously only minor pests such as armyworms, cabbage head caterpillar and cabbage webworms, all of which are now major pests in South-East Asia (Talekar, 1991). This left little choice but to change to IPM; if sole reliance on pesticides had been allowed to continue, the 'disaster phase' would have been inevitable. The IPM that has been developed successfully in several countries centers on several components. These are the use of egg, larval and pupal parasitoids, cultural practices, microbial insecticides (*Bacillus thuringiensis* [Bt]) and judicial use of insecticides only when necessary. The principal egg parasitoid used has been *Trichogrammatoidea bactrae*, and parasitoids attacking the larvae/pupae (*Diadegma semiclausum* and *Cotesia plutellae*, also *Diadromus collaris*) have been widely distributed.

In 1989, the Asian Vegetable Research and Development Centre (AVRDC) initiated an intercountry IPM program for DBM (AVNET). This program includes exchange, mass rearing and field establishment of important parasitoids of DBM, and also technician training.

In Malaysia, IPM for DBM in the Cameron highlands has been based on the conservation of the indigenous *C. plutellae*, together with the establishment of the exotic parasitoids *D. semiclausum* and *D. collaris* (Loke et al., 1992). Upon DBM reaching the economic threshold, the use of Bt is recommended. A similar strategy, involving conservation of *C. plutellae*, introduction of *D. semiclausum* and limited use of Bt, operates in the highlands of Taiwan. Here efforts are also being made to wean farmers away from chemicals (Talekar, Yang and Lee, 1992). Inundative releases of *C. plutellae* and the egg parasitoid *T. bactrae* have been largely successful in the Philippines (Morallo-Rejesus and Sayaboc, 1992). In Indonesia *D. semiclausum* effectively controls DBM in the highlands, but the parasitoid unfortunately cannot survive in the lowlands (Sastrosiswojo and Sastrodihardjo, 1986).

DBM adults are strongly attracted to the color yellow. In the lowlands of Thailand the combination of yellow sticky traps with releases of *T. bactrae* and *C. plutellae* has given successful control.

DBM is also attracted to mustard plants, and these are used as a trap crop in southern India as part of an IPM system. The trap mustard attracts 80–93% of DBM adults and almost all those of two other pests of cabbage, the leaf webber (*Crocidolomia binotalis*) and the stem borer (*Hellula undalis*). Other brassicas can be sprayed at the head-initiation stage with 4% neem kernel extract to check DBM. This IPM system has been adopted by many farmers in Karnataka State (Srinivasan, 1992).

With the aim of producing 'nonpolluted vegetables', IPM programs have been launched in China, and cover 200 cities in 22 provinces. These programs include resistant crop varieties, cultural controls and biological control, the latter focused on Bt and egg parasitoids of Lepidoptera. Pesticide use on vegetables has been restricted and residues are regularly monitored. About 85% of losses from pests and diseases can still be prevented with the IPM systems. Also, vegetables produced under IPM fetch at least a 10% premium in price (Di, 1990).

IPM programs for vegetables have therefore been launched successfully in China, India, Malaysia, Indonesia, Taiwan, Thailand, in the Philippines and also in Vietnam. However, many countries in the region, including Bangladesh, Sri Lanka and Pakistan, have yet to exploit nonchemical alternatives for pest control in vegetable crops. The need to adopt IPM in vegetables to replace sole reliance on pesticides is now well recognized in all the countries of Asia. For vegetables, perhaps more than for any other crop, IPM strategies must consider and modify the attitudes of consumers, since market demand has an immense influence on the acceptance of IPM by farmers.

It would appear that technologies for IPM on vegetables in Asia are now sufficiently available for the development of programs. What is now necessary is to extend such programs to far greater numbers of farmers.

Cotton

At least 12 countries in Asia grow cotton and, in 1989–90, 44% of the world's cotton was produced there. Cotton is the main cash crop of China, India and Pakistan, and is also important in Thailand and Myanmar. As in most regions where cotton is grown, insects are the major factor limiting production and the crop receives more pesticide than any other. Nevertheless, yields per hectare are still low and there are frequent, sometimes disastrous, pest outbreaks.

The potential high economic returns on the crop encourage farmers not only to use a great deal of pesticide (which may account for 40% of production costs), but also much fertilizer and irrigation. Pesticides tend to be overused and misused, leading to more pesticide-associated problems in cotton than in other crops. The development of resistance in several pests has put cotton farmers on the treadmill of increasing pesticide inputs and increasing loss of natural enemies.

The organochlorines, organophosphates and carbamates used in the 1960s and 1970s have, since 1980, been largely replaced by the synthetic pyrethroids. Of the many pests attacking cotton, particularly bollworm (*Heliothis armigera*) and the whitefly, *Bemisia tabaci*, now appear as typical resurgence problems associated with overuse of pesticides and elimination of natural enemies. As these two pests also move seasonally between cotton and other crops, the overuse of pesticide on cotton could make them harder to control on the other crops also. The adoption of IPM has therefore become imperative.

In India, although cotton accounts for only 5% of the area under cultivation, it accounts for half the pesticide use. Although use of pesticides has declined in food grains, on cotton it increased from 2.4 to 2.7 kg ha^{-1} between 1985 and 1989–90 (Anon., 1992). Eighty per cent of all pyrethroid use is on cotton, where *H. armigera* has now developed resistance.

Research has clearly shown that dependence on insecticides can be reduced considerably with an IPM package. This includes cultural controls, biological control by timely releases of parasitoids (primarily *Trichogramma* spp. egg parasitoids) and predators (*Chrysoperla* spp.), the use of nuclear polyhedrosis virus and judicious use of insecticides (Verma, Shenhmar and Gill, 1990; Sundramurthy and Chitra, 1992). IPM projects in some more progressive cotton-growing states, especially Tamil Nadu and Gujarat, have shown the benefits of IPM to farmers on their own fields. The high yields of cotton varieties and hybrids can be maintained, the abundance of natural enemies increased and insecticide costs reduced (Figure 6.4) (Anon., 1989).

Disasters in cotton, due to whitefly in 1983 and 1984, and due to bollworm in 1987 and 1988, occurred in Andhra Pradesh as well as in other cotton-growing states. These disasters have attracted considerable public attention, but they have been a blessing in disguise for IPM. They have made cotton growers in these states much more aware of the consequences of excessive pesticide use and thus more willing to accept IPM. Pest monitoring by both state and federal agencies has been strengthened, and the number of sprays applied has mostly been reduced from as many as 20 to as few as 3–6. The pesticide industry has itself become concerned about misuse of its products and pest resistance to them. The companies have therefore set up their own advisory and extension programs and are working with government agencies in setting up IPM demonstration trials for farmers.

In China, annual losses of cotton production due to pests are estimated to be 25–40% of potential yield. Several IPM programs have been initiated since 1985. One example is in the cotton-growing area

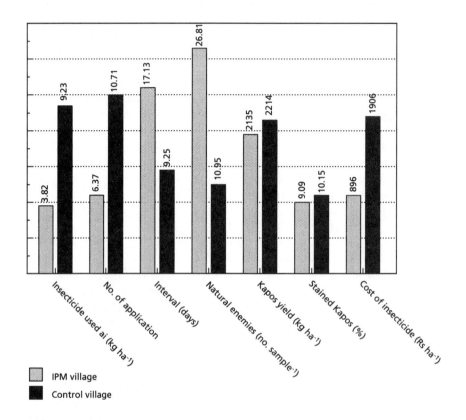

Figure 6.4 *Use of insecticides on cotton and yield of seed cotton; also abundance of natural enemies in villages practising or not practising ('control') IPM (1989 data of the Indian Council of Agricultural Research, New Delhi).*

north of Shihezi in North-West China. Here, the number of insecticide applications has been reduced from over 10 per year to an average of 1.5 by implementing IPM over a 10 year period. The crops are monitored for pests every 5 days and insecticides are applied in relation to locally devised thresholds, which take the abundance of natural enemies into account (Matthews, 1991).

The cotton situation in Pakistan shows every sign of deteriorating into the familiar pesticide treadmill. However, an IPM program was developed in the late 1970s and is being implemented through several cotton maximization programs. The Directorate of Pest Warning and Quality Control of Pesticides has a network of field workers to help cotton farmers. Many of these farmers have been trained in a pest scouting system (Ahmad, 1991).

In Thailand, cotton production has been in decline for some years, due to pest problems. Cotton IPM was initiated as part of a 5-year (1984–88) IPM project for nonsubsistence crops in cooperation between FAO and the Thai government. The economic thresholds for bollworms have been determined. *Heliothis armigera* is being controlled by nuclear polyhedrosis virus and the distribution in fields of eggs parasitized by *Trichogramma*. Cotton IPM programs have also been initiated in Bangladesh, Myanmar, Indonesia and the Philippines.

All cotton-growing countries in Asia now have policies to implement IPM in cotton. Most, however, do not have an institutional framework to achieve this. Although considerable research effort has led to IPM packages being available for cotton in most parts of the region, very few such packages have been implemented on a large scale.

Sugar cane

Sugar cane is an important crop in several Asian countries. The important insect pests include stem-boring caterpillars, leafhoppers (e.g. *Pyrilla*), whiteflies, scale insects, mealybugs and chafer grubs. Red rot disease can also be serious, as can rodents and grassy weeds. Control of many of the insects with sprays is made difficult by the density of the foliage or because they feed in concealed places or are in the soil. Nevertheless, insecticides have been used extensively and development of IPM has been slow.

Some alternative approaches to control, including cultural and biological methods, have been developed. For example, excellent control of stem borers has been achieved by release of egg parasitoids (*Trichogramma*). Sucking insects are kept in check by redistribution and conservation of natural enemies (Mohyuddin, 1991). Adoption of these successful methods is still only localized, with the result that pesticides continue to be used extensively.

However, recently control with pesticides has become less effective and, as a result, mill owners and farmers are showing interest in other methods. Prophylactic broad-spectrum treatments are slowly being replaced with more selective compounds applied based on need. In India and Thailand, attention is also being given to improved application and better timing.

There have been two notable successes with biological control in India. The borer *Scirpophaga excerptella* has been almost permanently controlled by an indigenous parasitoid (*Isotima javensis*) of the larvae, as the result of inoculative releases made 15 years ago (Solayappan, 1987). The second example is control of *Pyrilla* leafhopper with *Epipyrops melanoleuca* (Chaudhary and Sharma, 1988). This parasitoid has proved difficult to mass produce, so the approach has been to collect it in the field early in the season and redistribute it to new fields. The parasitoid is now established in most sugar-cane-growing areas of India. The technology for mass production of *Trichogramma* spp. is available in India, and inundative releases of *T. japonicum* have provided effective control of stem borers. Similarly in Pakistan, borers have been controlled by introducing several strains of *Apanteles flavipes* and augmentative releases of *T. japonicum*. *Pyrilla* has been checked by redistributing *E. melanoleuca* and by conservation of the egg parasitoid *Parachrysocharis javensis* (Mohyuddin, 1991).

IPM for sugar cane in Asia now needs more support in terms of infrastructure, particularly facilities for mass rearing of biological control agents. Sugar cane cultivation is controlled directly or indirectly by the sugar industry. Adoption of IPM should thus be easier, once the industry becomes convinced of the value of the technology. Moreover, the long season and dense canopy of sugar cane makes it an excellent ecosystem for exploiting biological control.

Horticultural and plantation crops

Horticultural crops such as citrus, mango, apple, banana, rubber, oil palm, cocoa, coffee and tea represent high-value cash crops, grown in Asia both in large estates and by smallholders. Most are grown as monocultures whose perennial nature sustains natural enemies that keep most pests at nondamaging levels. However, any breakdown of this equilibrium (for example, due to overuse of pesticides) can result in pest outbreaks. This has been appreciated in Malaysian plantations since the early 1970s and has led to the development of IPM. This has had a good success record there in oil palm, cocoa, rubber and coconut. The resultant minimal use of pesticide has provided good management of pests with improved economic benefit, greater safety for operators and less environmental damage.

IPM programs for most fruit crops exist in several Asian countries. Programs involving limited and highly selective pesticide applications for plantation crops such as oil palm and cocoa were developed several decades ago. These developments followed evidence that pesticide use in these crops had itself caused outbreaks of defoliating caterpillars by eliminating their natural enemies (Djamin, 1988; Ho, 1988; Liau, 1991).

In Indonesia, a 15 year master plan for IPM on estate crops has been underway since 1985, when a Directorate of Estate Crop Protection was established under the Ministry of Agriculture. In India, biological control has been researched for pests of fruits such as citrus, grapes, mango and apples (Singh, 1990).

Coconut palm is a particularly important crop in the Philippines, Indonesia, India, Sri Lanka, Thailand and Malaysia, and promising results in managing pests have been reported. The rhinoceros beetle (*Oryctes rhinoceros*) is a key pest combatted by combining insect pathogens (baculovirus and fungi) with cultural methods (Pillai, 1985; Kaske, 1988).

Successes with IPM in plantation and fruit crops have been mainly local, and pesticides are still widely used on these crops in the region. This is partly because there is currently no IPM for some major pest and disease problems, including bunchy top of bananas, the coconut beetle *Scapanes australis* and rodents in oil palm.

Africa

African agriculture is characterized by most small-scale farmers cultivating holdings of less than a few hectares, and often even less than one. Such small-scale farming not only dominates subsistence agriculture, but even cash crops. The crops are often cultivated as mixtures (intercropping, mixed cropping, see p. 119). In most countries farmers do not have title to their land, but only the right to cultivate it inherited through the maternal line. Some farmers are tenants or share crop land controlled by others (particularly in relation to cash crops such as cotton).

Who owns or has rights over land is extremely important in implementing IPM. Only those who will reap the fruits of their labor are likely to invest the necessary time and money. Thus although the women often do most of the actual field work, they have little incentive to do any extra IPM-related work since men are the legal owners of the harvest.

Only few of the cultivated crops Africa are indigenous. Exceptions are two of the most important subsistence crops, millet and sorghum, and coffee and cotton among the cash crops. Most of the other important crops were introduced to the continent a long time ago (e.g. rice), others with European colonizers from the sixteenth century onwards (e.g. cassava, maize, sweet potato and cocoa) and some much more recently (e.g. potato and wheat).

Unlike the crops, many pests are indigenous (e.g. most pests of millet, sorghum and coffee). However, there have also been many accidental pest and disease introductions from other continents. In recent years, the introductions of cassava mealybug, cassava green mite and the larger grain borer have resulted in major problems. Such introductions have stimulated considerable classical biological control activity, and several successes have resulted.

The knowledge of African farmers about pests is usually good. This is partly because cultural and mechanical pest control measures, which form an important part of African land management systems, have evolved over millennia. Over this time, selection of seed by farmers has incorporated much pest and disease resistance into the traditional land-race varieties.

Use of pesticides is limited by their high cost and by their nonavailability in most local markets. However, in commercial crops often under the control of cooperatives, pesticides are the major control method and are applied prophylactically on a calendar basis. Farmers of commercial crops are then likely also to treat any food crops they grow for home consumption. Insecticides also percolate down to some small farmers when substantial surplus insecticide is supplied by governments and donors for other purposes, e.g. locust control. Many governments also still subsidize pesticides, and this can lead to overuse. Another reason for overuse is that there is often an acute shortage of trained plant protection specialists. Also, the plant protection services are not well organized and extension is weak or absent.

Although pesticide use is still low in Africa when compared with other continents, it is increasing rapidly (WHO, 1990). The increase of 60% between 1983 and 1988 was the highest in the world during this period. Based on admittedly scanty data, a 200% increase was predicted between 1988 and 1993, compared with 40% in Latin America and only 25% in Asia. There is an urgent need for more accurate statistics about pesticide use in Africa.

A recent World Bank survey (Kiss and Meerman, 1991) points out that, although work on single components of IPM (e.g. biological control, host-plant resistance) is common throughout Africa, IPM projects are few (Table 6.3). Only Sudan has adopted IPM as the official national crop protection policy, although IPM is accepted as policy by some African growers' groups. Most IPM work in Africa has been through projects funded by international donors. One serious problem is the lack of education and training. Many researchers have been trained in the developed countries; their salaries are small and many have joined the international research institutes. The research staff remaining need more training as well as better access to literature. The need to make governments and donors aware of these problems is urgent, if over-reliance on pesticides is not to become a feature of Africa as it is elsewhere.

In broad outline, IPM in Africa should be targeted to two different situations: food crops, where no or little pesticide has yet been used, and commercial crops, where overuse of chemicals has occurred.

Case studies

Pearl millet in the Sahel

In the Sahel–Sudan region, millet forms the basic food for a large part of the population. It generally outyields other cereals on poor soils and shows great pH and salinity tolerance. It is cultivated both in monocultures and polycultures (with sorghum, leguminous plants or *Acacia albida* as a shade plant). In spite of its world importance, pearl millet has only recently been subjected to crop improvement by breeding (in Niger and at the

Table 6.3 Summary of IPM and IPM-related projects in Africa, November 1992 (compiled by Dr O. Zethner from FAO literature and individual project reports)

Contents of projects	Basic food crops	Basic food stores	Fruit	Beverages	Cotton	General
Phases						
Planning	1	0	1	0	0	1
Research	20	4	3	1	2	1
Development	15	5	4	1	3	1
Farmer active involvement	6	0	1	1	3	0
Coordination/organization	12	3	0	0	0	1
Components						
Resistance breeding	9	0	0	0	0	0
Biological control	8	3	5	2	2	0
Chemical control	7	3	2	2	3	0
Monitoring	5	0	1	2	1	0
Cultural control	7	0	1	1	1	0
Training of:						
researchers	11	3	0	1	1	0
extension staff	14	3	2	1	3	2
farmers	6	0	0	1	3	0
Total projects involved (total = 51)	32	5	6	2	4	2

International Crops Research Institute for the Semi-Arid Tropics [ICRISAT], India).

In spite of its relative tolerance to weed competition, millet still has weeds as its most severe problem, especially during drought. The parasitic weed *Striga* is becoming increasingly important in Africa.

A stem-borer complex, consisting of *Chilo* spp. in East Africa and *Sesamia calamistis* and *Coniesta ignefusalis* in West Africa, is found in most millet fields. *Coniesta ignefusalis* causes particularly heavy damage. Grasshoppers and armyworms (*Spodoptera* spp.) can cause serious defoliation, especially of young plants, and downy mildew (*Sclerospora graminicola*) stunts growth and causes deformed chlorotic leaves and tillers.

Grain-eating birds (*Ploceus, Passer* and *Quelea*) are the most damaging pests of the head. After drought hit the Sahel in the late 1960s, the millet spike borer (*Heliocheilus alpipunctella*) increased in importance. In some areas meloid blister beetles (*Psalydolytta*) often destroy entire fields by sucking grains in the milky stage. The smut fungus (*Tolyposporium penicillariae*) converts grain into dark sori, and downy mildew may, if attack is heavy, also destroy the head.

There is now sufficient knowledge for a provisional IPM program to be recommended for Sahelian conditions: millet varieties with tolerance or some resistance to birds, insects or diseases are recommended. Work on breeding such resistant varieties has been intensified, and varieties with long bristles on the head give some protection against birds and blister beetles. One of these varieties is the Gambian Souna bado laebi, and this has also shown good yields in spite of attack by spike borer and even locusts.

Crop rotation (often with groundnut in the Sahel) and removal of plant debris after harvest, a practice that reduces fungus and stem-borer attack, are recommended. Where millet stalks are used for construction (e.g. fencing), heating them before use is recommended to kill overwintering stem-borer larvae inside the stems.

Early sowing is recommended, especially where there is a risk of blister beetle attack. Seed should be dressed with fungicide, and sowing in rows by machine will simplify later weeding. Fertilizer should be applied close to the middle of the crop rows to benefit the millet and not the weeds. Weeding within the row, once only and before weeding between the rows, can increase yields by 25% (Zethner, 1987). Young *Striga* plants must be totally uprooted or killed before flowering, since their minute seeds can survive in the soil for more than 10 years and give rise to constant new infestations.

Plants infested with downy mildew or stem borer should be rogued to reduce later infestation of the head. A traditional control measure is lighting fires in the field; the smoke deters blister beetle from the millet heads particularly if groundnut shells are burnt.

Surveys during the IPM project of the Comité permanent Interétats de Lutte contre la Secheresse dans le Sahel (CILSS) revealed the presence in millet of several potential biocontrol agents. A high rate of parasitization of *Heliocheilus* by *Bracon hebetor* was recorded during the latter part of the growing season. That this parasitoid also attacks *Ephestia* moth in stored millet can be exploited while stored millet remains available, and the release of parasitized *Ephestia* larvae into millet fields in Senegal has given promising results, even if on a small scale (Bhatnagar, 1987).

As a result of cooperation between ICRISAT and the UK Overseas Development Administration (ODA), a highly efficient pheromone blend has been synthesized for *C. ignefusalis*. A trap that can be operated by farmers to control the borer directly has also been produced. This trap is also immediately useful for large-scale monitoring of the pest (ICRISAT, 1992).

Chemical insecticides raise yields far less effectively than fertilizer or the use of resistant varieties. Chemical control of the spike borer is probably always uneconomic, whereas chemical control of blister beetles and locusts may be warranted to prevent total crop destruction. Mostly economic thresholds, important prerequisites for minimum insecticide use, have yet to be determined. Unstressed millet has the potential to compensate for borer damage, with the result that equally sized populations may cause less damage in wetter than in drier years (Jago, 1991).

Sorghum in the Sahel

Sorghum originated in Africa, and is widely grown in semi-arid areas. Its pest spectrum is similar to that of pearl millet, although sorghum is more susceptible to weeds, especially under drought conditions. Most sorghum varieties are also more susceptible to *Striga* than any other cereals grown in the Sahel.

Besides many less economically important pests, four main insect pest groups occur. These are shoot fly (*Atherigona soccata* and other species), a stem-borer complex (including *Busseola fusca*, *Sesamia* and *Chilo diffusilineus*); gall midge (*Contarinia sorghicola*), causing malformation of the spikelets; and head bug (*Eurystylus* spp.), sucking on the developing grains. Smut fungi, destroying the grains, are the most important sorghum diseases in the Sahel, though grain molds are currently on the increase.

Lack of research has prevented comprehensive IPM recommendations to be formulated, but there has been some progress. ICRISAT has developed sorghum varieties with broad-spectrum resistance to *Striga*, though the resistance may be unstable with high weed pressure. There has also been a breakthrough with plant resistance to shoot fly (ICRISAT, 1992), but commercial resistant varieties are still some time away. Research at ICIPE in East Africa has revealed promise for resistance breeding against stem-borer species also present in West Africa.

Cultural control can be recommended as for pearl millet (see above), especially the removal of *Striga* and the destruction of crop residues and diseased plants.

Chemical control is not recommended apart from seed treatments against smuts and herbicides against *Striga* if a shortage of labor precludes hand-weeding.

Cassava

Cassava was introduced into Africa from South America by the Portuguese, probably late in the sixteenth century. It is now widely cultivated throughout tropical Africa from humid to semi-arid regions. Cassava is drought resistant (it can survive even 4–6 months of drought), performs well at low nutrient levels, requires few production skills and is easily adapted to the social framework of the farming community. It has played a vital role in alleviating famine conditions when other crops have failed.

Cassava is commonly intercropped by smallholders, though large commercial stands also exist. It has a long growing cycle (9 months to 2 years). When intercropped, the different crops usefully mature at different periods.

Cassava is normally grown on ridges that need to be preserved during weeding, which remains necessary until a good crop canopy has been formed. It has been relatively free of arthropod pests in Africa when compared with South America, its continent of origin. The large quantities of cyanogenic glucosides in the latex may have prevented attack by most indigenous herbivores. However, there are two accidentally introduced pests, cassava mealybug and cassava green mite, that avoid the chemical defense. These have spread in a surprisingly short time since introduction, to become the most severe pests of cassava in Africa.

Cassava mosaic virus in Africa is transmitted by the whitefly *Bemisia tabaci* and perhaps other insects, and is also carried in cuttings of infected plants. The virus can cause tuber losses of 20–60% (Hahn *et al.*, 1990). Bacterial blight caused by *Xanthomonas campestris* pv. *manihotis* is indigenous to Africa and has been reported from many countries. This disease may cause complete yield loss under conditions favorable to its development and spread. Movement of infected planting material and rain-splash are the most important agents of spread (Hahn *et al.*, 1990).

Biological control of cassava mealybug, which stunts the growing points and sometimes totally defoliates the plants, has already been referred to in Chapter 5. Losses of over 80% tuber yield were reported (Herren and Neuenschwander, 1991) as the pest spread rapidly across the continent at a rate of 300 km yr^{-1}. Effective control soon became a prerequisite for the cultivation of cassava in many countries. An International Institute of Tropical Agriculture (IITA) workshop in 1977 concluded that classical biological control and resistance breeding were both feasible options, and a biological control program was commenced in 1979. Ten years later it had grown into what is probably the world's largest ever biological control program, supported by many international donors and collaborating with several other international research organizations.

Following the discovery of the mealybug in Paraguay in 1980, 10 species of natural enemy were collected there, including the parasitoid *Epidinocarsis lopezi* (p. 94). Following screening at the International Institute of Biological Control (IIBC) in the UK and mass-rearing at IITA in Nigeria, ground and aerial releases were made. Monitoring showed that, 3 years after the first release, *E. lopezi* could be found in 70% of all cassava fields in south Nigeria. In 1985 the initial program was expanded into the Africa-wide Biological Control Programme (ABCP). By 1990, the parasitoid had become established in 25 countries where cassave is cultivated. There biological control has become self-perpetuating following a single input, and thus has shown the high benefit:cost ratio of 178 to 1, with a return of over US$20 million over the 20 years since the mealybug was first found in Zaire. National biological control programs and facilities had been established in 19 countries by 1987. This was a necessary development, since there were strong indications that the original stock liberated from the IITA center in Benin might not be sufficiently widely adapted. Adapted individuals do, however, appear in areas where mass release has taken place and these can be collected and mass-reared in national programs.

Cassava green mite (*Mononychellus tanajoa*) was also introduced accidentally from South America, first into Uganda in 1971 and later to most of the cassava belt in Africa. The spread appears to have been by aerial dispersal and the movement of infested plant material. Attacked leaves are deformed and become mottled and bronzed, ending in necrosis of stems and leaves downwards from the top of the plant. Cassava green mite is a less serious pest than the mealybug, but still can cause severe yield losses of tubers in susceptible cassava varieties. No satisfactory biological control has yet been found with the potential agents that have been introduced and released. Recently, most interest has become centered on predatory Phytoseiid mites, which can maintain control at lower pest densities of the green mite than can other predators.

Breeding for plant resistance against cassava mosaic virus was achieved by Nigerian researchers long before either mealybug or green mite had been introduced into Africa. One such 'old' resistant cassava clone (No. 58308) has been crossed with local cultivars and even a different cassava species. These new clones have maintained their resistance to the virus under different environmental conditions over the past 10 years.

Progress with resistance breeding to bacterial blight has not yet reached the stage of cultivar release.

Sources of plant resistance to the mealybug and green mite have been located in IITA breeding materials; with both pests resistance is associated with hair density on both leaf surfaces, the petioles and the stem tips. Small-scale cassava growers are willing to replace the older varieties with new resistant ones; where IITA distributed these varieties, about 90% of the surveyed villages planted them.

There are also cultural practices for the control of pests of cassava. Initially, selection of healthy planting material is very important, particularly with regard to virus and bacterial disease. Basal stem cuttings from 12–18-month-old plants give fastest growth and highest yields. Yields can be further improved by supply of nutrients through mulches or fertilizer and the use of nontillage systems. All these improvements also give the plants better recovery from insect attack. Intercropping would be expected to reduce the risk of insect or mite infestation, but no data are available.

However, without doubt the most important cultural crop protection measure is weeding. Unfortunately this is time-consuming. It has to be carried out 2–3 times per growing season, yet is essential to prevent weed competition drastically reducing yield. Labor is lacking on the farm at peak cultivation periods, so root crops tend to be given a low priority. Herbicides are becoming increasingly used in African agriculture, and not just by the large-scale cash-crop growers. Small-scale farmers are finding that herbicides may be cheaper than the labor for manual weeding. An integrated weed management system could combine low rates of pre-emergence herbicides with low-growing intercrops like melon under rapidly branching cassava cultivars that quickly shade the ground.

Rice, with special reference to Burkino Faso and Madagascar

Rice is cultivated in many parts of Africa, either under rainfed or irrigated conditions, and mostly by small-scale farmers. Its importance has increased due to a change in people's attitudes and eating habits, particularly in those African countries in receipt of large amounts of rice as food aid during emergencies such as periods of drought. Both the cultivation pattern and pest complex vary over the continent, with the different ecological conditions (e.g. dry land, swamp or irrigation) under which the rice is grown.

However, some serious problems are common to almost all the rice-growing areas (Hill and Waller, 1988). These include the lepidopterous stem borers in the genera *Maliarpha* and *Chilo*, rice blast (*Pyricularia oryzae*) and weeds. There are also bugs, gall midges, leaf-mining beetles, case-worms and armyworms, besides some serious diseases (brown spot, sheath blight, bacterial leaf blight) caused by fungi and bacteria. Crabs are serious pests in mango swamp rice in West Africa.

Two well-defined projects to introduce IPM in rice in Africa will be described here. One is the program 'Rational Pest Control in Irrigated Rice in Vallee du Kou, Burkino Faso', which has been supported by the Dutch government since 1985 (Meerman, 1991a). The other is 'IPM in Irrigated Rice in the Lac Alaotra Basin, Madagascar', supported by Switzerland since 1984 (Zahner, 1991).

Burkina Faso

The Vallee du Kou accounts for 25% of the total area of rice in Burkina Faso. Rice-growing here was started in the 1970s with Chinese assistance. Initial high yields then dropped dramatically in 1983 due to disintegration both of the irrigation system and the farmers' cooperative, and also because of problems with soil fertility.

The following new program has increased yields again by reorganization of the cooperative, restoration of the irrigation system, recovery of soil fertility and implementation of IPM.

The previous rice variety, which had become susceptible to rice blast with the appearance of a new physiological race of the disease, was replaced with resistant varieties developed at the West African Rice Development Association (WARDA) and other research stations (WARDA, 1990).

Specific weeding of *Echinochloa* spp., alternative hosts of *Chilo* borers, was introduced with fines for failure to remove the weed. Synchronized cropping was introduced to prevent a build-up of pest and disease organisms. The rice caseworm (*Nymphula depunctalis*) was effectively controlled by draining the paddies for 3 days, since the floating larvae cannot survive without water.

Chemical control had been 2–3 calendar-based sprays of pyrethroids, available through the cooperative, per season. Farmers were charged for the applications whether they used them or not. This was inappropriate as pest attack varies considerably between seasons; moreover, the health hazards were high since irrigation water was used for human consumption and fishing. A more rational use of insecticides to control stem borer was developed based on weekly monitoring by trained farmers and thresholds. This resulted in a 50% drop in insecticide use and net profit increases of around 10%. Farmers were highly motivated to adopt the new system. Incidence of pests and diseases in Vallee du Kou is now less than in other irrigated rice schemes in the country.

In West Africa as a whole, WARDA programs are promoting repeated tillage integrated with herbicide application (sometimes supplemented by manual weeding). These programs have eliminated problems with *Euphorbia heterophylla*, one of the most difficult weeds of upland rice. Annual weeds in irrigated systems can be shaded out with a layer of the free-floating pteridophyte *Azolla*, often also valued as a supplier of nitrogen. WARDA research teams are also developing resistant varieties, and biological and cultural measures, with the aim of developing IPM against blast, stem borers and land crabs.

Madagascar

The Lac Alaotra Basin, covering 60 000 ha cultivated by smallholders, was targeted by the government for a project to increase rice production by reducing the 30% losses from white rice stem borer (*Maliarpha seperatella*). Aerially applied chemical control was used in 1982–83, but was halted in recognition of inadequate basic information on its appropriateness or how to evaluate its efficacy. It was feared that a purely chemical approach could threaten the ecology of the rice ecosystem. Instead, a research and development project was initiated to assess the economic importance of the rice pests and to identify future alternative control methods (Zahner, 1991).

Weed control had previously been neglected, and mechanical or chemical methods raised yields by an average 20%, though irrigation was a prerequisite to realizing this benefit.

Chemical control of stem borer was shown to have no economic justification. No insecticide could be found that was both effective and acceptable from a toxicological and environmental point of view. Moreover, the borers inside the stem are difficult to kill. It was decided to accept losses of up to 30%, and the government ceased to provide free pesticide treatment of this pest. Farmers showed no interest in making their own investment in such control.

The other important pest, the leaf-defoliating beetle *Dicladispa gestroi*, has a high economic threshold because the plant can compensate for damage. Also, without insecticide, parasitoids recovered to levels adequate to keep the beetle below the threshold. It was therefore recommended that insecticides should only be used against the beetle in the nursery, and not in the open field. However, this recommendation may have to be reconsidered if the beetle is shown to be the vector of a new rice virus disease that has appeared.

Monitoring for stem borer started in 1982–93. The methods have been refined and developed into a full-scale scouting program, carried out by the plant protection service at no cost to the farmer. A training program is under way to transfer responsibility for scouting to the farmer.

Cotton, with special reference to the Sudan

Cotton is one of the major cash crops in Africa, and is cultivated in most parts of the continent, by both small- and large-scale farmers. In the Gezira (Sudan), where cotton has been grown under irrigation since 1867, the hectareage of the crop has increased substantially following the completion of the Senna Dam in 1975.

Diseases and insect pests attack cotton throughout its growth cycle, and have proved major obstacles to optimal yield. In Sudan, such yield losses have been estimated to range between 40 and 65% (El Amin, El Tigani and Amin, 1991).

Young cotton is severely affected by weed competition, especially when soil fertility and moisture are low. Weeding must be adequate and on time. Inter-row weeding should start about 2 weeks after seedling emergence, with the within-row weeds removed 1 week later at the time of thinning. Two or three more weedings are then required as the crop develops. With increasing costs of mechanical weeding, the use of herbicides is becoming more common.

The major pests of cotton today reflect the history of the crop in Sudan. Until the mid-1940s, leafhoppers were the only significant insect problem, with most others controlled by cultural practices or legislative controls.

The cotton system was gradually changed from a three- to an eight-crop rotation, to stop the build-up of bacterial blight. Farmers were required to burn cotton residues, not to grow ratoon cotton and to exercise a close season on some vegetable crops. These measures were to control shared problems such as bacterial blight, leaf curl virus and pink and spiny bollworm. Varieties resistant to bacterial blight and leaf curl virus were introduced, as well as hairy-leaved varieties resistant to leafhopper, but unfortunately particularly susceptible to whitefly.

Whitefly problems were thus aggravated, but other pests were increased by additional changes in cultural practices. Cultivation of cotton was expanded into more humid areas with more severe pest problems. Planting of groundnut and sorghum increased and this, with the introduction of earlier-fruiting cotton varieties, promoted bollworm in the early part of the season. The introduction of cotton varieties that responded vigorously to fertilizer led to increased use of the latter; this in turn stimulated whitefly (p. 117) and aphids. Whitefly has remained the most serious cotton pest in the Sudan, partly because the sticky honeydew (also produced by aphids) on the lint decreases its market value by up to 20%. Elimination of fodder crops from the rotation led to an extension of the growing season to the benefit of pests; it also removed at least one useful trap crop for bollworm. Policing of closed seasons and phytosanitary measures relaxed, enabling pest populations to build up the whole year round.

Because of this history, the major insect pests on cotton in Sudan today are whitefly (*Bemisia tabaci*), bollworm (*Heliothis armigera*), leafhopper (*Empoasca lybica*) and aphids (*Aphis gossypii*). The important diseases are wilt caused by *Fusarium oxysporum* and bacterial blight (*Xanthomonas malvacearum*).

From the mid-1940s, use of chemicals (first DDT) increased yields substantially, but led to the treadmill towards overuse without further yield increases. Over the 60 years between 1925 and 1985, average yields only increased by about 15% (El Amin, El Tigani and Amin, 1991). However, from 1945 to 1987 the number of sprays increased six- to nine-fold to reach 25% of production costs. The 'package deal' system, under which pesticide manufacturers were contracted to control cotton pests on a guaranteed yield basis at a fixed price per hectare, was abolished in 1981. This is perhaps because it became evident that whitefly populations were higher in areas taking part in the scheme than in those outside it!

Spraying over so many years had a disastrous effect on populations of natural enemies. Of 140 beneficial species recorded in the 1920s, only 40 could be found in cotton fields in the mid-1980s (Abdelrahman and Munir, 1989).

Returns from increasing pesticide inputs, that had become no longer affordable in Sudan, were anyhow decreasing. A project entitled 'Development and Implementation of Integrated Pest Control for Cotton and Rotational Food Crops in the Sudan' was therefore started in 1979. The project was funded by the Netherlands through FAO, in partnership with the IPM Unit of the Agricultural Research Corporation's Entomology Department of the Sudan (Abdelrahman and Munir, 1989; Schulten, 1989). Initially the project generated awareness of IPM, and carried out research on developing varieties resistant to whitefly, as well as on basic population dynamics of pests.

It became clear that there was danger in extrapolating the results of small-plot experiments. A second phase of the project therefore compared sprayed and unsprayed plots of 40 ha. The results showed 10–20% lower yields in the unsprayed plot. However, this reduction in yield was largely compensated for by the saving on spraying costs as well as by the absence of stickiness on the lint; natural biological control was holding whiteflies and aphids at acceptable levels. This phase also included trials with inundative releases of the egg parasitoid *Trichogramma* against bollworm.

In the present third phase, large-scale experiments and demonstrations are being continued. Although economic thresholds have been used for cotton in the Sudan since the 1950s, they have not been amended for reduced insecticide use. An important step has therefore been to evaluate new lower thresholds with the greater level of biological control now occurring in the fields. Cultural and biological control are used as much as possible, and recommendations for 'softer' pesticides, applied only when needed, are being developed. Not only are more selective compounds being tested, but also more selective methods of using intrinsically broad-spectrum compounds. Rotation of insecticides from different classes may counteract secondary pest outbreaks and the development of resistance. Also, biological control alternatives, such as more effective strains of the pathogen *Bacillus thuringiensis* and of baculoviruses, are actively being sought (Sechser, 1989).

Training of researchers, field staff and farmers has been an important part of the project. The large-scale trials have led to closer cooperation with farmers. Interest and confidence in IPM for cotton in Sudan is therefore increasing. As stated earlier, Sudan is the only country in Africa that has adopted IPM as an official policy.

There is still, however, much additional scope for IPM. New varieties of cotton highly resistant to whitefly have been bred, and should be used more widely. Cultural control methods are also available to make an important contribution to IPM; some of these predate the use of insecticides and could be resurrected. Deep ploughing can expose resting stages of several pests including *Heliothis* to unfavorable conditions or bury them. Sowing time can be chosen to escape some pests, and removal of weeds that are hosts of cotton pests is a further contribution to control. Proper fertilizer use encourages the compensatory ability of plants, though excessive use can promote sucking insects. Trap plants such as *Hibiscus* are known to reduce flea beetle attack on cotton seedlings, and re-introduction of a rotation (including a clean fallow) could reduce bollworm and whitefly.

The major constraints to IPM in cotton in Sudan are ones of manage-ment. It is still yields, and not gross margins, which motivate the Agricultural Production Corporation and crop protection field staff. Yield is also the basis on which cotton-producing tenants are paid. Costs of pesticides are not clearly separated from other inputs in the accounts for the tenants, who therefore do not realize how much they are paying for pest control. They therefore do not appreciate the value of cultural measures. The project in Sudan, with its collaboration with tenants through large field experiments, has started to overcome these difficulties.

Zimbabwe and Togo are facing similar problems in cotton to those in the Sudan, but solutions have not progressed as far; the Sudan has a much longer history of cotton as a major export crop. An approach to IPM is gradually building up in the commercial farms in Zimbabwe. Monitoring and scouting are well developed, but there is no real integration of chemical control with biological and cultural methods.

In Togo, almost all farmers are smallholders. Research has provided savings for farmers through reduced insecticide use. Baculovirus mixed with low-dose pyrethroid has shown promising results. Efforts to promote collaboration between cotton protection researchers are supported by the World Bank.

Coffee, with special reference to Kenya

Seventy per cent of Kenyan coffee is produced by smallholders on an average plot size of 0.5 ha. These smallholders have formed cooperatives, and grow subsistence crops. The remaining production is from estates varying in size from 4 to over 50 ha. Estates (yielding $870–1500$ kg ha^{-1}) greatly outyield the smallholders (average 600 kg ha^{-1}), though the more advanced smallholders may achieve considerably higher than average yields (Meerman, 1991b).

Yields are limited not only by pests and diseases, but also by poor canopy management and low use of fertilizer.

Some 30 pests of coffee are known in Kenya. The most important are leaf miners (*Leucoptera* spp.), several bugs and scales (e.g. the 'fried egg' scale, *Aspidiotus*, and the soft green scale, *Coccus alpina*), defoliators (especially loopers) and thrips. Since 1980 a previously rare indigenous scale (*Icerya pattersoni*) has attained pest status on large coffee estates in central Kenya. This new pest was not pesticide induced, but was probably promoted by changing irrigation and fertilizer practices. However, use of insecticides did exacerbate its outbreaks. The trial release of the coccinellid *Rodolia iceryae* has shown that natural enemies can bring this scale under control in areas where they have not been reduced by insecticides.

The coffee mealybug (*Planococcus kenyae*) was accidentally introduced to Kenya from Uganda in the early 1920s, and successfully controlled with two Ugandan parasitoids. This biological control saved the Kenyan coffee industry US$10 million in 1935–58. It also enabled smallholders to restart coffee culture, after having been forced by the pest to abandon the crop. That the parasitoids came from a neighboring country within Africa itself augers well for future biological control of major indigenous pests in Africa.

The principal coffee diseases are coffee berry disease (CBD, *Colletotrichum coffeanum*), which attacks the ripening berries, and coffee leaf rust (*Hemilea vastatrix*), which causes premature leaf fall and a reduction in flower number. A further important disease is bacterial blight of coffee (*Pseudomonas syringae*).

Most smallholders skimp on the fungicide recommendations of the Coffee Research Foundation and apply the fungicide incorrectly. By contrast, they apply 2–3 times more broad-spectrum insecticide than recommended (see below).

The Kenya Coffee Research Foundation recommends an IPM approach (Meerman, 1991b) involving Ruiru 11, which is a coffee variety possessing high resistance to CBD and leaf rust. However, bean production by this variety is limited and new high-yielding resistant varieties are really required. Cultural practices include pruning, to reduce humidity and therefore CBD, as well as reducing antestia bug (*Antestiopsis* spp.). Mulching reduces the number of thrips that successfully reach the soil for pupation, but does allow pupation of leaf miners and, unfortunately, increases their incidence.

The approach to spraying shows great variations. Some farmers use no insecticides at all, and generally benefit from adequate control by natural enemies except for short-lived leaf-miner outbreaks. Others, both smallholders and estate farmers, spray using the threshold levels recommended by the Coffee Research Foundation. Others, again mostly smallholders, spray much more than recommended (in ignorance, spraying even against natural enemies) and some large estates tolerate far less insect damage than the recommended thresholds. This overuse of insecticides leads to outbreaks of secondary pests and to an increasing economic burden, since pesticides have doubled in price.

The recommended thresholds for both pests and diseases really need revision and another urgent need is proper training materials for all farmers, but especially for the smallholders who grow so much of Kenya's coffee.

Livestock and tsetse fly

Tsetse flies (*Glossina* spp.) are the principal vectors of trypanosomiasis (known as 'nagana') of livestock. The disease is caused by protozoan parasites (*Trypanosoma* spp.) and is the most important disease of cattle in East, West and South Africa.

Since the beginning of this century, there have been major efforts to combat tsetse and the disease. Livestock have been moved out of tsetse-infested areas, bush has been cleared to try to create environments unsuitable for tsetse, and wildlife thought to act as disease reservoirs have been killed. In addition, frequent chemical applications have been made against tsetse, and livestock have been treated with chemical drugs (Odhiambo, 1990).

Some control has been effected thereby, but usually only for short periods, particularly since tsetse will reinvade cleared areas. The efforts have been costly and have not been sustainable, neither for the people nor for governments. It has been clear for some time that other control approaches, which are more economic and of less risk to the environment, are required.

Although chemical control programs are still used, efforts have been directed to identifying races of cattle tolerant to the trypanosome. The African Trypanotolerant Livestock Network illustrates these efforts, based on cattle in the Gambia (the small N'Dama breed) tolerant to trypanosomiasis (ILRAD, 1988; ILCA, 1989). This breed appears to produce more milk than previously thought (Agyemang *et al.*, 1991) and is a candidate for livestock development in Africa. There is the potential to use crosses between tryptotolerant breeds in areas with a high trypanosome challenge. Units of the network are also studying the relationship between nutrition and the disease, since well-fed animals are better able to withstand the disease than poorly fed ones.

Trapping of tsetse to reduce tsetse numbers gan with ICIPE's Tsetse Research Programme in the mid-1970s, and the technique is now used in several countries (Odhiambo, 1990; Shereni, 1990). The biconical trap developed at ICIPE (Figure 6.5) is covered with blue cloth. Baits of cow urine proved very effective in a pilot project at Nguruman in the Rift Valley, Kenya, and resulted in reasonably long-lasting suppression of the tsetse population. The trap is now often known as the NGU trap, from the first three letters of the location. The Masai were very satisfied that animals did not need pharmaceuticals. The pilot project showed that trapping could form the basis for an IPM strategy for tsetse control, on a sustainable basis and with community participation (Odhiambo, 1990).

Also since the mid-1970s, the sterile-male technique has been tried (p. 132). After irradiation of pupae, sterile males are released in a number much greater than the number of wild males in the locality. However, this technique has not achieved the same success as trapping.

An integrated approach to tsetse control should benefit all partners in the community. Cattle herding communities have to accept a reduction in stocking levels to the carrying capacity of the land, so that the animals remain well fed. Many scientists feel that trying to eradicate species of tsetse and trypanosomes from Africa is both expensive and not feasible. It would be more sensible to live with the vector by using methods such as trapping in small-scale control campaigns involving also tryptotolerant breeds of livestock and some chemical prophylaxis (Hardouin, 1987).

Figure 6.5 *The ICIPE biconical tsetse trap.*

South America

The subcontinent of South America comprises 12 independent countries and two European colonies. It is a diverse region, running from tropical lowland in the north (from 13° N) to the temperate tip at 55° S. East to west differences are mainly a consequence of altitude. The east is more uniform and rarely rises to above 1000 m above sea-level. The west ranges from a coastal Mediterranean type of climate to the cold inland on the Andes mountain chain, where altitudes can exceed 6000 m. The total area of the subcontinent, at 17.8 million km², is about twice that of the USA or about 12% of the world land area. Roughly 47% is occupied by one country, Brazil.

There is a problem in South America in trying to assess the amount of arable land. Little is known about areas with low or no habitation. However, it is estimated that only about 5% of the total area is cultivated, whereas 20% is used for livestock production. Ninety per cent of the total area could probably be used for either purpose.

Parts of this large subcontinent provide a large proportion of the world's needs of some important foods. There is considerable untapped potential for increasing this contribution of South America in the world agricultural market, especially in relation to crops particularly well adapted to the region. IPM will have a major role to play in allowing the full potential of each crop to be realized without endangering the natural yield capacity of the corresponding ecosystems.

The rural population of South America ranges from 14% of the total population in Argentina to about 53% in Brazil. Income from agricultural activities varies from 9% of gross income in Brazil to over 27% in Paraguay (IICA, 1991). In the subcontinent, there is considerable cultural diversity, resulting from an influx of people of many different origins. These people domesticated important indigenous crop plants adapted to the abiotic factors in their region, and imported other crops suitable for the local conditions from around the world. Many of the indigenous crops are now grown world-wide.

Easy communications and the efforts of international organizations have led agricultural development in many South American countries to follow a similar pattern. This has been prompted by the need to feed a growing population and to comply with repayment of foreign debts.

The principal cash crops grown are coffee, maize, wheat, soybean, sugar cane, cotton and fruit. Cattle, goats and sheep are also important in most countries. Until recently, coffee was by far the most important crop for at least two countries, Brazil and Colombia. However, low international prices have recently reduced the importance of this commodity significantly.

The formation of regional commercial groups of countries is an important development that is leading to considerable changes, particularly in the relative importance in different parts of South America of different crops and livestock. With custom tariffs between the members of each group progressively decreasing, competition will alter the spatial and temporal distribution of crops and livestock. Similar changes have occurred in other parts of the world where such trading groups of nations have been formed.

The first trading group, the Andean Pact involving Bolivia, Colombia, Ecuador, Peru and Venezuela, formed a free commerce zone in 1992 leading to unified customs from 1993 onwards. The second group, Mercosul, involves Argentina, Brazil, Paraguay and Uruguay, and is due to start operating in January 1995. The formation of these groups should lead to increased demand for better quality agricultural products, including ones with lower pesticide residues or even free of chemicals. IPM is therefore bound to play an important role in the development of these trading groups.

Access to international markets, financial realities and development of infrastructure are among the more important variations between the countries that have resulted in an uneven adoption of modern technology. All countries have, to differing degrees, adopted the use of machinery and chemical inputs, especially for products for export. However, most countries have areas where agriculture is solely for subsistence and where cultural practices have remained unchanged for many years.

Considerable degradation of natural resources has resulted from both high- and low-input farming, leading to unsustainable situations. Of particular note is the recent widespread cultivation of soybean in southern South America. This immediately resulted in increased deforestation and soil erosion and had other undesirable consequences. Also, cultivation systems in general have become more intensive, and virtually zero pest levels are required for exported crops such as fruit. These factors have led to undesirable environmental pollution in Chile, Argentina and southern Brazil. Such developments in agriculture, once thought to be inevitable, are now being questioned by the public. In turn, this has influenced the way local governments direct agricultural development and take more account of the need to conserve natural resources. The impact of the recent United Nations sponsored meeting in Rio de Janeiro cannot be over-emphasized, especially on those working professionally in agricultural research and development. Local radio and television frequently present the need for environmental conservation to a large part of the population. However, little of all this reaches the largely illiterate rural population, most of whom struggle to survive and therefore often overexploit the local natural resources. Unfortunately, very few reliable data exist in the region regarding pollution and contamination levels and their effects on important biological processes.

The use of pesticides in South America has been considerable. The amounts used in different countries have been very variable depending on country size, types of crops grown and access to international markets. Pesticide sales more than doubled in the subcontinent between 1976 and 1980 (IRL, 1981). They now correspond to about 10% of the world market, with Brazil accounting for about half of the pesticide used in Latin America.

Unofficial movement of pesticides across borders makes the picture in some countries far from clear. In Bolivia (SEMTA, 1990), where pesticides have been used since the 1960s (coinciding with massive colonization of the Bolivian tropics after the Agrarian Reform), usage jumped from less than 200 to over 1300 tons, almost all as chlorinated hydrocarbons. In 1988 the official total of imported agrochemicals (mainly pesticides) was over 5600 tons (SEMTA, 1990). The extensive cultivation of soyabean in the past few years will probably have meant a considerable recent increase in the amount of pesticide used.

Pesticide use in Peru has shown only a very slight increase in the past decade, rising from 1700 tons of active ingredient in 1981 to 2100 tons in 1991 (G. Diaz, personal communication, 1992).

Until recently, Brazil was the fifth largest pesticide consumer in the

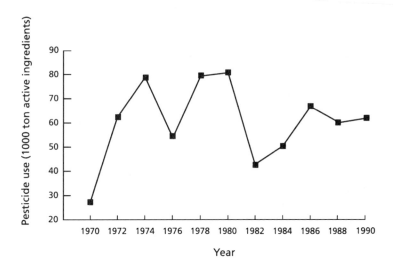

Figure 6.6 *Total pesticide use in Brazil, 1970–90 (data of Alves, A., Ciência (São Paulo), 4(22), 1986 and the Associação Nacional de Defensivos Agrícolas, São Paulo).*

world after the USA, Russia and the other members of the old Soviet bloc, France and Japan. Brazil has always been considered the largest potential world market for pesticides because of its large size and the importance of agriculture in its economy. However, the progression of pesticide use has changed dramatically in the past 20 years. While the economy was booming up to the early 1970s, annual pesticide use rose steeply (Figure 6.6). During the 1970s use remained high because of the incentive of government credits and subsidies. A period of economic crisis in the early 1980s caused a drop of about 50%. Since then a slight increase has occurred, but the level of use has tended to plateau out. However, recent changes in use have been very different for different groups of agrochemicals. The use of insecticides and fungicides has decreased in Brazil since 1986 (Figure 6.7), probably because of increased environmental consciousness, reduced subsidies and economic pressures on growers. The result has been the appearance of new alternative options. These have involved more limited use of chemicals or their replacement by other forms of pest control, including the development of resistant varieties and biological control (Flores, da Sá and de Moraes, 1992). Use of herbicides, however, has been increasing in Brazil for economic reasons. Labor for weeding has become expensive; moreover, there is a scarcity of workers in areas of agricultural expansion. This scarcity has been aggravated by new rural labor laws and social changes in Brazil that have led to a rural exodus since the 1970s. Soybean, fruit, sugar cane, wheat,

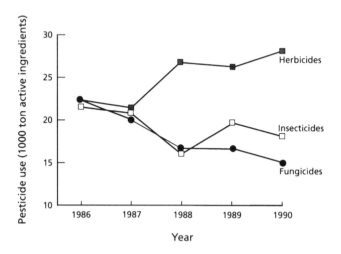

Figure 6.7 *Use of main classes of pesticides in Brazil, 1970–90 (data of Alves, A., Ciência (São Paulo), **4**(22), 1986 and the Associação Nacional de Defensivos Agrícolas, São Paulo).*

coffee and cotton are, in decreasing order, the crops with the largest demand for pesticide in Brazil (Thomas, 1988).

In Argentina, use of all three classes of pesticide has remained at the same level since 1987 (Figure 6.8).

In Colombia, pesticide sales have increased slightly in the past 15 years (Bellotti, Cardona and Lapointe, 1990). Insecticide sales accounted for about 60% of total sales in the 1970s and peaked at 15 000 tons of active ingredient in 1977. Sales of insecticide have since then decreased dramatically to only 20% of that peak following implementation of IPM with emphasis on biological control. This decrease in insecticide sales has, however, corresponded with increased sales of fungicides and herbicides.

Cotton is often regarded as the crop with the heaviest use of pesticide, but recent successful IPM programs have greatly reduced the demand in several countries in South America for cotton insecticides (Bellotti, Cardona and Lapointe, 1990). Today, soybean receives the largest amount of insecticide on the subcontinent. Large monocultures, especially of soybean, rice and sugar cane in Brazil, Argentina and Colombia, have resulted in increasing use of herbicides.

The Mercosul trade agreement referred to earlier will influence use of pesticide in the countries it involves. National regulations on registration and use of pesticides and biological products will have to be unified. It can be expected that the stricter national laws will be adopted by the whole region; in some countries, persistent organochlorines and organo-

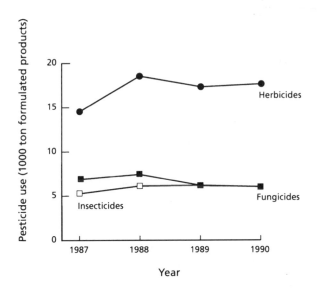

Figure 6.8 *Use of pesticide (formulated products) in Argentina, 1987–90 (data from the Camara de Sanidad Agropecuaria y Fertilizantes, 1990.* Boletin Infomativo CASAFE).

mercurials are still permitted, but are likely to be banned, or at least controlled very closely, under Mercosul.

In most South American countries, national research institutions generally dominate the development and implementation of IPM. Mostly they are large organizations with national mandates. Their involvement is indispensable given the ecological framework of IPM.

In those countries that have given most attention to agricultural research, a major role is being played by agricultural colleges, universities and other local government institutions. Thus in Brazil, several state universities have been actively engaged in IPM; the University of the State of Sao Paulo at Jaboticabal is the leading such institution. An Integrated Pest Management Center is attached to its campus. This center has been working mainly on IPM in cotton and citrus, and has played a major part in the implementation of IPM at the grower level (Gravena, Pazini and Fernandes, 1987).

Nongovernment organizations, including private companies, have played a progressively increasing role in the implementation of lower-input technology, including IPM. For example, the number of private laboratories in South America producing biological control agents has increased greatly in the past few years. Thus over 20 private laboratories produce parasitoids (especially *Trichogramma*) commercially in Colombia, and these are both sold internally and exported. These developments have been such that the Colombian government has established quality control rules through the Colombian Agricultural Institute (ICA) for the organisms produced (Garcia, 1990). In addition, most sugar mills maintain a parasitoid production facility for biological control of sugar-cane borer. Another important development is the laboratory production of *Diglyphus begini*, a parasitoid for use against leaf miners on various crops.

A few laboratories producing natural enemies are also found in Brazil. The oldest ones produced the fungus *Metarhizium anisopliae* for use against planthoppers in sugar cane. The parasitoid *Cotesia flavipes* is produced in most Brazilian sugar mills for the control of sugar-cane borer. For tomatoes in the north-east of Brazil, *Trichogramma* production is very important for control of leaf miner and fruit borer on thousands of hectares. Also in large-scale production is *Baculovirus anticarsia* against velvet bean caterpillar on soybean. Until recently this virus was only produced by the Brazilian Agricultural Research Corporation (Empresa Brasiliera de Pesquisa Agropecuária [EMBRAPA]) at its Centro Nacional de Pesquisa de Soja (CNPSo), but now four private companies produce the pathogen commercially by the same production methods. This is a good example of technology transfer from the public to the private sector; in this case, EMBRAPA receives royalties.

In Peru there are currently 23 laboratories that mass produce natural enemies, including native *Trichogramma* spp. among other parasitoids, predators and pathogens for field release.

Twelve commercial producers of beneficials operate in Venezuela to sell *Trichogramma* sp., *Telonemus remus*, *Bracon* sp., *Metagonistylum minense*, *Metarhizium anisopliae*, *Beauveria bassiana* and *Nomuraea rileyi*. One of these producers is attached to a growers' organization (Linares, 1990).

In Paraguay, growers' cooperatives are currently producing large quantities of *B. anticarsia* for use on soybean.

Many other examples of mass production of natural enemies for research or experimental trials could be cited; what is clear is that South America is extremely progressive in commercializing biological control agents and has a strong commitment to this approach to pest control.

Case studies

Soybean

Soybean cultivation has increased markedly in the past 30 years in southern Brazil, Argentina, Paraguay and Bolivia. Today, over 6 million ha are devoted to this crop in South America. It is used locally for the production of cooking oil; also the raw beans are exported.

In Brazil, commercial soybean production started in the early 1950s.

The crop is now grown on more than 8 million ha spread over both tropical and subtropical regions. Soybean exports account for nearly 30% of Brazilian agricultural exports.

Pest control is a major expenditure on this crop in Brazil, with each spraying representing about 10% of total production costs (Roessing, 1984). The main pests in Brazil are the velvet bean caterpillar (*Anticarsia gemmatalis*), which accounts for most of the insecticide used, and the stink-bug complex (*Nezara viridula, Piezodorus guildinii* and *Euschistus heros*), sucking the pods and seeds.

An IPM program was developed in response to overuse of insecticides. It was based on monitoring both pest and beneficial populations by counting insects knocked on to a cloth placed underneath the soybean plants. The program then involves the combined use (Iles and Sweetmore, 1991) of the following: early maturing varieties to avoid stink-bug damage; trap crops for stink bugs; early planting to avoid thrips, which vector virus diseases; soil management to reduce soil-borne fungal diseases; adequate tillage of the soil to reduce pests in the soil; use of varieties resistant to stink bugs, foliar diseases and nematodes; use of selected insecticides recommended by a design committee; and, for *A. gemmatalis*, the use of the nuclear polyhedrosis virus, *Baculovirus anticarsia*.

About 40% of soybean farmers have adopted this program. It has saved over US$ 200 million annually from reduced costs of pesticide, labor, machinery and fuel (Iles and Sweetmore, 1991).

The key to the program is the use of the *B. anticarsia* virus against the velvet bean caterpillar. This virus was identified in 1971 as an effective mortality factor and, as mentioned earlier, was initially mass produced by EMBRAPA/CNPSo. At present, nearly one million hectares of soybeans are sprayed annually with the virus (Moscardi and Sosa-Gómez, 1992). About 5 million ha have been treated since the beginning of the program in 1983; the savings are estimated at US$50 million. With the implementation of the IPM program, the use of insecticide has been completely abandoned in some areas. In most others, applications have been reduced to only one or two applications per year (Dossa, Avila and Contini, 1987). This compares with five annual applications in the early 1970s.

The success of this program revolves around the financial savings growers have achieved; most of them are high-income farmers. It involved a dramatic improvement in skills, with over 500 courses given in IPM, including insect recognition and the importance of natural enemies. Information on IPM technology was also provided by the extension services through weekly TV presentations, and these were an important element in the success of the program. Recently, there has been a development in the soybean IPM program in Brazil that has already resulted in a further reduction in the use of insecticide. This is the use of a parasitoid (*Trissolcus basalis*) on the eggs of stink bugs.

In Argentina the pest complex on soybean is similar to that in Brazil (defoliating caterpillars and stink bugs), and an IPM program was initiated in 1984. The major focus was on reintroduction as well as conservation of indigenous natural enemies. Implementation of the program reduced insecticide treatments per season from 2–3 to an average of only 0.3. In north-west Argentina, this corresponds to an annual saving of US$1.2 million.

In Colombia, research by ICA has shown that the major soybean pests (*A. gemmatalis, Omiodes indicata* and *Semiothisa abydata*) are significantly attacked by indigenous *Trichogramma* species. These, if occasionally supplemented by timely releases of laboratory-reared stock, can keep these pests below the economic threshold. Adoption of this practice has reduced the cost of pesticides to only 10–20% of what it was when there was total reliance on chemicals for pest control (Garcia, 1990).

Efforts to implement a soybean IPM program in Paraguay began in 1982 with international technical cooperation with Brazil. Today this program represents one of the most spectacular examples of the adoption of IPM technology. The key element was the control of velvet bean caterpillar, the most important soybean pest in Paraguay as well as in Brazil, by using *Baculovirus anticarsia*. This measure was adopted by the cooperative 'Colonia Unidas' in 1985 and in 1988–89 the cooperative used the virus on almost 19 000 ha. This reduced the number of insecticide applications by more than half without significant yield loss, and the cost of the virus treatment was only one-eighth of insecticide costs. The resulting saving was US$416 000. By the following year, the cooperative aimed to use the virus on half its soybean hectareage (35 000 ha).

Sugar cane

Sugar cane is a very important crop in tropical and subtropical parts of eastern South America. Here it has been used especially for alcohol production as well as for sugar. Throughout the subcontinent, the sugar-cane borer (*Diatraea saccharalis*) is the most serious pest of this crop.

In Brazil, where nearly 4 million ha of sugar cane are grown, annual losses caused by the borer have been more than US$100 million (Macedo and Bothelo, 1986). Nevertheless, insecticide is not used; instead, a successful biological program based on releases of parasitoids has been developed. The most effective such parasitoid is *Cotesia flavipes*, which has been mass produced and released in the field against the borer since 1975. Currently more than 30 laboratories produce *C. flavipes* in southern Brazil. The result between 1975 and 1990 has been a reduction in attacked sugar-cane internodes from 6.6 to 3.7%. Another serious pest of sugar cane in North-East Brazil is the sugar-cane hopper, *Mahanarva posticata*. The fungus *Metarhizium anisopliae* has been used as a successful control measure for almost 20 years. Since 1975 many sugar mills in Pernambuco State have produced their own *Metarrhizium*. Between 1970 and 1991, some 38 000 kg of conidia were produced by the Instituto do Açucar e Álcool (IAA) at Planalsucar, the State of Pernambuco Agricultural Research Institute (IPA) and private laboratories (Marques, 1992). This amount was sufficient to spray the fungus over 470 000 ha. From 1977 to 1987 this program reduced hopper infestation by 72%, and contributed to drastically decreased use of insecticides. Between 1985 and 1987 only 12 000 ha yr^{-1} required insecticide treatment, representing less than 10% of the area treated in 1971. The most recent figure is that the *Metarhizium* program now extends to 150 000 ha.

In the Valle del Cauca in Colombia, chemical control of pests in sugar cane has been completely replaced by an IPM program. This is based on periodic releases of parasitoids for control of sugar-cane borer, which is again the key pest. The parasitoids are: *Trichogramma pretiosum* against the eggs, the larval parasitoid tachinids *Parathesia claripalpis* and *Metagonistylum minense*, as well as the braconid *Cotesia flavipes*. These are all produced by the sugar mills themselves for use in their own cane fields. Borer infestation dropped from 10.6 to 2.9% between 1972 and 1985, despite increased planting of susceptible varieties (Escobar, 1986). Since 1976 the economic threshold has not been exceeded.

In Venezuela, as well as sugar-cane borer, the froghopper *Aeneolamia varia* is an important pest. Both insects have been controlled by an IPM program for over 5 years. The borer has been controlled by release of the introduced parasitoids *Metagonistylum minense* and *C.flavipes*. *Metagonistylum mimense* alone reduced damage by 50% before the introduction of *Cotesia* (Clausen, 1978). The froghopper has been controlled by periodic sprays of *Metarhizium anisopliae* in combination with weed control and post-harvest destruction of crop residues by burning. However, occasional pesticide applications remain necessary. In spite of its short history, this IPM program is already used on about 50 000 ha of sugar cane in Venezuela.

Laboratory mass rearing of natural enemies in Peru against crop pests was initiated by the Sugar Cane Grower Committee. *Trichogramma minutum* was produced in large numbers and gave excellent results against cane borer. This had considerable influence on several other South American countries. Since 1960, the Center for Introduction and Production of Beneficial Insects (CICIU) in Peru has promoted the development of laboratory facilities for multiplying natural enemies for use in sugar cane and several other crops. This has had considerable benefits to local growers (Beingolea, 1990).

Wheat

Wheat has long been an important crop in Argentina, which historically has supplied much of the international wheat market. Chile has also been an important wheat producer, although only for internal consumption, and wheat cultivation has recently expanded considerably in Brazil (see below).

In 1972, the aphids *Sitobion avenae* and *Metopolophium dirhodum* were detected on wheat in Chile. Insecticide was applied from the air on over 120 000 ha. In 1976, the Chilean National Institute for Agricultural Research (INIA), together with FAO, initiated an IPM program for wheat (Altieri *et al.*, 1989). Several natural enemies were introduced against the aphids; five were predators brought in from South Africa, Canada and Israel, and nine were species of parasitoids from Europe, California, Israel and Iran (Zúñiga, 1986). More than 300 000 coccinellids (in 1975) and more than 4 million parasitoids (from 1976 to 1981) were released in wheat fields. These introduced agents now keep the aphid populations below the economic threshold (Zúñiga, 1986).

Brazil was a major wheat importer for many years. However, because of government incentives and subsidies and the development of climatically adapted wheat varieties, the area of wheat cultivation has been expanded towards lower latitudes in Brazil. Again, introduced aphids have been the major pest problem. EMBRAPA, with the National Wheat Research Center (CNPT), initiated an IPM program in 1978 based mostly on natural enemy introductions from overseas. About 3.8 million parasitoids were released throughout the wheat-growing areas in the States of Rio Grande do Sul, Parana and Santa Catarina (Gassen and Tambasco, 1983). Whereas until 1977 almost all growers relied on insecticides for pest control, only 6% were still using them in 1982. As a result of the program, 1 million liters less of insecticide were used in the State of Rio Grande do Sul, and 1.6 million liters less in Parana State in 1989 than in 1977. The saving was more than US$15 million.

Cotton

Cotton is an important crop in many South American countries, though it has become less so in Brazil since the boll weevil (*Anthonomus grandis*) was accidentally introduced in 1983. Nevertheless, Brazil continues to be the largest cotton producer in South America and the sixth largest in the world. Traditionally, cotton is one of the crops accounting for the largest amounts of pesticide world-wide, and cotton in Brazil has been no exception. Until recently, up to 40 sprays per year were applied against the two key caterpillar pests, tobacco budworm (*Heliothis virescens*) and cotton leafworm (*Alabama argillacea*).

The first important IPM program for cotton in Brazil started in 1979.

It followed from the participation of research entomologists, extension agents and farmers in the First Brazilian Meeting on Cotton Pest Control (Ramalho, 1994).

An economic assessment of IPM in cotton showed that pesticide use could be reduced by half, achieving a 58% reduction in costs. However, the early IPM programs then had to be revised following the appearance of boll weevil. The one or two insecticide treatments required in the previous IPM recommendations had to be increased to four or five. Today's IPM involves the following components:

1. Use of rapid fruiting and early maturing varieties, particularly for boll weevil control; such varieties also escape heavy mid- to late-season pest populations.
2. Biological control with parasitoids and predators, but particularly exploiting naturally occurring insect diseases together with commercial microbial insecticides.
3. Uniform planting dates across a region with a synchronized cotton-free period.
4. Destruction of infested squares, bolls and alternate hosts; also early and synchronized stalk destruction after harvest.
5. Use of trap crops.
6. Crop rotation.
7. Control of boll weevil larvae by providing high soil temperature and low soil humidity during the growing season.
8. Judicious use of chemicals after monitoring for both pests and natural enemies in relation to economic thresholds; dose reduction and use of selective pesticides is recommended.

About 70% of cotton farmers in the main Brazilian cotton-growing State, Sao Paulo, apply the basic principles of IPM (Ramalho, 1994). However, most farmers do not scout their fields nor use economic thresholds as the basis for decisions on spraying; therefore much more extension effort is necessary. An important constraint on full IPM is that cotton is one of the most traditional Brazilian crops cultivated by very conservative farmers.

Heliothis virescens is also the main cotton pest in Colombia, where it is also important on several other crops. Cotton IPM in Colombia has centered on periodic releases of the commercially produced egg parasitoid *Trichogramma pretiosum*. Other IPM methods include post-harvest stalk destruction, a restriction of the permitted planting period and no use of insecticide for the first 70–80 days after planting. This program has greatly reduced pesticide applications on cotton. ICA reported that spraying frequency decreased from 20 to only two or three per year. Control costs are now only 3.5% of those when chemicals were the routine method (Garcia, 1990). A survey in Valle del Cauca in 1988 indicated that IPM was used in practically all cotton fields (26 000 ha). In one municipality (Zarzal) alone, with over 6000 ha of cotton fields, the number of sprays dropped from 20 to 12 in 1974–75, then to 1.2 in 1981 and to 0.8 in 1984.

Colombia has found that considerable ecological benefits have accrued from such reductions in spraying. Several species previously considered primary pests have diminished to secondary importance. However, the appearance of boll weevil in Valle del Cauca in April 1992 is a new challenge to the IPM program that will require modification of existing practices.

Boll weevil has also recently appeared in Paraguay. This country will similarly have to review its IPM strategy in the light of the large amounts of insecticide normally used to control this pest. Up to now, few annual insecticide applications have been needed on cotton because of the effectiveness of indigenous natural enemies. Normally only two treatments have been necessary.

Cotton in Peru is famous (or infamous) for the much-quoted Canete Valley as an example of the 'pesticide syndrome'. The 'disaster' phase was reached in the 1950s and was solved by what today would be called an IPM program, based on a package of legally enforced measures (p. 42). Yet ironically integration of cultural, biological, legal and chemical control for cotton had already been proposed in Peru as early as 1934.

In Venezuela the main pests of cotton are boll weevil, and the Lepidoptera *Sacadodes pyralis*, *Heliothis* sp. and *Spodoptera* sp. IPM consists in some cultural measures (control of alternative weed hosts, destruction of dropped squares and flowers as well as a restricted planting period), releases of parasitoids (*Trichogramma, Telonemus remus* and *Bracon*) and judicious use of insecticide against the boll weevil (Linares, 1990). Although this program is commonly used in Venezuela, no figures are available on its impact or the resulting savings.

Citrus

Citrus production in Brazil is concentrated especially in Sao Paulo State. Here, the area planted, almost 1 million ha, corresponds to about 80% of national citrus production (Pellegrini, 1990).

The most important pests are two species of mite, *Brevipalpus phoenicis*, which vectors the 'leprosis' virus, and the citrus rust mite, *Phyllocoptruta oleivora*. Other well-known citrus pests such as scale insects and fruit flies are widespread in occurrence, but only of local importance. IPM is based on economic thresholds for insecticide and acaricide sprays, with maintenance of weeds between the citrus rows. The latter measure provides refugiae for natural enemies. This program requires careful and frequent scouting, and the IPM Center at the University of Sao Paulo has provided considerable assistance to growers since 1986.

Twenty annual pesticide applications on citrus were normal in Brazil in the 1970s. In 170 orchards surveyed recently (S. Gravena, personal comminucation, 1992), the average number of annual sprays had fallen to 4.6. A further fall can be expected as more growers adopt the program, since mismanaged orchards can negatively affect IPM in adjacent ones. It seems likely that just one or two applications may eventually be necessary. Besides the savings on insecticides and acaricides, further savings are expected with reduction in herbicide use also. The goal of the program is to involve at least 6000 growers (30% of the total) within the next 4 years.

In Argentina an IPM program for citrus was initiated in 1977, based on introduction and conservation of natural enemies with selective pesticide use when economic thresholds are exceeded. Sprays against diaspid scales have been reduced to one or two per year, and more selective pesticides such as emulsion oil have been promoted. In north-west Argentina, savings on chemical treatment of US$2 million yr^{-1} have been generated.

Tomato

The spur for Brazilian IPM in tomato was the accidental introduction of the leaf miner/fruit borer *Scrobipalpuloides absoluta* to the north-east of the country in 1981. Insecticide applications every 3 days proved unsuccessful in controlling the pest, and losses exceeded 50% (Haji, 1992). The area planted to tomato, originally predicted to reach 15 000 ha by 1989, declined to only 5000 ha. Other Lepidoptera and mite pests also affect tomato, but their damage is minor compared with *S. absoluta*.

Since 1990, an IPM strategy has been adopted. The egg parasitoid *Trichogramma pretiosum* is released periodically, *Bacillus thuringiensis* rather than chemical pesticide is sprayed, there is a restricted planting period, and post-harvest destruction of plants is carried out. Also, better hygiene of containers and transport vehicles is maintained. The parasitoid releases have been the most effective element in the IPM program. As a result, three large laboratories for production of *Trichogramma* have been constructed to replace previous importation of parasitoid stocks from a commercial producer in Colombia.

Colombia itself, in Valle del Cauca, has been combining periodical releases of *Trichogramma pretiosum* with the action of the indigenous parasitoid, *Apanteles exiguum*. These two parasitoids, again together with sprays of *B. thuringiensis*, have been efficient enough to keep the same major tomato pest as in north-east Brazil, *S. absoluta*, under control. This program has almost eliminated the 40 annual pesticide sprays previously given, and has reduced control costs by about 70%. In 1988, it was concluded that 70% of tomato growers were no longer using any insecticide.

Alfalfa (lucerne) in Argentina

Alfalfa is the main leguminous crop of Argentina, and occupies some 5 million ha. IPM was stimulated by the severe damage following the accidental introduction in 1969 of the pea aphid, *Acyrthosiphum pisum*. The most important pests the program had to target were four aphid species, defoliating caterpillars and weevils.

Recognition of pests and natural enemies was emphasized in the IPM program. The control components were *Bacillus thuringiensis* (or reduced insecticide doses) against the caterpillars, and host-plant resistance together with introduction of parasitoids (*Aphidius* spp.) against the aphids.

Aphid-resistant alfalfa varieties can reduce insecticide against the aphids by two or three applications. The use of *Bacillus* and other selective products, or reduced dose of pyrethroids against caterpillars, has not damaged the diversity of natural enemies that help keep other pests under control (J.R. Aragon, personal communication, 1992).

Ornamentals in Colombia

This IPM success was developed by a private company, Flores del Cauca S.A. It is located 120 km from Cali and has almost 20 ha of plastic houses for chrysanthemum production. The key pest of this crop is leaf miner (*Liriomyza trifolii*) and other common pests are aphids (*Myzus persicae*), mites (*Tetranychus urticae*) and caterpillars (*Heliothis virescens* and *Pseudoplusia* sp.). IPM was based on biological control of leaf miner with periodic releases of the parasitoid *Diglyphus begini*. This was coupled with hand removal of mines, and capture of adult leaf miners with adhesive traps and nets.

Until 1985, 35 insecticide treatments over the 3 months of the crop were the normal control for leaf miner. Today, all-year-round chrysanthemum crops are grown without any insecticide against leaf miner. The biological control program has reduced the costs of pest control by 72.7% (Escobar, 1986).

Livestock

In South America, cattle are raised mainly in Argentina, Brazil and Uruguay to supply meat to local markets, but also for export of the surplus.

Horn fly (*Haematobia irritans*), ticks (*Boophilus microplus*) and the torsalo (*Dermatobia hominis*) are the main ectoparasites of cattle. Since the spread of horn fly throughout Brazil, particularly during the past 10 years, there has been a major effort to develop an IPM program for these ectoparasites. It consists in an integration of chemical, physical and biological measures (Honer and Gomes, 1992). The biological control is the recent introduction of the beetle *Onthophagus gazella*, which promotes more rapid recycling of cattle feces to inhibit the development of the flies. As yet, no economic evaluation of the impact of the program has been made.

Besides the IPM programs described above which have reached implementation, there are other IPM projects, still in the initial phase of development, in South America. These should make a significant contribution to better pest control in the near future.

One of these projects is the use of virus to control the most important maize pest, *Spodoptera frugiperda*. Promising results have been obtained on large trials in maize plantations in Brazil.

Another such development is the Sustainable Cassava Plant Protection Project, recently approved by the United Nations Development Programme (UNDP). It will be carried out in north-east Brazil as well as in four African countries (Benin, Cameroon, Ghana and Nigeria). The project will be based mainly on biological and cultural control methods, and planting systems known to result in reduced pest problems will be enforced.

Programs involving the use of insect pathogens have been sponsored by the Cooperative Programme for Agricultural Research in the Southern Cone: they are to be carried out in southern South America for the control of several lepidopteran pests.

Constraints on development and implementation of IPM in developing countries

The constraints on the development and implementation of IPM in developing countries are similar between continents. They are many and varied, but can be categorized as follows.

Technical

There is often a lack of basic studies on pests and their natural enemies, and individual components of IPM are usually researched in isolation. IPM developers then try to put together the information that has often been obtained by different specialists. This is done with poor knowledge of the possible positive or negative interactions between the components; also, undue emphasis is often given to just one component. Especially if the region is within the remit of an international agricultural research center, the emphasis is likely to be on pest-resistant varieties. In other situations, it may be on classical biological control. It is much less often that research results are available on improving cultural control or improving choice of pesticide and application method. Frequently also, economic thresholds have not been determined adequately for the pests.

Organizational

Even where economic thresholds have been established, there is far more difficulty in arranging satisfactory monitoring than in developed countries. Rarely are there adequately organized scouting services, or effective extension and training, for monitoring to become a farmer-operated tool. Both national research and extension services suffer from various financial, educational, organizational and administrative constraints. They are often sparsely manned (and even more sparsely wo-manned!), little trained in IPM, poorly equipped and with insufficient funds. The extension staff are often so few that they cannot serve more than a few farmers; in particular, their transport facilities are usually very poor. We recall a talk by an extension worker in Sri Lanka who had an enormous area over which to advise the farmers. All he found feasible was to cycle along any high ground and to scan the farmland though binoculars for flocks of birds. His warnings to farmers were based on his experience of which species of birds were often associated with particular species of insect pests!

The creation of effective national research and extension is one of the most urgent priorities for improving IPM in developing countries. There is often therefore little feedback from farmers to researchers to guide the latter in their choice of research priorities. It has to be added, however, that this problem does not always lie in the lack of an extension link. Nor is the problem only the lack of means or opportunity for researchers to visit farms and hopefully carry out some of their research there. Often researchers seem to prefer to stay and work within their research station without 'looking over the fence'.

Crop economics

Farmers may find the more selective pesticides often required for IPM are unavailable or too expensive. Also they fear that 'managing' pests (in contrast to attempting to maintain a totally clean crop) may lead to lower prices for a crop with some damage, even if this is only cosmetic. Also farmers may not be willing to take on the additional labor for weeding, scouting, etc. to reduce chemical control, if they have become accustomed to a less labor-intensive system with pesticides. Farmers may also find that using pesticides frees their time to earn extra money doing other things.

Social

At one end of the scale, farmers may be reluctant to depart from their traditional methods of pest control. At the other, increased cooperation by farmers and pressures from cooperatives can lead to above-optimum use of pesticides, with the lure of a temporarily high income; this is especially true in countries with considerable potential for agricultural expansion. Inputs become an 'assurance' and while this assurance is fulfilled, more complex methods for crop management are less attractive.

Legislative

Developing countries often have inadequate legislation governing product registration and use; moreover, enforcement of any legislation that does exist is frequently only partially effective. Farmers are unlikely to be influenced to change their attitude to chemical control for ideological reasons. Governments need to take the lead by introducing measures to make chemical control less attractive by legislation, more stringent registration, pricing without subsidies and even taxation. Concern continues to rise about the use of certain chemicals in developing countries. Thus, although several pesticides, often called the 'dirty dozen' (p. 43), have been banned in the developed world, they are still being produced and sold in Asia. However, it is necessary to be cautious in making this point. Better safety standards and better application equipment in the developed world make it possible to have the luxury of using environmentally safer products. However, these may be potentially more hazardous to the operator, e.g. organophosphate insecticides instead of organochlorines.

Educational

The prevalence of illiteracy in rural areas of many developing countries makes it difficult to distribute information to farmers in their learning phase about IPM. It also limits the introduction of techniques, including pesticides, where compliance with printed instructions is vital. Also, in many countries, adequate curricula for training in IPM do not exist. Neither are there mechanisms for 'snowballing' training from IPM workers to others and thus in stages to farmers in the community.

   Many of these issues are socio-economic rather than biological or chemical. Although the importance of ecology is so often stressed by biologists professing IPM, not all of them fully understand that the environment of IPM also includes social and economic environments. The socio-economic implications of IPM in developing countries are therefore covered in more detail in Chapter 7.

Conclusions

Pesticides were 'invented' in temperate countries for use in temperate countries and are perhaps less appropriate to the tropics with faster pest breeding rates and higher levels of indigenous natural biological control (p. 114). There has been rapid disillusionment with pesticides in many tropical developing countries, as they face the appearance of pest resistance to insecticide, pesticide-caused environmental damage, and escalating unsupportable costs of an ever-increasing number of pesticide applications.

It is therefore not surprising, however encouraging, that IPM is moving forwards so rapidly in developing countries, especially as international donor and technical support agencies have made considerable inputs to foster IPM there.

What is abundantly clear is that the developing countries have not had the option available to temperate agriculture of building IPM up slowly. They have not had the many man-years of applied research needed for a synthesis of target-specific control measures (p. 157). However, developing countries have, for some crops (a good example being cotton in Brazil), found it possible to adapt a more complex IPM package already developed elsewhere. Otherwise progress from scratch in IPM has had to be very rapid. Most of the case studies reported above have been based on more generally applicable ideas. The principal one has been to reduce insecticide pressure on the agroecosystem by introducing economic thresholds and more selective applications (e.g. insect pathogens, reduced doses, less persistent and more selective materials). Increased conservation of indigenous natural enemies then occurs, and the more biological control of one major pest results in other pests reverting to non-pest status. A second stage of IPM is the use of host-plant resistance and/or periodical release of mass-produced natural enemies against the major pest, to replace insecticide as far as possible. South America stands out as a region where such releases of biological control agents have been the main principle of IPM. The other regions have relied far more on conserving indigenous biological control. In this second stage of IPM, a variety of cultural controls have often been introduced or resurrected.

It is also true, and highly relevant to the IPM successes achieved in developing countries, that economic thresholds can be considerably higher than in developed countries. Farmers in the latter, particularly in temperate climates, have not faced as serious side-effects of pesticides and have become accustomed to maintaining high yields of blemish-free produce. It has often been said that, in developed countries, 95% of the pesticide is used to control the final 5% damage. This last fraction is often cosmetic or related to storage life of the produce. Such virtual total control of damage has been a luxury the developing world has not experienced and, as yet, has not aspired to in evaluating the efficacy of IPM programs.

References

Abdelrahman, A.A. and Munir, B. (1989) Sudanese experience in integrated pest management of cotton. *Insect Sci. Applic.*, **10**, 787–94.

Agyemang, K., Dwinger, R.H., Grieve, A.S. and Bah, M.L. (1991) Milk production characteristics and productivity of N'Dama cattle kept under village management in the Gambia. *J. Dairy Sci.*, **74**, 1599–608.

Ahmad, Z. (1991) Integrated pest control in cotton in Pakistan. Presented at Conference on IPM in the Asia–Pacific Region, Kuala Lumpur, September, 1991.

Altieri, M.A., Trujillo, J., Campos, S.L. *et al.* (1989) El control biologico classico en America Latin en su contexto historico. *Manejo Integrado de Plagas*, **12**, 82–107.

Anon. (1989) *Technology for Increasing Cotton Production in India*, Central Inst. for Cotton Res., Coimbatore.

Anon. (1991a) Rice – the path ahead. *Far Eastern Agric.*, **July–August**, 19–22.

Anon. (1991b) Draft Report of the Conference on IPM in the Asia–Pacific Region, Kuala Lumpur, September 1991.

Anon. (1992) *Technology Status Study on Pesticides*, Biotech Consortium India Ltd, New Delhi.

Asian Development Bank (1987) *Handbook on the Use of Pesticides in Asia Pacific Region*, Asian Development Bank, Manila.

Beingolea, G.O.D. (1990) Sinopis sobre el control biológico de plagas insectiles en el Perú, 1909–1990. *Rev. Peru. Entomol.*, **33**, 105–12.

Bellotti, A.C., Cardona, C. and Lapointe, S.L. (1990) Trends in pesticide use in Colombia and Brazil. *J. Agric. Entomol.*, **7**, 191–201.

Bhatnagar, V.S. (1987) Conservation and encouragement of natural enemies of insect pests in dryland subsistence farming: problems, progress and prospects in the Sahelian zone. *Insect Sci. Applic.*, **8**, 791–6.

Chaudhary, J.P. and Sharma, S.K. (1988) Biological control of *Pyrilla* using parasites and predators, in *Biological Technology for Sugarcane Pest Management*, (eds H. David and S. Easwaramoorthy), Sugar Cane Breeding Inst., Coimbatore, pp. 180–207.

Clausen, C.P. (1978) *Introduced Parasites and Predators of Arthropod Pests and Weeds: a World Review*, ARS/USDA, Washington, DC.

Cramer, H.H. (1967) *Plant Protection and World Crop Production*, Pflanzenschutz-Nachrichten, Bayer.

Di, Y.B. (1990) Status and control practices for vegetables in selected countries of Asia – Pacific region, in *Status and Management of Major Vegetable Pests in the Asia–Pacific Region*, Regional Office for Asia and Pacific, FAO, Bangkok, pp. 108–15.

Djamin, A. (1988) Crop protection and pest management of oil palm in Indonesia, in *Pesticide Management and Integrated Pest Management in Southeast Asia*, (eds P.S. Teng and K.L. Heong), Consortium for Int. Crop Prot., USAID, Maryland, pp. 205–12.

Dossa, D., Avila, D. and Contini, E. (1987) Alocaçâo de Recusos e Rentabilidade des Pesquisas Originadas do Centro Nacional de Pesquisa de Soja, Document 26, EMBRAPA-CNPSo, Londrino.

El Amin, A.M., El Tigani. M. and Amin, M.A. (1991) Strategies for integrated cotton pest control in the Sudan. 1. Cultural and legislative measures. *Insect Sci. Applic.*, **12**, 547–52.

Escobar, G.J.A. (1986) Criterios para la evaluacion de programas de manejo integrado de plagas, con referencia especial al caso en caña de azucar. *Conferencias Especiales 13, Congresso Socolen, Cali, July, 1986*, pp. 64–97.

FAO (1989) *International Code of Conduct on Distribution and Use of Pesticides: Analysis of Responses to the Questionnaire by Governments*, FAO, Rome.

Flores, M.X., da Sá, L.A.N. and de Moraes, J.G. (1992) Controle biológico: importância econômica e social. *A Lavoura* (Rio de Janeiro), Suppl. Sept./Oct., 6–9.

Garcia, F. (1990) Avances y perspectivas del control biologico en Colombia, *Colombia Ciencia Tecnol.*, **8**, 8–12.

Gassen, D.N. and Tambasco, F.J. (1983) Controle biológico dos pulgões do trigo no Brasil. *Inf. Agrop.*, **9** (104), 49–51.

Gravena, S., Pazini, W.C. and Fernandes, O.A. (1987) Centro de manejo integrado de pragas – CEMIP. *Laranja, Cordeirópolis*, **1**, 33–46.

Guan-Soon, L. (1990) Overview of vegetable IPM in Asia. *FAO Plant Prot. Bull.*, **38**, 73–87.

Hahn, S.K., Caveness, F.E., Lema, K.M. and Theberge, R.L. (1990) Breeding cassava and sweet potato for pest and disease resistance in Africa, in *Integrated Pest Management for Tropical Root and Tuber Crops*, (eds S.K. Hahn and F.E. Caveness), IITA, Ibadan, pp. 66–72.

Haji, F.N.P. (1992) Histórico e situação atual de traça do tomateiroo nos perímetros irrigados de submédio São Francisco. *Anais 3. Sympósio de Controle Biológico, Águas de Lindóia, August, 1992*, pp. 57–9.

Hardouin, J. (1987) *Let us live with tsetse flies.* Proc. CEC International Symposium, Ispra, 1987, pp. 183–6.

Heinrichs, E.A. (1988) Role of insect-resistant varieties in rice IPM systems, in *Pesticide Management and Integrated Pest Management in Southeast Asia*, (eds P.S. Teng and K.L. Heong), Consortium for Int. Crop Prot., USAID, Maryland, pp. 43–54.

Herren, H.R. and Neuenschwander, P. (1991) Biological control of cassava pests in Africa. *Ann. Rev. Entomol.*, **36**, 257–83.

Hill, D. S. and Waller, J.M. (1988) *Pests and Diseases of Tropical Crops*. Vol. 2, *Handbook of Pests and Diseases*, Longman, New York.

Ho, C. T. (1988) Pest management on cocoa in Malaysia, in *Pesticide Management and Integrated Pest Management in Southeast Asia*, (eds P.S. Teng and K.L. Heong), Consortium for Int. Crop Prot., USAID, Maryland, pp. 193–203.

Honer, M.R. and Gomes, A. (1992) O manejo integrado de mosca dos chifres, berne e carrapato em gado de corte. *EMBRAPA-CNPGC Circular Técnica, no.22*, pp. 1–60.

ICRISAT (1992) *Annual Report for 1991*, ICRISAT, Patancheru.

IICA (1991) *El Cono sur en Gráficos*, CONASUR/IICA, Buenos Aires.

ILCA (1989) *ILCA Trypanotolerance Trust*, International Livestock Centre for Africa, Addis Ababa.

Iles, M.J. and Sweetmore, A. (eds) (1991) *Constraints on the Adoption of IPM in Developing Countries – a Survey*, NRI, Chatham.

ILRAD (1988) *Annual Report, 1987*, International Laboratory for Research on Animal Diseases, Nairobi.

IRL (1981) *The Evolution of the Latin American Pesticides Industry and the Impact on Supply Patterns*, Information Research Ltd, London.

IRRI (1989) *IRRI Toward 2000 and Beyond*, IRRI, Los Banos.

IRRI (1991) *IRRI 1990–1991: A Continuing Adventure in Rice Research*, IRRI, Los Banos.

Jago, N.D. (1991) Integrated pest management for rainfed millet in northwest Mali, in *Integrated Pest Management and African Agriculture*, (eds A. Kiss and F. Meerman), World Bank, Washington, DC, pp. 25–33.

Johnson, E.L. (1991) Pesticide regulation in developing countries of the Asia–Pacific region, in *Regulation of Agrochemicals*, American Chemical Society, Washington, DC, pp. 55–71.

Kaske, R. (1988) IPM activities in coconut in Southeast Asia, in *Pesticide Management and Integrated Pest Management in Southeast Asia*, (eds P.S. Teng and K.L. Heong), Consortium for Int. Crop Prot., USAID, Maryland, pp. 181–6.

Kiss, A. and Meerman, F. (eds) (1991) *Integrated Pest Management and African Agriculture*, World Bank, Washington, DC.

Liau, S.S. (1991) IPM in plantation crops. *Proc. Conf. on IPM in the Asia–Pacific Region, Kuala Lumpur, September, 1991*, pp. 187–92.

Linares, F.B.A. (1990) Incidencia del MIP en la agricultura campesina venezolana. Presented at Consulta Sudamericana sobre Manejo Integrado de Plagas en la Agricultura Campesina, Santiago de Chile, November 1990.

Litsinger, J.A., Canape, B.L., Bandong, J.P. *et al.* (1987) Rice crop loss from insect pests in wetland and dryland environments of Asia, Report, IRRI, Los Banos.

Loke, W.H., Lim, G.S., Sayed, A.R. *et al.* (1992) Management of diamondback moth in Malaysia: development, implementation and impact. *Proc. 2nd Int. Workshop on Diamondback Moth and other Crucifer Pests, Tainan, 1990*, AVRDC, Taiwan, pp. 529–40.

Macedo, N. and Bothelo, P.S.M. (1986) Controle integrado de broca da cana-de-açúcar *Diatraea saccharalis* (Fabr., 17940) (Lepidoptera, Pyralidae). *Brazil Açucareiro*, **106** (2), 2–14.

Marques, E.J. (1992) Controle microbiano de cigarrinhas (Hemiptera: Cercopidae) com *Metarhizium anisopliae* (Metsch.) Sorok.: eficiência e limitações. *Anais 3. Simpósio de Controle Biológico, Águas de Lindóia, October, 1992*, pp. 73–8.

Matthews, G.A. (1991) Cotton growing and IPM in China and Egypt. *Crop Prot.*, **10**, 83–4.

Meerman, F. (1991a) Rational pest control in rice in Burkino Faso, in *Integrated Pest Management and African Agriculture*, (eds A. Kiss and F. Meerman), World Bank, Washington, DC, pp. 77–81.

Meerman, F. (1991b) Pest and disease control in coffee in Kenya, in *Integrated Pest Management and African Agriculture*, (eds A. Kiss and F. Meerman), World Bank, Washington, DC, pp. 83–8.

Mohyuddin, A.I. (1991) Utilisation of natural enemies for the control of insect pests of sugarcane. *Insect Sci. Applic.*, **12**, 19–26.

Morallo-Rejesus, B. and Sayaboc, A.S. (1992) Management of diamondback moth with *Cotesia plutellae*: Prospects in the Philippines. *Proc. 2nd Int. Workshop on Diamondback Moth and other Crucifer Pests, Tainan, 1990*, AVRDC, Taiwan, pp. 279–86.

Moscardi, F. and Sosa-Gómez, D.R. (1992) Controle microbiano de pragas: possibilidade de integração na América Latina. *Anais 3. Simpósio de Controle Biológico, Águas de Lindóia, August, 1992*, pp. 126–7.

Norton, G.A. and Way, M.J. (1990) Rice pest management systems – past and future, in *Pest Management in Rice*, (eds B.T. Grayson, M.B. Green and L.G. Copping), Elsevier Applied Science, New York, pp. 1–14.

Odhiambo, T.R. (1990) Keynote address. Proc. Int. Study Workshop on Tsetse Population and Behaviour. *Insect Sci. Applic.*, **11**, 259–63.

Pellegrini, R.M.P. (1990) Desempenho da agricultura paulista em 1988/89. *Inf. Econ.*, São Paulo, **20** (4), 43–7.

Pfuhl, E.H. (1988) Radio-based communication campaigns: a strategy for training farmers in IPM in the Philippines, in *Pesticide Management and Integrated Pest Management in Southeast Asia*, (eds P.S. Teng and K.L. Heong), Consortium for Int. Crop Prot., USAID, Maryland, pp. 251–5.

Pillai, G.B. (1985) Coconut pests of national importance, in *Integrated Pest and Disease Management*, (ed. S. Jajaraj), Tamil Nadu Agricultural University, Coimbatore, pp. 166–73.

Ramalho, F.S. (1994) Cotton pest management: part IV. A Brazilian perspective. *Ann. Rev. Entomol.*, **39**, 563–78.

Roessing, A.C. (1984) Taxa Interna da Retorno dos Investmentos em Pesquisa de soja, Document 6, EMBRAPA-CNPSo, Londrino.

Saha, P.K. (1991) Overview on pest control in Asia. Presented at Seminar on Pest Control, Asian Productivity Association, Tokyo.

Sastrosiswojo, S. and Saastrodihardjo, S. (1986) Status of biological control of diamondback moth by introduction of the parasitoid *Diadegma eucerophaga* in Indonesia. *Proc. 1st Int. Workshop on Diamondback Moth and other Crucifer Pests, Tainan, 1990*, AVRDC, Taiwan, pp. 185–94.

Schulten, G.G.M. (1989) The role of FAO in IPM in Africa. *Insect Sci. Applic.*, **10**, 795–807.

Sechser, B. (1989) New developments in pesticides for IPM in Africa with special reference to cotton pests. *Insect Sci. Applic.*, **10**, 815–20.

SEMTA (1990) *Situación Atuel de Plaguicidas en Bolivia.* Consulta Latinoamericana sobre Manejo Integrado de Plagas en la Agriculturura Campesina, Algarrobo, 1990, Servicios Múltiples de Tecnologias Apropriados, La Paz.

Shepard, B.M. (1990) Integrated pest management in rice: present status and future prospects in Southeast Asia, in *Pest Management in Rice*, (eds B.T. Grayson, M.B. Green and L.G. Copping). Elsevier Applied Science, New York, pp. 258–68.

Shereni, W. (1990) Strategic and tactical developments in tsetse control in Zimbabwe (1981–1989). *Insect Sci. Applic.*, **11**, 399–410.

Singh, S.P. (1990) Biological suppression of pests in fruit crops. *Proc. Indo-USSR Joint Workshop on Problems and Potentials of Biocontrol of Pests and Diseases, Bangalore, June, 1990*, pp. 91–165.

Solayappan, A. (1987) Biological control of sugarcane pests in India: recent developments, in *Platinum Jubilee Souvenir*, Sugarcane Breeding Inst., Coimbatore, pp. 158–60.

Sri-Aruntoi, S. (1988) The organisation and implementation of the surveillance and early warning system in Thailand, in *Pesticide Management and Integrated Pest Management in Southeast Asia*, (eds P.S. Teng and K.L. Heong), Consortium for Int. Crop Prot., USAID, Maryland, pp. 241–50.

Srinivasan, K. (1992) Integrated pest management in cruciferous and solanaceous crops, in *Summer Institute on Integrated Pest Management in Horticultural Crops*, Indian Inst. of Hortic. Res., Bangalore, pp. 43–58.

Sundramurthy, V.T. and Chitra, K. (1992) Integrated pest management in cotton. *Indian J. Plant Prot.*, **20**, 1–17.

Talekar, N.S. (1991) Integrated management of diamondback moth: a collaborative approach in Southeast Asia, Report, AVRDC, Taiwan.

Talekar, N.S., Yang, J.C. and Lee, S.T. (1992) *Introduction of* Diadegma semiclausum *to Control Diamondback Moth and Other Crucifer Pests*. Proc. 2nd Int. Workshop on Diamondback Moth and other Crucifer Pests, Tainan, 1990, AVRDC, Taiwan, pp. 263–70.

Thomas, M.S. (1988) The pesticide market in Brazil. *Chem. Ind.*, **6**, 179–84.

Verma, G.C., Shenhmar, M. and Gill, J.S. (1990) Biological control of cotton pests. *Proc. Indo-USSR Joint Workshop on Problems and Potentials of Biocontrol of Pests and Diseases, Bangalore, June, 1990*, pp. 57–76.

Waibel, H. (1990) Pesticide subsidies and the diffusiõn of IPM in rice in southeast Asia: The case of Thailand. *FAO Plant Prot. Bull.*, **38**, 105–11.

Waibel, H. and Meenakanit, L. (1988) Economics of integrated pest control in rice in Southeast Asia, in *Pesticide Management and Integrated Pest Management in Southeast Asia*, (eds P.S. Teng and K.L. Heong), Consortium for Int. Crop Prot., USAID, Maryland, pp. 103–11.

WARDA (1990) *Annual Report, 1989*, West African Rice Development Association, Bouake.

Wardhani, M.A. (1991) Developments in IPM: The Indonesian case. Presented at Conference on IPM in the Asia–Pacific Region, Kuala Lumpur, September.

WHO (1990) *Public Health Impact of Pesticides Used in Agriculture*, WHO, Geneva.

Zahner, P. (1991) Integrated pest management in irrigated rice in Madagascar, in *Integrated Pest Management and African Agriculture*, (eds A. Kiss and F. Meerman), World Bank, Washington, DC, pp. 95–102.

Zethner, O. (1987) Integrated Pest Management Project for Basic Food Crops in the Sahel, Report, FAO, Rome.

Zúñiga, E. (1986) Control biológico de los áfidos de los cereales en Chile. I. Revisión histórica y líneas de trabajo. *Agric. Tecnica, Chile*, **46**, 475–7.

CHAPTER 7

Social and economic aspects of integrated pest management in developing countries

The objectives and methodology of IPM raise the socio-economic issues of: integration within the whole social or health system; who should monitor pest populations and what is an acceptable threshold; the high information and technical requirements (including cost:benefit and cost:efficiency assessment); and the attitude and involvement of institutions. The last named are identified as governments, national agricultural research centers, farmers' organizations and the private sector. Community-based tsetse control with the ICIPE trap, the IPM training program in Indonesia and ICIPE's project in western Kenya are cited as case studies.

Introduction

It is evident from the preceding chapters that there is considerable information on the components of IPM, particularly on the four main approaches of chemical control, biological control, cultural control and host-plant resistance. It is also evident that much of the research contributing to IPM has been on the control of a single pest organism, and it is a large step, in socio-economic terms, to integrate such research into IPM on the farm.

This chapter focuses on how success in adopting IPM technologies has been achieved in developing countries. It also considers how constraints can be overcome to implement IPM to a greater extent in the future. The discussion will be limited to those socio-economic issues that stem directly from the objectives and methodology of IPM as opposed to the more general topic of pest and vector control. These issues are identified as:

1. the basic objective of IPM that the various methodologies should be integrated compatibly, both with each other and within the cropping or health system involved;
2. the principle that pest populations should be monitored and kept below damaging levels;
3. that IPM is both information and management intensive;
4. that IPM must be cost-effective;
5. that institutional factors affect IPM technologies and success in implementation.

Integration of IPM into agricultural and health systems

One basic objective of IPM is integration at three levels. First, IPM should, ideally, integrate the various control technologies for the various classes of damaging organisms (pests, diseases and weeds) (Smith, Apple and Bottrell, 1976). Secondly, IPM has to integrate these technologies into the farming system, which has many constraints and challenges and requires many decisions other than those related to pest control. Thirdly, there has to be integration of a farming system involving crops and/or livestock with people, with nontarget organisms such as fish and beneficial insects, with other economic enterprises and with the environment in general.

The first integration challenge is both conceptual and methodological. As pointed out above, researchers in scientific institutions typically focus on a single pest problem. The user such as the farmer, by contrast, handles a complex of damaging organisms in relation to his system. This system often involves people and animals as well as crops. More often than not he is presented with just one component of IPM. This is especially true with biological control, which is often applied independently of other components and in which introductions are often made independently of the user (Beirne, 1980). Yet the only person who works with the entire system is the farmer. One way these challenges have been tackled successfully by researchers has been through the production of IPM menus comprising several flexible components (p. 156), from which the user can select the items that fit into his system. IPM packages, where the user must follow preset instructions (p. 158), perhaps present a more difficult challenge in terms of socio-economic integration.

An example of a multicomponent IPM approach integrated with local farming systems is the recently completed ICIPE Oyugis–Kendu Bay Crop Pest Research Project funded by the Belgian government through the African Development Bank. This project combined sound agronomic practices with pest-resistant varieties of cowpea, sorghum and maize, and with intercropping of some of these crops. Low-cost inputs of ox ploughs, maize shellers and improved granaries to reduce post-harvest losses were also incorporated. The increase in yield that resulted from the project was 40%. Farmers rejected some cereal varieties as not fitting in well with their production systems, so the researchers carried out further on-station trials to meet farmers' specifications. The ox ploughs had, in fact, been brought in specifically at the request of the farmers.

At a later stage the farmers requested tsetse fly control in the project. This request followed an outbreak of nagana and death of some of their cattle. Although ICIPE had developed a suitable community-based tsetse trap, the request could not be met because the whole project was approaching its end. This illustrates how system-orientated users can make demands that the single-pest-orientated researcher does not envisage and can rarely meet.

The project area is also a producer of sweet potatoes, cassava and other crops. Tick-borne diseases of livestock are endemic there. The international agricultural research centers specializing in IPM components for all these crops and livestock face a challenge. Can they join forces, and produce an integrated pest and vector management program incorporating all the components already researched and available? Next is the challenge to unravel the best pest and vector suppression mechanisms already embedded in the traditional systems and to use them as the basis for improving IPM. For example, shifting cultivation reduces pests through burning, rotation, fallowing, intercropping and shading. Much biological control occurs naturally and is sometimes enhanced by traditional cultural measures. Host-plant resistance will often have been selected 'naturally' because of pest pressure or by seed selection over centuries.

The realization that there is a rich pest control potential in traditional agricultural and health practices has led to the establishment of related specialist institutions and publications. Although, ironically, mainly sited in developed countries, they are playing a crucial role in promoting the integration of IPM with traditional indigenous health and agricultural systems in developing countries. Notable centers promoting indigenous knowledge and practices as bases for IPM include the Center for Indigenous Knowledge for Agriculture and Rural Development (CIKARD) at Iowa State University in the USA, the Information Centre for Low External Input Agriculture (ILEIA) at Leusden in the Netherlands and ICIPE in Nairobi, Kenya. The first two of these publish *CICKARD News* and the *ILEA Newsletter*, respectively. The last named is devoted to the study of insect science and its application to pest and vector control using IPM strategies.

Sautier and Amaral (1989) have exemplified the kind of contribution that traditional agriculture can make to pest management and how IPM can be integrated. They have documented how some farmers in central Brazil abandoned synthetic organic fungicides and turned to the older Bordeaux mixture. This has some fertilizing properties in addition to its fungicidal ones. There were, however, problems that forced a decline in the use of Bordeaux mixture. Farmers changed again, this time to composting for improving balance in the soil and promoting plant health and plant resistance to disease in a more sustainable way. Sautier and Amaral (1989) remark: 'Probably, in the field of plant protection, farmers' knowledge and practices are the most rich and diverse, but also the most unknown and often undervalued.'

Veterinary science, previously slow to recognize the potential of traditional husbandry, has produced a spate of publications in the past 10 years on new possibilities for promoting animal health. These developments involve many disciplines (including veterinary medicine, immunology, microbiology, parasitology, pharmacology and botany), which are combined with social science in the study of traditional veterinary knowledge and practices. This approach is known as ethnoveterinary medicine or ethnoveterinary science. It has had some successes in identifying and improving on traditional methods of controlling vectors using low-cost, environmentally safe materials. For example, when researchers at ICIPE learnt that farmers in central Kenya were using a natural product from the tobacco plant as a tick repellent, they tested it in the laboratory. This confirmed that it was active against the tick *Rhipicephalus appendiculatus*. The chemical(s) responsible is now being identified with a view to improving the product. Similarly, researchers at Iowa State University learnt from Peruvian farmers, and then confirmed in laboratory tests, that the use of a wild tobacco reduces the number of external parasites on livestock by more than 90%. This is better control than achieved by anything synthesized by the chemical industry (Goodell et al., 1992).

A final issue for integration is that farming and health systems in developing countries are not static. They are continually subject to change, which in turn results in changes in the pest problems (Norton and Heong, 1988). Reliable information on changes taking place in farming and livestock systems in developing countries is fragmentary. This makes it hard to identify which technologies for integrating in IPM would be likely to be viable and durable. Pest problems therefore need to be redefined periodically.

Kiss and Meerman (1991) identify three broad categories of farming system and then prescribe appropriate IPM objectives for each. For subsistence systems, IPM should aim to increase the level and reliability of production. For intensive systems with high inputs, IPM should seek primarily to reduce costs, health hazards, ecological disruption and environmental damage. For transitional systems between these two extremes, IPM should seek to balance yield increase with avoiding overuse of pesticide.

The message of these objectives is as follows. High-input farmers using chemical pesticides have frequently run into the associated problems and have been the first to turn to IPM. However, the philosophy of IPM is equally suitable for raising yields at the other end of the agricultural spectrum and allowing the economic and judicious use of some pesticide without the danger of the system entering the 'pesticide treadmill'.

Monitoring pest populations

The major socio-economic challenge of this principle of IPM is the question of who is to monitor the insect populations. Theoretically, there are four options: researchers alone, extension staff alone, farmers alone or a combination of two of or all these options.

Within the context of the developing world, the first option (researchers alone) is unrealistic because there are far too few such people and large areas of countries without any at all. The second option (extension workers) is impractical for much the same reason, but additionally because the extension workforce is largely unfamiliar with IPM concepts and strategies. The adoption of the third option (farmers) is hampered by the following two constraints:

First, monitoring relies on highly technical methods and skills that require trained personnel; researchers on IPM could rarely produce protocols for monitoring which can easily be used by farmers in developing countries. Both the establishment and use of economic thresholds for pests of crops or livestock require a high level of technical skill. It is thus not surprising that, even where thresholds are available, farmers in developing countries frequently spray when they see any pests or damage from them. Some scientists are critical of the economic threshold concept for much of

228 SOCIAL AND ECONOMIC ASPECTS OF IPM IN DEVELOPING COUNTRIES

the developing world. They have concluded, usually from field evidence, that economic thresholds are inappropriate there. This leads them to suggest that the research effort devoted to thresholds should instead be directed at finding better pest management methodologies (Mumford and Norton, 1987). However, most scientists working in developing countries would argue strongly for economic thresholds and (as described in the previous chapter and again below), many farmers there are using thresholds successfully.

The second constraint is how far thresholds based purely on an objective scientific assessment of a 'number against effect' relationship, isolated from social considerations, can be an appropriate decision model. For example, some people would argue that the threshold for insect vectors of human disease is virtually zero if an infinite value is set on a human life. However, for the future, eradication of certain diseases is not a practical option, through lack of funds, if nothing else. Thus we are faced with the problem, what is an acceptable level? The massive program in the Volta River Basin of West Africa is not expected to eradicate river blindness (onchocerciasis). However, it should reduce the population of the Simuliid (blackfly) vectors to a low level. Although some transmission will persist, the threat of blindness would be sufficiently reduced that communities will return to the fertile valleys. Similarly, an efficient malaria control program should aim at reducing transmission to a point where malaria is no longer a major health problem, and the risks of controlling the disease are acceptable. What is 'acceptable' is a socio-economic, rather than an experimentally based decision; resources are finite while the needs of expanding populations are not. Thus Baldry (1983) considered that the primary objective of tsetse management is to reduce the population of vectors to the point where transmission and the resultant socio-economic impediment are negligible, i.e. where the disease situation can adequately and economically be dealt with by normal medical and veterinary services.

The high information and management requirement inherent in IPM

The previous sections of this chapter have already explained that IPM is both information and management intensive. Pimentel (1987) has pointed out that lack of success of IPM has been due partly to its requirement for far more basic information than is currently available. Also, it is much more sophisticated than the routine use of pesticides and requires trained workers. As pointed out earlier, farmers using pesticides tend to spray to a fixed schedule since this avoids the chore of monitoring the crop. Similarly, although traps provide the most efficient, cost-effective and environmentally friendly form of tsetse control, trapping is not really as easy as it sounds; it requires a good knowledge of tsetse fly biology, ecology and behavior and also entails practical difficulties. Very often the local people are initially very enthusiastic about the traps, but interest in maintaining them wanes as tsetse populations decline, with predictable consequences (Laveissière *et al.*, 1990). This shows inadequate understanding of tsetse, particularly of the ability of the fly to rebuild its population after repressive measures by man.

This example of tsetse illustrates an important feature of the community-based approach to IPM; members of the community need to have raised their understanding of problems and their causes. This has important implications that are rarely considered. Where understanding the problem requires scientific knowledge, there are formidable challenges. These lie particularly in the selection of the science, in translating scientific concepts and information into local languages (usually for the first time) and in choosing and preparing communication techniques (such as audio and visual aids) when some members of the community are illiterate. Meeting these challenges requires input from experts in several different disciplines: biologists, social scientists, linguists, graphic artists, etc. Cost is another element that is often ignored. Community participation in IPM entails financial costs in, for example, production of training materials, paying and accommodating the experts, and their transportation. Three case studies where attempts have been made to tackle these challenges are reported below.

Case studies

Community-based tsetse control with the ICIPE trap

Tsetse flies are visually attracted to the ICIPE biconical trap (Figure 6.5) by the black, white and blue colors and olfactorily by cow urine. They enter the trap but are prevented from leaving and die of exposure to the sun. The trap can reduce tsetse numbers by 99.9% as well as being cheap and easy to maintain (Brightwell *et al.*, 1987).

Encouraged by the success of the trap in a pilot project at Nguruman in the Rift Valley, Kenya, with participation by the local nomadic Masai (p. 199), ICIPE has launched a new community-based project in the Lambwe Valley in South-West Kenya, among the Luo people who mainly grow subsistence crops but also tend sizeable herds of cattle, goats and sheep. The costs of community participation and collaboration with Kenya National Agricultural Research Services (NARS) are included in the budget.

To raise the understanding within the community of tsetse and trypanosomiasis, a week of hands-on instruction for a core number of trainers (selected from the community itself) was organized in November 1992. Topics covered included: the history of tsetse invasion and human and animal trypanosomiasis in the region, effects of tsetse infestation, the biology and ecology of the fly, trapping and other methods of tsetse control. Instruction was also given in effective organization and management of community-based tsetse control based on the ICIPE trap. Participants on the course raised several questions. One was how spraying affects pupating tsetse under the ground, and this sparked off a lively discussion among the scientists themselves. Another participant wondered whether wildlife and people would also be sprayed. Others questioned the economic value of the Ruma National Park, which is the major habitat of tsetse in the region.

At the end of the course farmers were asked to present a talk in written form with visual aids and were promised the help of ICIPE in the preparation. One farmer, however, preferred to work with a village artist near his home. Together they produced a painting depicting four scenes. In the first, a group of people were mourning a relative who had just died of sleeping sickness. An approaching funeral procession was being thrown into confusion as a bull at the head of the procession collapsed and died of nagana. Near the mourners, another bull was being slaughtered. In the community, death of a relative is a double blow. Not only do the mourners have to be fed for up to a week, but it is up to the bereaved family to provide the first animal for slaughter. The second scene showed a group of people migrating with their livestock because of trypanosomiasis, leaving behind them lush crops of maize, sorghum, sunflower, etc. The third scene showed an aeroplane spraying not only tsetse, but also the mourners, the migrating villagers, others clearing vegetation, fleeing wildlife and the bull being slaughtered. This one painting encapsulated the farmer's topic: an assessment of tsetse and trypanosomiasis in the Lambwe Valley, including a lesson for economists in how local people view the costs of tsetse infestation and control.

Indonesia's IPM training program

The Indonesian agricultural reforms of the mid-1960s aimed at intensifying production through increased use of pesticide as a key component. Pesticide-induced problems on rice that resulted have already been described in Chapter 6 (p. 175), as have the IPM measures that were introduced to restabilize the situation.

The presidential decree in 1986 referred to in that account read very much like an extract from an IPM manual and proclaimed three major measures:

1. Pesticides were only to be used when other methods of pest control had been proved ineffectual.
2. Types of pesticides and methods of application would take into account the need to conserve natural enemies.
3. Pesticides that might cause pest resurgence and resistance or other damaging side-effects were made illegal.

Top priority was given to changing the behavior of people in relation to agricultural practices. Measures were therefore taken to improve the technical knowledge, organizational ability and management skills of farmers, pest and disease observers, extension personnel, bureaucrats in key ministries and the public. The program was nothing less than an endeavor to reorientate the nation towards making manpower development, environmental preservation and human health the bases for agricultural development (Goodell *et al.*, 1992).

The training of farmers in pest control and management skills is being carried out mainly through 'farmers' schools without walls'. One thousand out of the envisaged 1800 have already been organized. These 'schools' operate for 10–12 weeks during a crop season, and each school operates IPM on a 1000 m^2 'learning field' of rice. Farmers are trained in ecosystem analysis involving skills in insect and disease surveillance, plant health, weed density, water management and weather. They also receive training in organization and management skills. The methods emphasize letting the individual discover for himself and recognizing the farmer as an expert. With the training team, the farmers produce training materials from their agroecosystem analysis (i.e. insect collections, field trials, posters, workbooks, etc).

A training target set for 3 years starting in 1989 was 100 000 farmers, which represented just 0.65% of the total number of rice farmers in Indonesia. One source estimated that 45 000 farmers were trained in rice IPM during the first year. This estimate excludes *ad hoc* training in response to particular pest outbreaks. For example, 75 000 farmers were trained to form the core of what has been described as the largest mobilization in history – 80 000 schoolchildren and 350 000 farmers collected egg masses, set traps and reared and released parasitoids to arrest an outbreak of white stem borer in 1990–91. The successful outcome of this exercise was a reduction of damage by 93%.

Two thousand of the 35 000 extension workers and 1000 of the 30 000 pest and disease observers were targeted to receive three seasons of IPM training and one season of university training. Each trainee is expected to grow several rice fields of his own and to perform all the necessary tasks. Courses are given in economic analysis, including economic thresholds, statistics and ecosystem analysis. The trainees also receive instruction in group dynamics, horizontal communication and group cooperation Finally, they receive training in IPM not only in rice but also in associated crops such as maize, sorghum, etc.

Since the key decision makers are in higher echelons of various ministries, the trainers organize field days each season. These provide an opportunity for senior personnel to learn about IPM and the methods used to train field staff and farmers. They can also see the results at field level for themselves. For the public, the IPM program has sought to influence the media by organizing field visits, and also by providing information to the media networks, journalists, nongovernment organizations (NGOs) and consumer associations. Materials have also been provided for video and press dissemination. These aim not only to counter years of marketing and promotion of pesticides but, more important, to put across the IPM message in a positive way.

The result of the Indonesian training program, as already mentioned on p. 176, has been a considerable reduction in pests. This has been accompanied by an increase in rice production and a dramatic financial saving on pesticide use. However, there have also been changes that cannot be quantified in terms of money. Environmental health has been improved through less exposure of consumers to pesticide residues; farmers are now aware of the dangers. Extension staff and other pest scouts have gained confidence in their ability to manage pest problems. Also, bureaucrats at all levels have now given full moral and financial support to the IPM program.

Case study – rice in the Philippines

A remarkable recent phenomenon was that the rice yields of the top third of farmers surrounding the International Rice Research Institute (IRRI) increased between 1966 and 1987, yet at IRRI itself (during the same period) there was a decline. The yield of the highest yielding variety in long-term fertility trials fell by over 1.25% per annum, and yields at Philippine research stations remained stagnant or decreased by between 0.4 and 1% per year. The farmers around IRRI began with an average yield more than 2 tons below the IRRI highest yielding variety. However, by the late 1980s their yields were 1 ton ha^{-1} higher than at IRRI (Pingali et al., cited in Goodell et al., 1992). Pingali et al. attribute this to training and skilful management. The farmers 'were able to learn about new technologies, discriminate among those offered to them by the research stations, adapt them to their particular environmental conditions and provide supervision input to ensure the appropriate application of the technology'.

Finally, there is a useful catalog (van Alebeek, 1989) listing 627 sources of information on training and extension materials; of these, 252 are listed as suitable for farmers.

Economic viability of IPM

There are two prerequisites for a strict economic analysis of IPM. First, IPM should only be introduced if losses of crops or livestock represent a significant constraint to production, when compared with other factors that compete for investment (Kiss and Meerman, 1991). Secondly, the objectives and effects of the envisaged IPM measures need to be established. After these preliminaries, the IPM technology can be subjected to several economic analyses, of which the most important are described below. A case study illustrating the application of some of these analytical procedures comes at the end of this section.

Cost:benefit analysis

The main task of this analysis is to estimate likely costs and benefits and determine prices of resources on an annual basis. Incidentally, although the analysis is always under the name 'cost:benefit', the calculated ratio is more often than not expressed in 'benefit:cost' terms.

Benefits are of several kinds: readily quantifiable ones such as yield and monetary income or indirect ones such as benefit to the environment, reduced health hazards, equitable income distribution and human nutrition. In the same way as the easily quantifiable benefits, so the indirect ones also need to be tabulated on an annual basis. Similarly, costs are not only the obvious ones; there are hidden costs such as subsidies, environmental damage, human health hazard, etc. Estimating both benefits and costs on an annual basis is complicated by the delayed nature of some costs (e.g. pollution). There is the additional complication that benefits may increment over a long time, whereas some costs (e.g. of classical biological control, host-plant resistance) are of much shorter duration. Such difficulties make cost:benefit analysis a specialized and demanding task, but it remains an important aid to both users and policy makers. For example, the cost:benefit assessment that control of mealybug in Kenya cost about US$75 000 but would yield benefits of US$42.5 million in 35 years must have reassured both the Kenya government and taxpayers that the money invested in mealybug control was well spent.

Pimentel (1987) attempted to include all indirect costs and benefits of IPM for the USA, and found that this reduced the benefit:cost ratio from 4:1 to 3:1. His analysis was, however, based on research station data, and benefits in the field would probably have been less.

Cost:efficiency assessment

Cost:efficiency assessment is carried out along similar lines to cost:benefit analysis, but with the aim of deciding the cheapest or most cost-effective way of achieving a given objective. For IPM, the costs of reducing pests to a specified threshold are related to the resultant benefits. Because of its focus on a single threshold value, cost:efficiency assessment presents fewer problems than cost:benefit analysis, and has been completed successfully on several IPM projects in developing countries.

A crucial component of efficiency is 'effective demand'. This is a measure of the ability of users to pay for the IPM technology; not all technologies that have favorable cost:benefit ratios can be afforded by the poor. Information on effective demand can be obtained relatively easily from statistics on income and expenditure of the intended users of the technology. In spite of this, 'effective demand' calculations rarely seem to figure in IPM projects. Related to 'effective demand' is 'willingness to pay', i.e. do users put value on the technology, even if they could afford it? If they are not convinced that they will be better off, they are unlikely to use the technology. That target users are not convinced of the benefit is one of the reasons often put forward to explain disappointing implementation of IPM (Norton and Heong, 1988). The remedy is to show that the technology is profitable under the users' field conditions. On-farm trials can provide the necessary assurance (Reichelderfer, Carlson and Norton, 1987), if conducted with the farmers' own management and financial resources, and much progress has been achieved in this way.

Case study – ICIPE's project in western Kenya

This IPM project on the 'Reduction of Food Losses Through Insect Pest Management and Use of Small-Scale Low-Cost Farm Equipment' in Kendu Bay and Oyugis Divisions, western Kenya, aspects of which have been discussed earlier in this chapter, had the following objectives:

1. To reduce food and crop losses due to insect pests through IPM and to provide the participating farmers with low-cost small-scale farm implements.
2. To test the IPM components developed at ICIPE on farmers' own fields and under their management, for economic viability and social acceptability.

The IPM components of the project included host-plant resistance to insect pests, and cultural practices such as intercropping, adjustment of planting time and destruction of crop residues. These components were expected to be socially acceptable, economically viable and effective enough to produce benefits measurable by increased crop yields of 'resource-poor farmers'. The IPM measures were therefore aimed at improving food supply. Pest populations were to be kept below thresholds that would be regarded by the farmers as 'tolerable', the essence of a practical 'economic threshold' (p. 73).

The IPM components were introduced to randomly selected farmers in 1986. As part of the project, an economic analysis of the impact of the technology was carried out in 1990.

Economic and social costs of IPM are dynamic and are often long-term in their impact. There may frequently also be unquantifiable multiplier effects with time. For the analysis of this project, costs and benefits were classified as attributable either to the project or to the farmer. The marginal rate of return (MRR) for the IPM technology was then derived.

Major fixed costs charged to the project were ox ploughs, drafts, yokes, hand tools, maize shellers and grain storage structures. The variable costs to the project were seed, fertilizer, personnel in the supervision and training of farmers, office space, office equipment and stationery. Participating farmers were expected to meet labor costs as well as rent for land, basket work for the improved granaries and stands for the maize shellers.

The ultimate benefit of the project was to be that small-scale poor farmers would be provided with simple equipment and a farmer-acceptable, practical and cost-effective IPM technology. However, 'benefit' in the analysis was measured simply in terms of increased crop yields.

MRRs were computed for the different IPM menus introduced both at Oyugis and Kendu Bay during the two rainy seasons (long and short) in 1988. The analysis of the sorghum-based cropping system showed that the ICIPE pest control menu yielded greater MRRs than did the farmers' own systems at both sites. In Oyugis, the MRRs from introducing the sorghum cultivars LRB5 and LRB8 intercropped with ICV2 cowpea were 62.3 and 72.2% for the two sorghum cultivars respectively. In Kendu Bay, these values were 382 and 304%. Labor costs were higher in Oyugis that year due to heavier rains that demanded harrowing and weeding twice rather than the once in Kendu Bay.

In both long and short rainy seasons, the maize-based cropping system showed that MRRs for both V-37 and KRN1 varieties were also higher than those on the farmers' own systems. However, they were lower than for commercial maize hybrids. The results were similar whether V-37 and KRN1 were monocrops or intercropped with ICV2.

The report on the experiment concluded: 'The introduction of the ICIPE IPM technology in Oyugis and Kendu Bay, on balance, yielded significant net benefits to farmers in 1988 as shown by the analysis of Marginal Rates of Return.'

This case study illustrates that an economic assessment which is useful for policy makers can be made, even where many relevant measurements discussed earlier have to be omitted. Thus benefits were narrowly defined in terms of increased yield, although there were also benefits of other kinds. These included the provision of ox ploughs and improved storage facilities.

Institutional factors affecting IPM

The achievement of IPM objectives is largely dependent on the attitudes and involvement of institutions. Of these, the four most important kinds are government, national agricultural research services (NARS), farmers' organizations and the private sector (in the production and distribution of agrochemicals).

Government

Governments exert the single most important influence on IPM in developing countries. This is because they not only set national policies and strategies, but usually are also the single most important investor, employer and controller of resources. They also create and control other institutions important to IPM, such as NARS and, particularly in developing countries, farmers' organizations.

Brader's (1979) generalization, that developing countries give a low priority to crop protection, is still valid. A study by FAO (FAO, 1989) revealed that in 1986 80% of governments in Africa and the Asian–South Pacific region did not promote IPM. This contrasted with only just over 10% of developed countries. The respective comparative percentages in relation to not having strategies to counter pesticide resistance are 78% (Africa), 60% (Asia–South Pacific) and 5% (developed countries) (Johnson, 1991).

Government policies are also crucial in the distribution and regulation of use of chemical pesticides. A publication about the Indonesian IPM program referred to earlier in this chapter is indicative of the importance given to pesticides in developing countries when it states: 'Since the introduction of the Green Revolution technologies, insecticide has been packaged as a product component along with fertilizers, irrigation, credit and hybrid seeds.' Most developing countries do not believe that increased output of crops and livestock can be achieved without insecticides. As Carter, cited in Goodell et al.(1992), puts it: 'Apparently in peasant farming areas the pressure to boost food production and turn a profit means a shift toward chemically intensive practices.' This belief has, at times, been held and promoted by otherwise well-informed and powerful forces. Thus in 1978, scientists, consulting for organizations as prestigious as the World Bank and IRRI, reported that Indonesia's biggest problem in rice pest control was to satisfy the demand for more insecticide. Indonesia duly bought in more insecticide, but the result was the second worst year on record for pests in rice. The Indonesian government has since, however, strongly challenged the link between pesticide use and increased yield. It has claimed that 25 years of field research has never shown that insecticides contribute to tropical rice production (Indonesian National

IPM Programme, undated). Indeed, as described earlier, increased rice yields have been obtained through a combination of IPM with a dramatic decrease in insecticides.

The many examples from developing countries showing that high pesticide use and high yields are not linked (Chapter 6) are bound to influence the policies of many of these and other developing countries. Already, several governments in Asia, including those of Thailand, Vietnam, Malaysia and Sri Lanka are trying to follow Indonesia's lead through the Inter-country IPM program involving nine Asian countries.

Governments also greatly influence IPM by their policies on pricing, marketing and seed multiplication. The most obvious examples are the provision or removal of subsidies on pesticides, and restrictions on their use. Thus the Malaysian government's combined policies of guaranteed prices, subsidies and irrigation have removed much of the risk factor in rice production. Simultaneously, the potential for productivity has been increased (Norton and Heong, 1988). However, this is a net effect; at one extreme, unfortunately, there has emerged the phenomenon of the 'weekend farmer' who sees routine chemical pest control as the only acceptable strategy.

Developing countries need to increase their surveillance of chemical pesticide use in the interests of user safety. The 1986 FAO study (FAO, 1989; referred to earlier in relation to the attitudes of governments in promoting IPM) also found that, with 78% of the governments in Africa and 62% of those in the Asia–South Pacific region, environmental effects of pesticides were not investigated. Furthermore, only 57 and 64%, respectively, of these governments observed the International Code of Conduct on the Distribution and Use of Pesticides (p. 44), although this had been adopted by all FAO members in 1985. As a result, a great volume of obsolete pesticides still finds its way into developing countries. In many such countries, governments still have to develop and apply rigorous monitoring of agricultural inputs, including seeds. Where the government has the monopoly on the availability and certification of pesticides, the responsibility becomes more onerous. The findings of a survey 10 years ago (Goodell, 1984) still apply in many developing countries. Out of 100 farmers who believed they were planting pest-resistant rice varieties, 40 were actually growing highly susceptible ones; furthermore, 70% of bottles of pesticides purchased from retailers in the same survey contained adulteration to more than twice the acceptable level.

Finally, governments are the best placed organizations to set up mechanisms for coordinating the activities of the complex of governmental, nongovernmental and international organizations that participate in IPM programs. Governments must recognize the need to involve such a complex of agencies in IPM programs. An excellent example is provided by the Indonesian program already referred to. A National Steering

Committee was set up by BAPPENAS (the development planning agency) to implement the IPM training development program. BAPPENAS coordinated the activities of the Ministries of Agriculture, Education and Culture, Health, and Population and Environment. BAPPENAS also collaborated with both FAO, which provided training, and USAID, which provided funding.

National agricultural research services (NARS)

A recent study of these services in Asia outlined their functions as follows: policy formation, research coordination, priority and program definition, program planning, program monitoring, program evaluation, development of infrastructure, human development, funding, distribution of information and technology transfer (Senanayake, 1990). A very similar list has been given for the functions of NARS in sub-Saharan Africa (Taylor, 1991).

Partly because of the complexity inherent in NARS, their effectiveness differs both between and within continents and countries. In Asia, NARS enjoy more autonomy and have been given greater support from national governments than is true in Africa. The Council system developed in the 1920s in India, and later adopted by Bangladesh, Pakistan, Sri Lanka and the Philippines, played a large role in the uptake of the Green Revolution. With this was associated the development and distribution of pest-resistant rice varieties. The autonomous NARS model developed in Indonesia enabled NARS to assume a pioneering role in shaping IPM policy. This model was later adopted by Malaysia, South Korea and some South Pacific countries. Asian NARS are almost exclusively staffed by a rapidly expanding workforce of nationals who are generally well trained. For example, between 1975 and 1987, the number of agricultural scientists in India grew from 5666 to 33 357. During the same period, the number grew from 635 to 1600 in Bangladesh and from 463 to 2000 in Indonesia (Senanayake, 1990). As stated earlier, IPM requires sound scientific knowledge, making a trained workforce a prerequisite.

In Asia, NARS have made good use of the results of the international agricultural research centers in the region, notably IRRI and ICRISAT. They have also forged useful links with other international organizations engaged in IPM, in particular FAO and the Pesticide Action Network/ International Organization of Consumer Unions.

The picture in Africa is more one of future potential. NARS are more recent and, especially in anglophone Africa, still reflect the colonial heritage and need to be restructured to meet the challenges of post-independence development (Taylor, 1991). They also rely more heavily than in Asia on outside funds and staffing. Thus, in sub-Saharan Africa expatriates still make up 25% of the agricultural researchers at NARS

(Jain, 1990), and the expatriate proportion among socio-economic researchers at NARS is even higher (Sutherland, 1987). Although the 38% of researchers at NARS in sub-Saharan Africa with postgraduate training is a similar proportion to that in Asia, in absolute terms Asia has far more; moreover, sub-Saharan Africa is losing qualified personnel at the rate of 7% per annum (Jain, 1990).

Research linkages between different regions in Africa have been progressively severed. Previously, in anglophone West Africa alone, there were seven inter-territorial regional research organizations linking Ghana, Nigeria, Sierra Leone and the Gambia. Similarly in East Africa, seven regional research centers had been developed to serve Uganda, Kenya and Tanzania, and in southern Africa the Central African Research Organization served Malawi, Zambia and Zimbabwe. All this regional organization was broken up in the 1960s and 1970s. Thus ended an era of cost-effective collaborative research and instead began the more isolated national research services of today (Taylor, 1991).

The implications for IPM of all the above factors in sub-Saharan Africa are predictable. A small workforce, with a small proportion of staff with postgraduate training, will need to be boosted to carry out the basic research required for the development of IPM. Weak links with extension services and the farmers need to be strengthened to meet the demands for IPM and to develop the capacity to communicate results to users. A dearth of socio-economists makes it difficult to incorporate their input at an early phase of the research, as at on-station and on-farm but scientist-managed trials. Such weaknesses are already reflected in the poor dissemination of information and poor coordination of IPM components. Yet suitable components are being proposed by the international centers (IITA, the International Livestock Centre for Africa [ILCA], WARDA, ICIPE and others) located within Africa, and having programs in the various countries of the continent. The picture is one of unrealized potential. Perhaps more importantly it is also one of an expectation that the investment in the agricultural research system that has occurred in the 1970s–1980s will 'deliver the goods' in the next decade. Already there are signs that NARS in Africa are facing the challenges. For example, NARS in Kenya, Zambia, Rwanda, Somalia and Ethiopia have recently entered into collaboration with ICIPE to promote IPM in their countries. Kenya NARS have developed several pest-resistant varieties of coffee and cowpea.

Farmers' organizations

Farmers' organizations are not new to the developing world. Traditional associations of farmers dating back several hundred years and based on irrigation needs have been documented for Asia (Coward, 1980), Africa and Latin America (Goodell et al., 1992). Other kinds of production-based associations of farmers have been common in Africa, and many of them are still in operation.

It is a serious failure of the colonial and post-independence regimes that these traditional associations have never been recognized, let alone transformed into viable production units. By contrast, the government of Japan saw the potential of its traditional irrigation associations. It promoted their modernization, and made them the foundation of nationally orientated production units that spearheaded development even beyond purely agricultural aspects.

In Africa, two kinds of farmers' organizations have played an important role in agriculture since the colonial era.

First, there have been the 'elite farmer' organizations. Most of these are derived from the earlier colonial plantation-farmer organizations, whose principal objective was to maintain the privileged position of ex-colonial farmers. Many of these organizations are based on single commodities, e.g. coffee, tea, pyrethrum. The farmers in these organizations rely mainly on chemicals for pest control. In Kenya, for instance, the organizations run shops that stock agrochemicals, including banned brands. This group of farmers has, in other continents than Africa, been the first to adopt IPM technology to reduce their input costs; they are therefore potential customers for IPM technology.

Secondly there are the cooperative associations that most small-scale farmers have been encouraged to join. The primary objective of these associations has been to facilitate the marketing of crops (particularly cash crops) and livestock products, especially to overseas markets (Widstrad, 1972).

The introduction in the 1970s of the 'bottom-up' approach to agricultural research and development has promoted the active involvement of farmers in generating new agricultural technology, especially for use by the resource-poor farmer in developing countries. In relation to IPM, such a group approach to farmer participation is not only desirable but also necessary, for several reasons. For example, many important pests in Africa, Asia and Latin America are both gregarious and migratory. As an example, tsetse flies are distributed over an estimated 10 million km^2 in Africa. The pest makes it virtually impossible to rear livestock and sometimes even to grow crops in areas of high abundance. IPM technologies for managing such pests are likely to be more cost-effective if they can be group-orientated; indeed, the most promising technology for controlling tsetse has proved to be community-based traps (Laveissière et al., 1990). Even those technologies that can be operated by an individual farmer can be diffused more rapidly and cost-effectively if the farmers are organized.

The contribution that farmers' organizations can make to IPM is of several kinds:

1. In developing countries, social scientists play a 'watchdog' role in protecting and articulating the interests of farmers. Once farmers are organized effectively and adequately trained in IPM, they can, and should, take over from the social scientists and articulate their interests for themselves. This is the concept of 'empowerment' which underlies farmer participation.
2. Once farmers' organizations are formed from the grass roots and attain national status, they should be fully involved and form part of NARS and IPM programs. This would add a user's perspective to the research planning process, the setting of priorities and the allocation of resources.
3. At lower levels of IPM development, farmers' groups strengthen and facilitate links between researchers, extension workers and the farmers themselves. Many researchers in IPM would confirm that farmers' group meetings have enabled them to get first-hand information on the socio-economic and ecological impacts of the design, implementation and adoption of their IPM technologies. Some researchers have themselves organized farmers' groups specifically for obtaining 'feedback'. This saves a great deal of time in assessing socio-economic constraints and reaching large numbers of farmers.
4. Users' organizations have also proved of value in the development of IPM training materials. The ICIPE and Indonesian case studies (see earlier) showed this very clearly. Farmers drew maps for a variety of purposes including agronomic monitoring and environmental mapping (Gupta, cited in Goodell et al., 1992). In Bangladesh, farmers who were involved in IPM trials presented the successful innovations to fellow farmers, extension workers and researchers through farmer-led workshops (Abedin and Haque, cited in Goodell et al., 1992).

The private sector

The role of the private sector in crop and livestock production is most obvious in research and development of new pesticides and in pesticide distribution, since this is the source of their financial profit. Until recently, they were therefore reluctant to participate in the research and development of IPM, and are still reluctant in situations where there is no pressure to reduce pesticide use. Classical biological control, for example, tends to be free to the public and users, except perhaps indirectly through general taxation. Another disincentive for the private sector is that initial investment in IPM has to be on a long-term basis, though the investment is low when compared with the many million dollars needed to develop a new pesticide. However, partly because of the increasing rarity of new active compounds, and partly because future sales of agrochemicals may increasingly depend on integration with IPM, many agrochemical companies have become more interested in more selective pesticides, such as cuticle inhibitors. Some 'selectivity' of a chemical has now become an asset rather than the death knell of a new product. Many international companies have also diversified into host-plant resistance, both by purchasing or investing in seed companies, or by involvement in genetic engineering of new crop varieties. Other multinational companies are seeking to increase the proportion of their activity in what they call 'biologicals'. At present, this term usually means insect pathogens marketed as a cocktail with an insecticide. The move towards IPM in many parts of the world has sparked off a new private industry in the production of biological control agents as, for example, in Latin America (Chapter 6). Of course, the private sector has always been directly involved in certain aspects of IPM, such as the production of vaccines to replace vector control.

Even purely in relation to traditional chemical pesticides, chemical companies have had to take account of the growing campaign for environmental and health safety. Some prices of compounds have been lowered, and the true effects of pesticides a little disguised by referring to them in promotion literature as 'preventive medicine'. Another strategy has been to try to counteract reduced sales by giving monetary incentives to sales agents and retailers who, in turn, offer favorable credit terms to their customers. This is particularly attractive for rural users who have no alternative source of credit with which to purchase their crop protection requirements. For such users, the idea of ceasing to use chemicals is tantamount to ceasing to protect their crops or livestock. The chemical companies and their retailers also often achieve sales by offering services, such as instructions on their products in the user's language. Most scientists would agree that it is counterproductive to mount an all-out crusade for the total abolition of chemical pesticides and, in contrast to *Silent Spring*, this is perhaps the most important message of the present book in relation to pest control. As has already been stated so often here, pesticides are valuable. They may be one secret of success of IPM, if ways can be found of using them compatibly or even in positive synergy with other components.

Most large chemical companies are showing an increasingly responsible attitude to the use of their products, even if this is not always borne out by the actions of their local representatives. It has, after all, always been a golden rule of the private sector that 'the customer is always right'. Change has, and will increasingly, come about through the combined efforts of governments to regulate the conduct of the private sector in relation to pesticides, and the conduct of the users themselves. No private producer can afford to ignore the product specifications demanded by the client. They may even have been imposed on him.

References

Baldry, D.A.T. (1983) Control of tsetse flies, in *Pest and Vector Management in the Tropics*, (eds A. Youdeowei and M.W. Service), Longman, London, pp. 290–9.

Beirne, B.P. (1980) *Biological Control: Benefits and Opportunities in Commonwealth Agricultural Bureaux*, Unwin, Old Woking.

Brader, L. (1979) Integrated pest control in the developing world. *Ann. Rev. Entomol.*, **24**, 225–54.

Brightwell, R., Dransfield, R.D., Kyorku, L. *et al.* (1987) A new trap for Glossina pallidipes. *Trop. Pest Management*, **33**, 151–9.

Coward, E.W., Jr (1980) *Irrigation and Agricultural Development in Asia: Perspectives from the Social Sciences*, Cornell University Press, Ithaca.

FAO (1989) *International Code of Conduct on Distribution and Use of Pesticides: Analysis of Responses to the Questionnaire by Governments*, FAO, Rome.

Goodell, G.E. (1984) Challenges to international pest management research and extension in the Third World. Do we want IPM to work?. *Bull. Entomol. Soc. Am.*, **30** (3), 18–26.

Goodell, G.E., Ssesnnyonga, J.W., Lako, G.T. and Tadla, S. (1992). *Social and Economic Aspects of Integrated Pest Management*, Review Paper, ICIPE, Nairobi.

Jain, H.K. (1990) *Organisation and Management of Agricultural Research in Sub-Saharan Africa. Recent Experience and Future Direction*, Working Paper No. 33, International Service for National Agricultural Research, The Hague.

Johnson, E.L. (1991) Pesticide regulation in developing countries of the Asia–Pacific region, in *Regulation of Agrochemicals*, American Chemical Society, Washington, DC, pp. 55–71.

Kiss, A. and Meerman, F. (eds) (1991) *Integrated Pest Management and African Agriculture*, World Bank, Washington, DC.

Laveissière, C.L., Eouzan, J.-P., Grèbart, P. and Lemasson, J.-J. (1990) The control of riverine tsetse. *Insect Sci. Applic.*, **11**, 427–41.

Mumford, J. and Norton, G.A. (1987) Economics of integrated pest control, in *Crop Loss Assessment and Pest Management*, (ed. P.S. Teng), American Phytopathological Society, St. Paul, pp. 191–200.

Norton, G.A. and Heong, K.L. (1988) *An Approach to Improving Pest Management: Rice in Malaysia, Crop Protection*, Butterworth, London, vol. 7.

Pimentel, D. (1987) Is Silent Spring behind us?, in *Silent Spring Revisited*, (eds G.T. Marco, R. Hollingworth and W. Durham), American Chemical Society, Washington, DC, pp. 175–87.

Reichelderfer, K.H., Carlson, G.A. and Norton, G.A. (1987) *Economic Guidelines for Crop Pest Control*, FAO, Rome.

Sautier, D. and Amaral, B. (1989) Integrated pest management or integrated system management? *ILEIA Newsletter*, **October**.

Senanayake, Y.D. (1990) *Overview of the Organisation and Structure of NARS in Asia*, Working Paper No. 32, International Service for National Agricultural Research, The Hague.

Smith, R.F., Apple, J.L. and Botrell, D.G. (1976) The origins of integrated pest management concepts for agricultural crops, in *Integrated Pest Management*, (eds J.L. Apple and R.F. Smith), Plenum, New York, pp. 1–16.

Sutherland, A.J. (1987) *Sociology in Farming Systems Research*, Russell Press, Nottingham.

Taylor, T.A. (1991) *Organisation and Structure of NARS in Anglophone Sub-Saharan Africa*, Working Paper No. 38, International Service for National Agricultural Research, The Hague.

van Alebeek, F.A. (1989) *Integrated Pest Management. A Catalogue of Training and Extension Materials*, Dept. Entomology, Wageningen Agricultural University, Wageningen.

Widstrad, C.G. (ed.) (1972) *African Cooperatives and Efficiency*, Scandinavian Institute of African Studies, Uppsala.

CHAPTER 8

Hazard assessment of persistent chemicals in the environment

The occurrence and effects of the persistent polychlorinated aromatic compounds are considered. The assessment of the effects of PCBs in terms of dioxin equivalents is discussed. The use of biological change – biomarkers – in hazard assessment is considered. The concept of the 'polluter pays' and the environmental and economic costs of pesticide use are discussed. National schemes to reduce pesticide usage are considered.

Introduction

The presence of synthetic chemicals in the environment is a matter of great public concern. We would like to put this concern into perspective with four initial points:

1. There is no doubt that the media overplay the environmental (and human health) dangers involved. Bad news unfortunately makes more compelling reading or viewing than good news! A headline in *The Sunday Times* read 'Toxic cocktail is poisoning Arctic wildlife'. The article went on to say 'before long, vast areas of the Arctic will be left devoid of wildlife unless pollution is controlled' and 'ultimately entire populations may be wiped out'. That a wide variety of pollutants are found in Arctic wildlife, thousands of miles from their source, is a cause for concern. However, mere presence is no evidence of adverse effects. In fact, populations of most of the species cited in the article – polar bears, seals and snow geese – are on the increase. In Canada, the polar bear population started to increase after hunting quotas were established in the 1970s, and the population of seals increased rapidly once commercial hunting stopped in the mid-1980s. Now the Canadian government is faced with protests from fishermen who claim the rapid increase in the seal population is affecting the fish stock.

    Even when the facts are available, interpretations may go far beyond them. In discussing the sex reversal found in alligators in Florida, a BBC *Horizon* program included the statement 'everything that we're seeing in wildlife has an implication for humans. I believe that we have the potential to have major human reproductive problems.' Yet sex reversal in reptiles can occur when the temperature changes. Extrapolation of wildlife findings to humans should be viewed with great caution.

    There is usually an outcry when a chemical company has withdrawn a pesticide only a few years after release, because of the discovery of an environmental hazard. We would argue that such a report is almost good news. Since industry begins environmental testing some years before a product is marketed, a longer-term hazard will show well before an equivalent length of commercial use. The report therefore means the system works.

2. The tone of both sides in debates of this sort tends to be extreme. The comments of the environmentalists: 'Goodness knows what harm this may not be doing us,' are in strong contrast with the reassurance given (at least publicly) by many government officials or representatives of the chemical industry: 'There is absolutely no evidence this is doing us any harm.' Yet both sides of the argument simply translate

as: 'We don't know.' What is often missing from both statements is a scientific assessment of how adequate the research into answering the question has been. Thus, although it is very difficult to be certain that the small residues of pesticide most of us take in with our food are totally innocuous, it is some reassurance that people have lived with these residues for 50 years. Also, strenuous efforts by many scientists to correlate human death or ailments with residues have failed for most pesticides.

3. Progress nearly always entails some risk. We live with the motor car in spite of the death toll it engenders. We continue to develop and use medicines in spite of the tragic side-effects many of them have later shown. We do not argue for a ban on electricity because of the many lethal electrocutions that occur annually. Nor do we give up our use of energy because coal-mining and oil extraction are hazardous processes. The problem with chemicals and the environment is the selfish aspect – the dangers of the car, thalidomide, electrocution and mining are not contagious. By contrast, none of us can escape from the environment. This is not to argue against the maximum reduction of risks; the point is rather that we must accept that human innovation has to be a step into the unknown. No one can conceive the inconceivable.

4. The environmental danger of chemicals has probably never been their use *per se*, but their usefulness. Mr Benz would never, in his wildest dreams, have envisaged that exhaust emissions from motor cars could ever become a problem. Engineers would probably have never built sewage outlets into the sea if they had designed the large coastal cities of today from scratch. A commercial failure is unlikely ever to become a serious environmental pollutant.

Polychlorinated aromatic hydrocarbons

DDT

Much of *Silent Spring* is given over to a discussion of the organochlorine pesticides, especially DDT. Then, the amount of DDT produced in the United States was about 60 000 tons yr^{-1}, and world-wide production was probably twice that amount. Although the translocation of DDT from the original spray site was already well known, the truly global nature of contamination by DDT (and its principal metabolite, dichlorodiphenyl-dichloroethylene [DDE]) was only demonstrated when these compounds were found in the tissue of penguins and a crab-eater seal in the Antarctic

(Sladen, Menzie and Reichel, 1966). Neither of these species normally ranges beyond the limits of the pack ice of the region. A subsequent investigation showed the presence of other organochlorine pesticides, viz. BHC, heptachlor epoxide and dieldrin in addition to DDT and its metabolites, in both fish and birds collected in Antarctica (Tatton and Ruzicka, 1967). The levels found were low; DDE (the highest of any of the organochlorines measured) did not exceed 100 p.p.b. in any of the samples analyzed by Sladen and co-workers. While no one suggested that organochlorines at this level caused any toxicological effects, the fact that they could be transported to the most remote parts of the world was considered remarkable at that time. These measurements can be considered as a triumph for analytical chemistry, but the undoubted ability of the analytical chemist to determine pesticide residues has been a mixed blessing. While the importance of determining residue levels accurately is not disputed, the analysis has often become an end in itself. We know far more about what is where, than what a particular degree of contamination means to the organisms living in that environment.

Hazard is a function of exposure and toxicity. If there is no exposure to a toxic compound, there is no hazard. Equally, if there is no toxicity there is no hazard in exposure. However, this is too simplistic because, certainly for DDE, residues can be found in almost every sample from the environment, and any compound is toxic if the dose is large enough. Nevertheless what is important to keep in mind is that one needs to know both the exposure and the toxicity that this exposure can cause. Far too many major programs measure residue levels but do not address the more difficult, but more important, question of the effect that these residues are likely to have.

Polychlorinated biphenyls

In 1966, the presence of polychlorinated biphenyls (PCBs) was discovered in the Baltic (Jensen, 1966). Within a short time, PCBs were found to be as ubiquitous as DDE (Risebrough et al., 1968). In some ways the discovery in the environment of PCBs, polychlorinated dibenzofurans (PCDFs, or furans) and polychlorinated dibenzodioxins (PCDDs, or dioxins), realized the worst fears raised in the chapter in Silent Spring entitled 'Elixirs of death'. Each of these groups of compounds contains many individual compounds. Some, especially the PCBs, are globally distributed in the environment, and some, like the dioxins, are highly toxic.

The basic formulae and the number of individual compounds in each family are given in Figure 8.1. It will be seen that the formulae are quite similar and indeed there is a fair degree of interrelationship between them. Neither furans nor dioxins are manufactured as such; furans occur as impurities in commercial PCBs, both furans and dioxins are impurities in

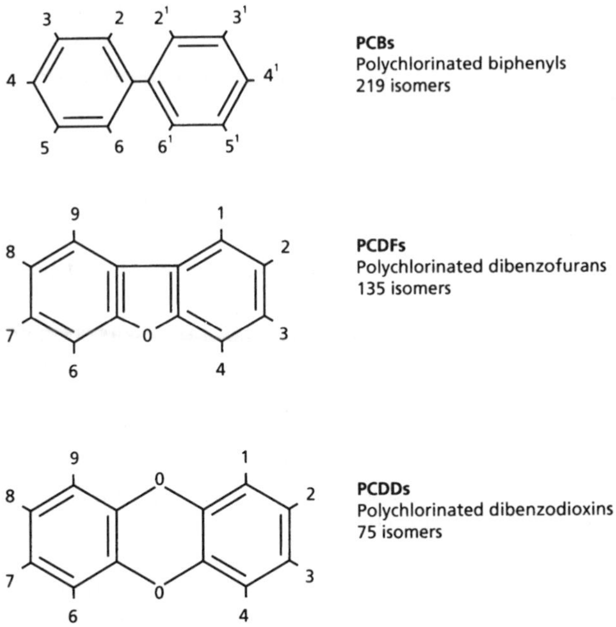

Figure 8.1 *Structure and number of congeners of PCBs, PCDFs and PCDDs. The molecules may have chlorine atoms at any or all of the numbered positions.*

the manufacture of chlorophenols and both are formed in combustion of chlorinated waste.

One problem that had to be solved immediately after the discovery of PCBs in the environment was the extent to which measurements of DDT were in error because of the presence of PCBs. Although one PCB could be, and had been, mistaken for DDT, it was soon found that the more important compound, DDE (the major metabolite of DDT), did not have a PCB that interfered with its analysis (Reynolds, 1969).

Environmental effects of DDT

The discovery of PCBs in the environment came at the time when moves were being made to ban DDT in the United States and other Western countries. Thus there was concern whether some effects ascribed to DDT were actually due to PCBs. Environmentally, the most important adverse effects of DDT were mortality of fish and effects on reproduction by some classes of birds. The case of fish mortality that had the greatest impact followed forest spraying in eastern Canada and the effect on the important salmon fisheries. These effects are described in *Silent Spring*. Because of the economic importance of these losses, the use of DDT was phased out in the programs to control spruce budworm in the mid-1960s. The earliest known effect of DDT on birds was the mortality of songbirds (such as American robins). This occurred when DDT was used at extremely high doses in attempts to halt the spread of Dutch elm disease. This phenomenon was reported at some length in *Silent Spring*. However, at normal operational doses, DDT does not cause avian mortality. Indeed, the switch from DDT to phosphamidon was to cause massive losses of songbirds. The most serious effect of DDT on birds was eggshell thinning, which caused reproductive failure of many species of raptorial birds throughout the northern hemisphere. This phenomenon of eggshell thinning was discovered in Great Britain in 1967 (Ratcliffe, 1967) and was soon shown to be global in character. By the early 1970s it had been linked to DDT. Experimental studies showed that PCBs had no such effect. Eggshell thinning and its effect on reproduction form one of the best case studies of environmental investigation. The dosage response of the degree of eggshell thinning to the residue of DDE in the egg has been established for many species. The relationship between the degree of eggshell thinning and population decline has also been established. It was found that thinning in excess of 18% was associated with declining populations. Some populations of raptorial birds were even eliminated, for example the peregrine falcon in eastern North America. Since the banning of DDT in many countries, eggshells have become thicker and most populations have recovered, at least to some extent. The history of the DDE-induced eggshell thinning of raptorial birds has been reviewed recently (Peakall, 1993).

Other pesticides in the environment

DDT illustrates several points made in the introduction to this chapter. When the chemical first became available for pest control half a century ago, no one could have conceived the variety of environmental effects a persistent pesticide could have. Nor could the eventual scale of use have been envisaged. Even so, it is surprising, given the overuse of DDT for so many years, that so little irreversible damage was done. Following the DDT experience, the agrochemical industry now tests for environmental effects of candidate pesticides on the soil microflora and microfauna. Effects on other nontarget taxa, such as bees, birds and insect natural enemies, are also researched. The fate in the crop plant is also followed in detail through the metabolic breakdown pathway, and the hazard of any breakdown products is assessed.

Modern testing procedures would no longer allow a chemical with the properties of DDT to be marketed. The pesticides in use today pose far less of an environmental threat than the earlier organochlorines. Much of the pesticide that lands on the plant surface is subject to oxidation catalyzed by sunlight (photochemical oxidation). A great deal more also enters the air, in the form of the tiny drops that have never contacted plants. Another fraction is evaporated from crop plants and weeds through transpiration. The roots of plants take up pesticide; this eventually adds to the aerial 'contamination', as does evaporation of pesticide from the soil. However, the kind of long-distance transport of pesticides in the air, which may have occurred with the organochlorines, is unlikely to occur with modern pesticides. For these, the atmosphere is a useful destination, since it acts as a very effective incinerator, again through photochemical oxidation.

Even pesticide which reaches the soil is not necessarily an environmental hazard. Provided a soil is not sandy, much pesticide will be firmly adsorbed on to mineral and clay particles. Soil microbes then slowly degrade the pesticide into harmless products, using the carbon and other elements. For modern pesticides, this process takes less than a year, and it appears that microbes can deal with amounts of pesticides considerably greater than those used in agriculture.

The aspect of pesticide pollution which remains potentially serious is the fraction that ends in surface-water run-off, and enters rivers and lakes. Of course, it will suffer considerable dilution, and breakdown by hydrolysis will occur. Water-borne microbes will also degrade the pesticide in a similar way to that in soil. However, pesticide in the surface water, but adsorbed on to soil particles, will be protected from breakdown, and will sediment to the bottom of water courses with the soil particles. However, here anaerobic micro-organisms will then destroy the pesticide rapidly, even the more persistent compounds. Before this happens, fish and other aquatic organisms may have accumulated pesticide. This phenomenon led to the 'biomagnification' up the aquatic food chain reported for organochlorines, but is much less likely to occur with modern compounds. Only a few parts per billion (one part being a thimbleful in an Olympic-size swimming pool) of modern pesticides can be detected in fresh water; traces in oceans are extremely hard to find.

The kill of fish is still an important environmental impact, even with modern pesticides, and particularly with the synthetic pyrethroids. Pimentel et al.(1993) quoted an estimate by the USEPA that between 6 million and 14 million fish were killed annually between 1977 and 1987 by pesticides in the USA.

As far as direct effects on other animals are concerned, the present environmental impact on birds is considered later in this chapter. Otherwise, problems seem considerably less than with the older organochlorines. However, some organophosphate and pyrethroid insecticides have been reported to reduce sperm production in rabbits. Also, some pesticides are clearly toxic to earthworms; fungicide against apple scab can affect them even by the residues remaining on leaves when they fall to the ground (Stringer and Lyons, 1974).

Another important impact of pesticides (particularly herbicides) is on biodiversity, both directly and by killing plants that provide food, harbor prey or provide cover. This has already been discussed (p. 107 and 112).

Hazard assessment of complex mixtures

With DDT, we are only dealing with the effect of two compounds (DDT and its stable metabolite DDE). However, with the PCBs, polychlorinated dibenzofurans (PCDFs) and polychlorinated dibenzodioxins (PCDDs), there are over 400 different compounds.

The first stage in developing a cost:benefit equation is assessing the damage being done by the pollutant. For PCBs, furans and dioxins, this is a formidable task. There are two major complications in the toxicology of these compounds. First, the toxicity of individual compounds varies greatly. For example, the 2,3,7,8-tetrachlorodioxin is 10 000 times more toxic than the 1,2,3,8-tetrachlorodioxin, although the difference is only the position of one chlorine atom (Rappe, cited in Eisler, 1986). Secondly, the toxicity of individual compounds to animals varies greatly from species to species. The acute toxicity of the 2,3,7,8 compound is over 1000 times greater to the guinea-pig than it is to the hamster. In making an assessment for risk to humans it is necessary to know if man is a 'guinea pig' or a 'hamster'.

Determination of dioxin equivalents

In practice we are always dealing with a complex mixture of compounds that varies from place to place and time to time. Fortunately a system has been devised to sum the effects of these mixtures. The breakthrough came from studies by molecular biologists on the isolation and characterization of receptors. Receptors are highly specific and are the cell's principal means of reacting to its environment. In the mid-1970s, Poland and co-workers identified the Ah receptor by its strong specific binding to 2,3,7,8-tetrachlorodibenzodioxin (TCDD). This was a key finding in bringing molecular biology into the realm of toxicology (Poland and Knutson, 1982). The Ah receptor is responsible for the control of the mixed function oxidase enzyme system, the key system in the defense of the body against foreign compounds. From this point on progress was rapid. Studies were broadened to cover many specific PCB, furan and dioxin compounds.

When a molecule binds to a receptor, its exact shape is important. It was found that the ability of individual compounds to induce the mixed function oxidase system is influenced greatly by the degree of chlorination and the chlorine substitution pattern. The most toxic PCBs are those that have no chlorine atoms next to the central bond, allowing the molecule to rotate and fit into the receptor. The most toxic PCDFs and PCDDs are those substituted in the 2, 3, 7 and 8 positions. Studies on the structure-activity relationships of many organochlorines show that the toxic effects and strength of binding to the Ah receptor are related.

This complex biochemistry has been applied to field investigations in providing a means of expressing the complex mixtures of organochlorines as 'dioxin equivalents'. The concept is based on the established correlation between the concentration required to induce a specific mixed-function oxidase enzyme (alkyl hydrocarbon hydroxylase) and the concentration required for toxic effects for many PCBs, PCDFs and PCDDs. In 'dioxin equivalents', the activity of the most powerful compound (2,3,7,8-TCDD) is considered to have a toxic equivalence factor of one. The potencies of the other compounds are then calculated from the correlation. This number can then be multiplied by the concentration, and the equivalence expressed in terms of dioxin calculated. An example is shown in Table 8.1. Although the potencies of other compounds, such as individual PCBs, are lower than for dioxin, their concentrations are often much higher. They therefore frequently contribute more to the total 'dioxin equivalent' than dioxin itself. For example, recent studies in Lake Michigan suggest that 90% of the dioxin equivalents in the eggs of fish-eating birds are caused by two specific PCBs. The approach is now also used in reverse. A measure of the degree to which the activity of the enzyme is increased is then converted into dioxin equivalents. This bioassay approach is rapid and inexpensive compared to the conventional chemical analysis by gas chromatography and mass spectrometry.

The dioxin equivalents of egg samples of fish-eating birds collected from 41 colonies on the North American Great Lakes were determined (Tillitt et al., 1992). The relative ranking of colonies correlated well with known areas of contamination. Material from Green Bay and Saginaw Bay gave the highest and those on Lake Superior the lowest values. When the overall reproductive success of cormorant and tern colonies was plotted against dioxin equivalents of eggs from each colony, a high degree of correlation was found (Figure 8.2). This strong correlation suggests that

Table 8.1 Calculation of dioxin equivalents (data from Ankley G. T., Niemi, G. J., Lodge, K. B. et al. (1993) Arch. Environ. Contamin. Toxicol., **24**, 332–44; Jones P. D., Giesy, J. P., Newsted, J. L. et al. (1993) Arch. Environ. Contamin. Toxicol. **24**, 345–54)

Compound	Toxic equivalent factor (TEF)	Concn (n g^{-1})	Dioxin equivalents (TCDD-EQ)
2,3,7,8-TCDD	1	0.8	0.8
3,3′,4,4′-5 PCB	2.2×10^{-2}	5 460	12.0 (126)
3,3′,4,4′-PCB	1.8×10^{-5}	30 700	0.5 (77)
2,3,3′,4,4′-PCB	8×10^{-6}	398 000	3.18 (105)

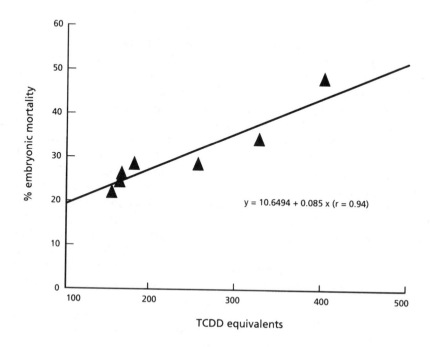

Figure 8.2 *Relationship of embryonic mortality of Caspian tern and dioxin equivalents of the egg contents (from Peakall, D.B., Animal Biomarkers as Pollution Indicators, Chapman & Hall, London, 1992, with permission).*

the abnormally low reproductive success of fish-eating birds in some areas of the Great Lakes is caused by sublethal levels of PCBs.

  Toxic equivalent factors (TEFs) are now being included in regulations. In several European countries and Canada, guidelines for emissions from municipal waste incinerations now use these factors (Fiedler, Hutzinger and Timms, 1990). Recommended levels in soil have also been expressed in TEFs in Germany and the United States.

Environmental estrogens

The issue of the *New Scientist* dated September 25, 1993, under the heading 'Pesticides linked to breast cancer', gave an advance report of work to be published in *Environmental Health Perspectives* (a journal put out by the US National Institutes of Environmental Health Sciences), which claimed that DDT and PCBs can act as weak estrogens and have been correlated with an increasing incidence of breast cancer. These chemicals are considered 'preventable causes of breast cancer'. A program by the BBC in November 1993 linked these weak estrogens to breast and testicular cancer, and also to the decreasing sperm counts of males in the Western world. There is, however, far from a consensus on these links. Other scientists quoted in the *New Scientist* article point out that the concentrations of these chemicals do not fit well with the epidemiology of the disease. They conclude that 'environmental estrogens can have only a small effect, if any.' The director of the US National Cancer Institute is quoted as saying: 'The environmental hypothesis is extremely interesting and scientifically important and we're going to pursue it.'

The most widely cited wildlife examples of endocrine dysfunction are the feminization of male alligators and the occurrence of masculinized female fish in Florida. Also widely cited is the finding that male fish have high levels of vitellogin (a protein used in egg formation) in several rivers in Britain, and that gulls in California lay super-normal clutches of eggs. These, and other findings, are detailed in a volume in the series *Advances in Modern Environmental Toxicology* entitled 'Chemically-induced alterations in sexual and functional development: the wildlife/human connection' (Colborne and Clement, 1992). One cannot dispute the consensus report in this volume that man-made chemicals have the potential to disrupt the endocrine system of animals. However, the connection made between this finding and humans, particularly the claim that human sperm levels have halved (Sharpe and Skakkebaeck, 1993), is far less clear.

Many different chemicals have been put forward as the cause of these endocrine disruptions. DDT, PCBs, effluent from paper mills, nonylphenols, phthalates and breakdown products from human birth-control pills have all been blamed. A valid approach would seem to be establishing the estrogen equivalents of these various compounds by determining their estrogenic activity (based on estrogen itself with the value 1), and multiplying these values by the concentration. This would parallel the use of dioxin equivalents already described.

These stories in the 1990s echo the earlier concerns raised in *Silent Spring*. On p. 235 of the latter, it is stated:

> A substance that is not a carcinogen in the ordinary sense may disturb the normal functioning of some part of the body in such a way that malignancy results. Important examples are the cancers, especially of the reproductive system, that appear to be linked with disturbances of the balance of sex hormones . . . the chlorinated hydrocarbons are precisely the kind of agent that can bring about this kind of indirect carcinogenesis.

Polluter pays?

The fundamental question that has to be answered is 'How serious is the damage being caused?' Clean-up of the environment – from a wide variety of pollutants – will be a major task of the next decade. The costs involved in this process are immense, making it essential to find the most efficient use of the measures available. A decision to do too little may cause irreversible damage to the environment, while a decision to do more than is necessary will waste resources better spent elsewhere.

The slogan 'the polluter pays' is, like most slogans, highly simplistic. In the case of past errors, it may be difficult to decide who the polluter was and, even more difficult, and expensive, to extract payment. Sometimes it is impossible; the company has gone out of business and there are no longer assets that can be used to pay for cleaning up the pollution. Another case that causes difficulties is where there are a number of different companies involved. There can be formidable legal problems, and costs, in deciding the degree of responsibility of each company.

Even when there are not complications, one has to realize that costs are passed on. Let us define the cost of producing a product as X, and the cost of pollution controls associated with the manufacture of the product as Y. The cost to the consumer becomes X + Y plus, of course, the profit to the manufacturer. Society has to decide if it wants pollution controls enough to increase the price from X to X + Y. If so, the price of chemicals has to include the cost of not allowing unacceptable pollution to occur.

Another problem is the inconsistency in environmental standards in different parts of the world. A manufacturer may well decide to locate his plant in a country where environmental legislation is weak or nonexistent. Then he avoids costs Y – to say nothing of reduced labor costs – and thus has the competitive edge on companies manufacturing in countries with strong environmental legislation. For example, the *New Scientist* (February 15, 1992) reported that some major multinational companies in the Netherlands were threatening to move abroad if the Dutch government went ahead with plans to impose environmental and energy levies ahead of the EEC's proposed carbon tax.

Enforcement of legislation, rather than the lack of legislation, is the greatest problem. Most countries have environmental legislation. The Ministry of Health of Tanzania defines a hazardous chemical substance as 'one which a) does not lend itself easily to chemical changes caused by natural effects and is also easily accumulated in biological organisms, b) is suspected of harming human health when ingested continuously, c) through chemical changes caused by natural effects, produces substances corresponding to a) or b).' However, the ability of governments in developing countries to protect their citizens from such hazards is very limited. The former Soviet Union had some of the strictest environmental legislation in the world, yet eastern Europe is now one of the most polluted areas of the globe. Even within the EEC, there has been criticism that environmental standards are more strongly enforced in the north than the south. Even within the EEC there is therefore not a level playing field.

There are, of course, some cases where the principle of the 'polluter pays' does work. For example, the farmer who illegally dumps pesticides can be fined. Even in the 'Alice in Wonderland' world of farm subsidies, the farmer has to bear the cost himself. However, most of the costs of meeting overall water standards are going to be borne by the water authorities and therefore passed on to the consumer.

It is important that pollution standards should be realistic. Recent water quality regulations issued by the EEC have been criticized as too stringent and being set on a poor scientific basis. This is not the forum to debate this issue but, even in the most affluent country, it does not make sense to spend large sums of money to clean water beyond what is necessary. It can be a difficult balance. There is the obvious tendency to make regulations very tough, to avoid being proved wrong from the viewpoint of safety to human or environmental health. However, the wasting of resources can also have a negative impact on human and environmental health, because funds are then no longer available for more worthwhile programs. Realistic pollution standards that can be met world-wide, and be enforceable, should be the objective.

Economic and environmental costs of pesticides

Attempts have been made in recent years to cost environmental damage: while the concept of a realistic cost–benefit analysis is appealing, in practice it is very difficult to avoid using arbitrary valuations and making assumptions that have a wide range of error.

An attempt to estimate the environmental and social costs of pesticide use in the USA has been made by Pimentel et al. (1993). This estimate is US$8.1 thousand million per year. Their paper categorizes the costs under 10 headings. The underlying difficulties can be illustrated by examining two of these categories, bird losses and public health impacts.

Bird losses, the largest single category, are put at US$2.1 thousand million. The estimate is based on the following equation:

cropland area $\times$ number of birds ha^{-1} $\times$ fraction killed $\times$ value of each bird.

The first figure, the area of cropland, can be obtained quite accurately and was set at 160 million ha. The number of birds per unit area of cropland is more difficult, but there is a large body of data suggesting that the figure used of 4.2 birds ha^{-1} is probably reasonable. There is much more uncertainty about the percentage of dead birds that were killed by pesticide. Pimentel *et al.* use a value of 10%, which they state to be at the lower end of the range cited in a review by Mineau (1988). However, recent work by Knapton and Mineau (1994), using banded birds, failed to find any losses in cropland treated with OPs. Even if one accepts that 67 million birds are killed, no basis is given for the completely arbitrary value of US$30 put on each individual bird. Some dead birds will be of species considered as pests. Even for waterfowl, US$30 seems a high price for a duck! If one substituted the value of US$1.70 that is used for a fish, the cost of US$2.1 thousand million falls to US$114 million.

The public health impacts are estimated at US$786.5 million. By far the largest part of this total (US$707 million) comes from pesticide cancers. The cost of treatment (US$70 700) is probably reasonable, but the estimate of the number of cancers (10 000) is far from firm. Pimentel and co-workers state: 'A realistic estimate of the number of US cases of cancer in humans due to pesticides is given by D. Schottenfeld (University of Michigan, private communication, 1991), who estimated that less than 1% of the nation's cancer cases are by exposure to pesticides.' There are two major difficulties with this statement. First, the study has not been published, so there is no way that the basis of the conclusion can be examined. Secondly, Dr Schottenfeld is quoted as saying 'less than 1%' of cancer cases are caused by pesticides. How much less? At 0.1% the cost estimate falls to US$71 million and the total cost to US$150 million.

Reduction in pesticide usage

Whether it is possible to calculate the environmental cost of using pesticides is open to doubt. Nevertheless there is no disagreement with the proposition that, if a reduction of pesticide usage can be achieved without increasing crop losses, this would be highly desirable. The sources of pressures to move towards IPM were discussed in Chapter 4.

In several countries, formal targets to reduce the amount of pesticide applied have been passed into law. Sweden was first, with a 5 year program (introduced in 1986) setting the target of reducing by half the weight of the active ingredients used. The base line for this halving was the average for the period 1981–85. Table 8.2 shows the reductions achieved, by categories of pesticides.

It is claimed that this goal has been achieved without serious expense or losses in yield. There are, of course, many factors that affect yields. However, in two important crops, spring barley and winter wheat, the average yields were higher in 1990 than the average for 1981–85.

In 1990 the Swedish parliament passed further regulations to reduce by another 50% the amounts of pesticides used by 1996. To achieve this

Table 8.2 Sales of pesticides in Sweden (from Emmermann, A. (1992) *Rep. Swedish Board of Agriculture*, with permission)

	Weight of active ingredient (tons)		
	1981–85	1990	Reduction (%)
Seed dressing	161	97	39.8
Fungicides	599	608	− 1.5
Herbicides	3536	1658	53.1
Insecticides	150	38	74.7
Growth regulators	82	49	40.2
Total	4528	2450	45.9

goal a reduction of some 1200 tons of active ingredients will be required. A quick look at Table 8.2 makes it clear that this must come largely from reductions in the amounts of herbicides and fungicides used.

One problem with a legal requirement to reduce the weight of active ingredient is that it exerts pressure to use compounds for which the rate of application is lower. However, there is no meaningful correlation between application rate and environmental damage across different chemicals. Forest spraying in eastern Canada, with phosphamidon at 140 g ha^{-1}, was responsible for major kills of songbirds. Yet fenitrothion, at twice the rate (280 g ha^{-1}), had a much smaller impact. Obviously it would not be desirable to substitute phosphamidon for fenitrothion to achieve a 50% reduction in the active ingredient. A shift from organo-phosphates to pyrethroids makes for a marked reduction in the weight of active ingredient used per hectare. For terrestrial vertebrates this may be a good move, but not necessarily for aquatic ones (the problems associated with the pyrethroids were discussed in more detail in Chapter 2). Clearly such differences need to be borne in mind, and indeed the Swedish National Chemicals Inspectorate has put forward 'Principles for identifying unacceptable pesticides' (Andersson *et al.*, 1992).

A readable account of the program to reduce the amount of pesticides used in the Province of Ontario in Canada is given by Surgeoner and Roberts (1993). They state: 'It would be a triumph of virtue if one could say that the program to reduce pesticides by 50% in the Province of Ontario was based on a consultative process between farmers, agriculture researchers, politicians and the general public. As in many endeavors the truth is more revealing.' The truth was that, because of public concern about the human and environmental health impacts of pesticides, the concept of a 50% reduction was included in the platform of one of the political parties. The first ever win in a provincial election of the Social Democrats in 1987 resulted in the 50% reduction being put into law. The reduction was to take place over 15 years. It has been suggested that the period of 15 years means that those who put the plan into being will not be held accountable if the program fails.

One point of the program was to make Canadian agriculture more competitive with that of the United States. Because of added regulatory costs and smaller markets, Canadian farmers typically pay considerably more for pesticides. Perhaps GATT and the North America Free Trade Agreement will also address this problem. As already discussed in Chapter 1, the GATT agreement is likely to have considerable environmental ramifications.

The exemplary reduction in pesticide use in Indonesia was described in Chapter 6, with its origin in brown planthopper (BPH) outbreaks, caused at least partly by overuse of chemicals. In 1986, a bold initiative was taken by the government. This was the issue of Presidential Decree no.3 of that year. It banned the use of 57 previously registered broad-spectrum insecticides on rice, and only a few insecticides with a narrower spectrum were permitted. This decree halted one of Asia's most serious environmental crises. Subsidies were decreased to 70–75% of the original level in 1986, to 40–45% in 1987, and finally completely withdrawn in January 1989.

The decree displayed a strong commitment and will of government to maintain rice self-sufficiency in a situation when overuse of insecticides was the cause of reductions in rice yields. Simultaneously, of course, it furthered the protection of the environment and human health. Pesticide use has fallen dramatically, and the production of rice is now greater than ever before (see p. 176).

Thus, in the world's fifth most populated nation, well-considered macroeconomic decisions have benefited farmers, consumers and the government, and have drastically reduced environmental damage. The National IPM programme in Indonesia is little short of a social movement (Wardhani, 1991). It links scientific development of ecological concepts with intensive farmer training in sound management techniques in the field. It represents one of the first large-scale examples of what might be called 'second-generation green revolution technology'.

Use of biomarkers in environmental assessment

An approach that is being used increasingly is the extent and severity of biological change, 'biomarkers', as indicators of the effect of pollution. The reality is that we cannot return the world to a pristine condition. The contamination of the planet ranges from highly industrial areas, profoundly altered by man, to wilderness areas such as Antarctica. Even here the blubber of the penguins and seals contains small amounts of PCBs and other contaminants. Is this contamination important? One possible approach to answering this question, using biomarkers, is to see if the organisms in a specific area are physiologically normal. An analogy to human medicine may be made. The approach is equivalent to a checkup involving the examination of the patient and having a battery of laboratory tests performed. If all the results are within normal limits, then it can be assumed that the patient is healthy and that no further action is needed.

Can this idea of physiological normality be universally applied? There one runs into difficulties, as some areas are already profoundly altered. For example, one cannot expect the entire area of an industrial port to maintain the same populations of the same animals under the same physiological conditions as a pristine area. Nor, on the other hand, is it acceptable for the port to be allowed to pollute the entire estuary and nearby lake or ocean.

The following criterion has been proposed for action: 'that the physiological functions of organisms, outside the exclusion zone, should be within normal limits' (Peakall, 1992). This gives what is, by environmental standards, a rigorous and reasonably practical endpoint. The concept of an 'exclusion zone' acknowledges that there are some areas so altered that remedial action is not feasible. Certainly the concept is tricky. It is difficult to see how an impartial scientific basis for the zone could be established, but it is equally difficult to see how pollution control can otherwise proceed.

The major advantages of the approach are:

1. It is, at least initially, independent of the pollutants involved and thus avoids the problem of mixtures and unknown substances. Only if the studies outside the exclusion area reveal abnormalities, would detailed investigations be needed.
2. Proof is not required, again at least initially, that any observed effect is deleterious. If the area over which the effect is seen is large, it may be argued that such proof is necessary. However, it is not a basic precondition.
3. Philosophically it is a defensible position. Analytical chemistry is now so precise that zero cannot be an objective. Although pollutants are present, the functioning of the animals living in the area may be normal.

There are, naturally, limitations in this approach. The most important are:

1. It implies that we have a battery of tests sufficiently good to give confidence in whether the physiology of the animals is indeed normal. Also, it implies that this battery of tests covers effects caused by all the major classes of pollutants.

2. A difficult question to answer is which, and how many, species should be tested. However, the same problem of interspecies differences exists with all other approaches.
3. It does not tackle the question: 'Is harm caused by the abnormal physiological state?'

The criterion for what is physiologically normal is a critical part of the approach. Controls should come from an environment as physiologically similar as possible to the area being studied. So far as contamination is concerned, not only is it not feasible to find control material with zero contamination but this may not be even desirable. In practical terms, broad areas of low-level contamination are now the norm. The argument can be made that, since this is the best that we can expect in an industrialized world, such areas should be the controls. In using such controls, an acceptance of this level of pollution is implied. Pragmatically, there seems little other option.

The exclusion zone/physiological normality approach is a mixture of politics and science. This mixture, although difficult, is an inevitable part of finding the solution to environmental problems. An obvious problem occurs if biomarkers are outside normal limits for a wide area. Once the area required to be cleaned up becomes large, then the pressure to question the harm caused by the physiological changes increases. At the moment we do not have the body of data on biomarkers, systematically collected over wide areas, that will answer this question. Studies, such as those being undertaken on the North American Great Lakes, may provide the answer. If it is found that biomarkers, over a range of species, are outside the normal biological variation over wide areas, we shall have to rethink the approach.

Biomarkers have sparked considerable interest in recent years. A wide range of biomarkers is now available (Huggett *et al.*, 1992; Peakall, 1992), covering effects caused by all the most common classes of pollutants. The strategy of using biomarkers has been discussed at a recent NATO workshop (Peakall and Shugart, 1993). While biomarkers have not yet been used in environmental legislation, it seems that we are rapidly approaching this point.

Control strategies for polyaromatic hydrocarbons

Control strategies for the different polychlorinated aromatic hydrocarbons have varied. PCBs – and for that matter DDT – were synthesized and used in that form, whereas the furans and dioxins are by-products of other reactions.

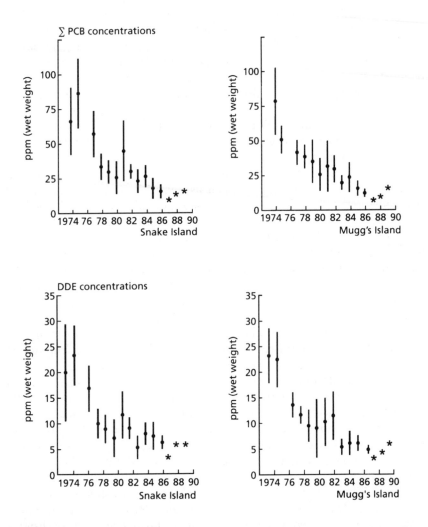

Figure 8.3 *Temporal trends in the concentration of PCBs and DDE in herring gull eggs from two colonies in Lake Ontario, Canada. Means with standard deviation are based on samples of 9–11 eggs; means denoted by * are based on a pool of 10 eggs (from* Toxic Chemicals in the Great Lakes and Associated Effects. Vol. 1, Constant Levels and Trends, Report of Environment Canada, Toronto, *with permission).*

In the case of PCBs, Monsanto (the sole US manufacturer) made a voluntary reduction of sales for 'open-circuit uses' in September 1970, a move subsequently followed by manufacturers in Europe. 'Closed-circuit' uses, such as closed-system electrical and heat transfer uses, continued for some time, but these involve far less loss to the environment. The action on PCBs was in marked contrast to that on DDT. Here, the final ban in the US was produced only after years of bitter court actions, although in other countries, such as Sweden, bans were readily put in place earlier. The environmental effect of these restrictions on DDT can be illustrated using data from the Canadian Wildlife Service Program on levels in herring gull eggs collected from the Great Lakes (Figure 8.3). At first, levels of DDT dropped rapidly; they have now levelled out.

The highly chlorinated PCBs are very stable and cycle almost continuously through the environment. The slow leaching of these compounds out of dumps means that there is still some new input from earlier uses.

Dioxins are not manufactured as such but are by-products of several industrial activities. They are present as impurities in some commercial herbicides (especially 2,4,5-T) and chlorophenols, and are formed by combustion, including that taking place in commercial incinerators. The explosion at the chemical plant at Seveso, Italy in 1976 resulted in the death of small animals in the surrounding area. Levels in domestic animals were sufficiently elevated to be considered a risk to human health and the animals were destroyed. Despite some human illness, no death of humans could be definitely attributed to this accident. Application of waste oil contaminated with dioxin to control road dust in Missouri caused the death of many horses and other animals near the treated areas. In one case, the town of Times Beach was purchased by the US Environmental Protection Agency and permanently evacuated in 1982.

The degree of contamination of herbicides by dioxins varies a great deal. The notorious Agent Orange, used by the military in Vietnam, contained up to 5% (Eisler, 1986). Other formulations contained amounts that were much lower. Regulations have now been put in place to reduce the amount of dioxin contamination. The herbicide 2,4-D also contains appreciable concentrations of dioxins, but most of these are the less toxic isomers.

In 1977, both dioxins and furans were found in fly ash from municipal incinerators (Olie, Vermeulen and Hutzinger, 1977). Since then, many investigations have shown the emission of dioxins and furans from a variety of combustion sources (including fossil fuels, wood and cigarettes). Most of the work has focused on municipal incinerators. The amounts produced vary greatly with both the type of material burnt and the operating conditions. Nevertheless, combustion is clearly an important source to the environment of dioxins and furans.

Another major source of dioxins is from the use of chlorinated phenols, especially the pentachlorophenols, as wood preservative. In Canada, the use of chlorophenols has been discontinued for short-term protection of wood, where the potential for loss to the environment is high. However, they are still used in treatment facilities for the long-term preservation of wooden structures.

The furans have similarities to the dioxins in that they are not deliberately manufactured and that there are many different compounds with a wide range of toxicities. While not as extensively studied as the dioxins, they are widely distributed at similar or higher concentrations to the dioxins. The main sources of furans are as impurities in PCBs and chlorinated phenols and as products of combustion.

Although PCBs are no longer manufactured, substantial amounts still enter the environment when waste materials are burned at low temperatures. Fires in buildings that contain PCB-electrical equipment are of particular concern, especially since the highly toxic 2,3,7,8 isomer is then formed. Chlorinated phenols are another source, although the isomers involved do not include 2,3,7,8 in appreciable concentrations. Combustion also produces the furans in amounts equal to, or higher than, dioxin levels.

Since dioxins and furans are not manufactured as such, they cannot be controlled by direct regulation as has been done for DDT and PCBs. Efforts to reduce the amounts in the environment have had to focus on activities such as the cleanup of herbicides and control of incineration conditions.

Conclusion

Could the approach, now being undertaken by the OECD, of producing enough data for a preliminary environmental hazard assessment of high-volume chemicals, be expected to solve the problems that have occurred with persistent chemicals? The answer is not a simple 'yes' or 'no'. It should have been able to handle the problems caused by PCBs. Their stability and high octanol:water ratio should have alerted us to potential problems. Dioxins, on the other hand, are not purposefully produced, and it would require detailed studies of manufacturing processes to show the sources of these highly toxic materials.

The position is similar with the heavy metals. Obviously the elements themselves are immortal and their direct uses, such as toxic mercurials as fungicides, could have been expected to cause problems. The increases of mercury caused by acidification and flooding would, however, not be detected by examining production processes.

It is to be hoped that the proactive approach will become the dominant approach to preventing problems with toxic chemicals. Nevertheless, the reactive approach, based both on formal monitoring programs and field observation, will still be needed as a safety net.

References

Andersson L., Gabring, S., Hammar, J. and Melsater, B. (1992) Principles for identifying unacceptable pesticides, *KEMI Rep.*, 4/92.

Colborne, T. and Clement, C. (eds) (1992) *Chemically-Induced Alterations in Sexual and Functional Development: The Wildlife/Human Connection*, Princeton Scientific, Princeton.

Eisler, R. (1986) Dioxin hazards to fish, wildlife, and invertebrates: A synoptic review. *U.S. Fish Wildl. Serv. Biol. Rep, no.85.*

Fiedler, H., Hutzinger, O. and Timms, C.W. (1990) Dioxins: sources of environmental load and human exposure. *Toxicol. Environ. Chem.*, **29**, 157–234.

Huggett R.J., Kimerle, R.A., Mehrle, P.M., Jr and Bergman, H.L. (eds) (1992) *Biomarkers. Biochemical, Physiological, and Histological Markers of Anthropogenic Stress*, Lewis, Boca Raton.

Jensen, S. (1966) Report of a new chemical hazard. *New Scientist*, **32**, 612.

Knapton, R.W. and Mineau, P. (1994) Effects of granular formulations of turbufos and fonofos applied to cornfields on mortality and reproductive success of songbirds. *Ecotoxicol.*, **4**, 138–53.

Mineau, P. (1988) Avian mortality in agro-ecosystems. 1. The case against granular insecticides in Canada. *BCPC Monograph*, **40**, 3–12.

Olie, K., Vermeulen, P.L. and Hutzinger, O. (1977) Chlorodibenzo-*p*-dioxins and chlorodibenzofurans are trace components of fly ash and flu gas of some municipal incinerators in the Netherlands. *Chemosphere*, **8**, 455–9.

Peakall, D.B. (1992) *Animal Biomarkers as Pollution Indicators*, Chapman & Hall, London.

Peakall, D.B. (1993) DDE-induced eggshell thinning: an environmental detective story. *Environ. Rev.*, **1**, 13–20.

Peakall, D.B. and Shugart, L.R. (eds) (1993) *Biomarkers, Research and Application in the Assessment of Environmental Health*, Springer Verlag, Berlin.

Pimentel, D., Acquay, H., Biltonen, M. *et al.* (1993) Assessment of environmental and economic impacts of pesticide use, in *The Pesticide Question. Environment, Economics, and Ethics*, (eds D. Pimentel and H. Lehman), Chapman & Hall, New York, pp. 47–84.

Poland, A. and Knutson, J.C. (1982) 2,3,7,8-tetrachlorodibenzo-*p*-dioxin and related halogenated aromatic hydrocarbons: Examination of the mechanism of toxicity. *Ann. Rev. Pharmacol. Toxicol.*, **22**, 517–54.

Ratcliffe, D.A. (1967) Decrease in eggshell weight in certain birds of prey. *Nature, Lond.*, **215**, 208–10.

Reynolds, L.M. (1969) Polychlorinated biphenyls (PCBs) and their interference with pesticide residue analysis. *Bull. Environ. Contamin. Toxicol.*, **4**, 128–43.

Risebrough, R.W., Reiche, P., Herman, S.G. *et al.* (1968) Polychlorinated biphenyls in the global ecosystem. *Nature, Lond.*, **220**, 1098–102.

Sharpe, R.M. and Skakkebaeck, N.E. (1993) Are oestrogens involved in falling sperm counts and disorders of the male reproductive tract? *Lancet*, **341**, 1392–5.

Sladen, W.J.L., Menzie, C.M. and Reichel, W.L. (1966) DDT residues in adelie penguins and a crabeater seal from Antarctica. *Nature, Lond.*, **210**, 670–3.

Stringer, A. and Lyons, C. (1974) The effect of benomyl and thiophanate-methyl on earthworm populations in apple orchards. *Pesticide Sci.*, **30**, 189–96.

Surgeoner, G.A. and Roberts, W. (1993) Reducing pesticide use by 50% in the Province of Ontario: challenges and progress, in *The Pesticide Question. Environment, Economics, and Ethics*, (eds D. Pimentel and H. Lehman), Chapman & Hall, New York, pp. 206–22.

Tatton, J.O'G. and Ruzucka, J.H.A. (1967) Organochlorine pesticides in Antarctica. *Nature, Lond.*, **215**, 346–8.

Tillitt, D.E., Ankley, G.T., Giesy, J.P. *et al.* (1992) Polychlorinated biphenyl residues and egg mortality in double-crested cormorants from the Great Lakes. *Environ. Toxicol. Chem.*, **11**, 1281–8.

Wardhani, M.A. (1991) Developments in IPM: The Indonesian case. Presented at a conference on IPM in the Asia–Pacific Region, Kualar Lumpur, September.

CHAPTER 9

Nonpesticidal chemicals that have an impact on agriculture

Three broad areas of environmental concern – acid rain, climate change and the hole in the ozone layer – all have implications for agriculture and forestry. The impact of SO_x on human health and the acidification of aquatic systems are discussed briefly and the controversy over the role of acid rain in the decline of forests is examined in more detail. The possible impact of climate change on agriculture and forestry is discussed and the role of methane in global warming is considered. The possible impact of ozone depletion on photosynthesis is considered. The international aspects of these problems and the steps that have been taken to solve them are described.

Introduction

The nature of pollution problems has changed over the decades. We have moved from the obvious and local to the subtle and widespread. At the beginning of the 1960s we had the obvious problem of dead pigeons following the use of dieldrin as a seeddressing. This was followed, by the end of that decade, by problems of avian reproduction, such as of the peregrine. By the early 1970s, the problems that man was causing in large bodies of water, such as the Great Lakes of North America and the Baltic, were becoming apparent. Later that decade the regional problem of acid rain became a prime concern. Lastly, by the end of the 1980s, clearly changes were occurring in the atmosphere that were truly global. Two changes – global warming and ozone depletion – have come to the fore. These are issues so widespread and so serious that they can only be tackled internationally at the highest level. These changes are made more complex by the fact that the component effects are interrelated. In this chapter we look at the impacts of these effects (acid rain, climate and ozone depletion) on agriculture.

Acid rain

A world map of estimated annual deposition of sulfur has been produced by Rodhe, Galloway and Dianwu (1992). In addition to SO_x, the map includes dimethyl sulfide (mainly from oceans) and sulfates. Problems for terrestrial systems occur when high deposition and high sensitivity of soil (i.e. a low buffering capacity) coincide. The map (Figure 9.1) shows three regions where high deposition (1 g m^{-2}yr^{-1}) coincides with soil sensitive to acidification; these are eastern North America, Europe and South-East Asia. The only region in the southern hemisphere that approaches these levels is an area of southern Africa with deposition rates up to 0.7 g m^{-2}yr^{-1}. Here, however, the soils can absorb acid. Rodhe *et al.* point out that just a modest increase of sulfur emissions in China would greatly increase the area of high deposition in South-East Asia. At present the industrialization of China is based on energy from coal. Since much of the area of South-East Asia has soil sensitive to acidification, serious problems stemming from China can be expected. The World Bank has recently given a grant of US$1 million to map the ecological impact of acid rain, and to help countries in devising strategies for emission reduction.

The adverse effects of SO_x and NO_x can be considered under the three main headings of: the effects of these gases on human health in cities and near to industrial sites, the acidification of bodies of water, and the effects

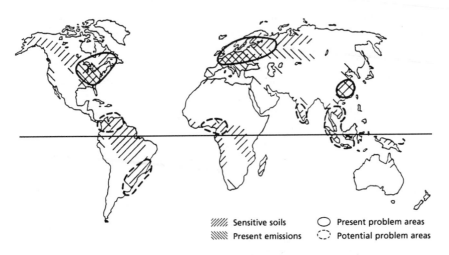

Figure 9.1 *Mapping of present and predicted sulfur deposition and sensitive soils to locate present and potential problem areas (from Rodhe, H. and Herrera, R. (eds), Acidification in tropical countries, SCOPE, **36**, Wiley, Chichester, 1988, with permission).*

of acid rain on forests. Although all these aspects of the problem are mentioned below, only the last will be considered in any detail.

Much of the 'local' impact of SO_x from power stations and other industrial sources has been reduced by discharging waste gases through tall chimneys. This has spread the gases over wide areas and the problems caused by this approach are considered later in this chapter.

Since nearly half the world's population lives in urban areas, the quality of air within these areas is a major human health concern. The air pollutants that occur widely in cities world-wide are SO_x, NO_x, carbon monoxide (CO), ozone (O_3), lead and suspended particulate matter. A detailed report on the air quality of 'megacities' – those with a population of over 10 million – has recently been issued (WHO/UNEP, 1992). Some solutions (changing from dependence on coal, using unleaded petrol) reduce the overall air pollution, but others (e.g. high stacks and resiting power plants) merely shift the problem elsewhere. Problems are caused by SO_x and NO_x to both aquatic and terrestrial environments; these two are considered under separate headings.

Aquatic environments

The effects of acidification are better understood in the aquatic than in the terrestrial environment. There is no doubt that acidic deposition can cause surface water acidification. A recent survey in acid-sensitive areas of the United States (Baker *et al.*, 1991) found that atmospheric deposition was the dominant source of acid anions for 75% of the 1180 lakes and for 47% of the 4670 streams included in the National Surface Water Survey. Detailed studies on experimentally acidified lakes have been carried out (Schindler *et al.*, 1985). These workers found some dramatic changes in the lake food web when the pH was changed from 6.8 to 5.0 over 8 years. There were changes in phytoplankton species, cessation of fish reproduction, disappearance of benthic crustaceans and the appearance of filamentous algae. However, no changes in primary productivity, rates of decomposition or nutrient concentrations were observed. Key changes in the food web had already been noted when the pH dropped to 5.8.

The decline and loss of fisheries caused by acidification has been documented in the north-eastern United States and Scandinavia. In the north-eastern United States fishless acidified lakes have been found, especially in the Adirondack Mountains of New York State. Some streams in Pennsylvania and Massachusetts are also considered to be affected by acidification, but Haines and Baker (1986) concluded that 'the extent of damage to date appears small relative to the total resource.'

In Norway a survey was made in 1986 of 1000 lakes to compare with an earlier one in 1974–75 (Henriksen *et al.*, 1989). It was found in southern Norway that the number of barren lakes had doubled. The chemical changes were characterized by a decrease of calcium and an increase of aluminum rather than by a marked decrease in pH. In 1986, 52% of the lakes surveyed were considered as endangered. Models suggest that a reduction of 30% in the loading would lead to recovery in 28% of these lakes, and a reduction of 50% would ensure viable conditions for fish in 40%. The total land affected by acidification in Norway has increased from 33 000 km^2 in 1974–79 to 36 000 km^2 in 1986.

The effect of acidification on metal bioavailability has been examined. Aluminum and mercury are the two most critical metals; effects of others such as cadmium and lead seem less important. The impact in both cases is serious, but quite different. The effect of aluminum, which has been documented in both laboratory and field studies, is toxicity to the fish themselves, including acute effects on the gills and effects on reproduction. So far as fish populations are concerned, the latter effect is the more important, with reproductive failure leading to the decline and extinction of populations. There is considerable variation in the sensitivity of individual fish species to pH. Some, such as the fathead minnow, are affected at pH 5.6 whereas others, such as yellow perch, can survive pH values as low as 4.5. In contrast, although fish in acidic waters do accumulate increased levels of mercury, these are well below toxic levels. The problem is that piscivorous birds and mammals are likely to be at risk (Scheuhammer, 1991). This risk to predators of fish is reflected in the 'fish advisories' that have been issued. In Canada, fish consumption restrictions have been recommended for 90% of the walleye and 60% of the lake trout populations. This followed a survey of over 1000 lakes in Ontario. Similar restrictions have been made in the north-eastern United States and in Sweden. Although the number of lakes in Sweden for which advisories have been issued is only 250, it has been estimated (Hakanson, Nilsson and Andersson, 1988) that as many as 10 000 lakes may be involved. Limited data suggest that liming of lakes can improve this situation.

A similar problem, again not involving anthropogenic sources of mercury, but through an increase in bioavailability caused by human activity, has occurred when large areas of land have been flooded during hydroelectric projects. One such project was the Churchill–Nelson River hydroelectric development project in Manitoba, Canada. Levels of mercury in fish increased soon after impoundment. In predatory fish, they increased from *c.* 0.5 p.p.m. to *c.* 2.5 p.p.m. due to mobilization of mercury from the soil and subsequent methylation by bacteria (Bodaly *et al.*, 1984). Mercury levels in fish showed no decline 5 years after impoundment. Elevated levels were noted in fish-eating mammals, although no toxic effects were noted. Many local people had blood levels above 20 p.p.b., but no adverse effects were detected.

A current project that is causing considerable concern and controversy is the James Bay II project, in Canada. The original James Bay project was begun in 1973 to generate hydroelectric power. It involved the diversion of three rivers and involved the flooding of 10 000 km^2 of land. The new project would involve flooding an additional 4000 km^2 of land in which the soil is rich in mercury. The area affected is equivalent to that of France. There is considerable scientific debate about the time-scale involved for the rise and subsequent decline of mercury levels in fish. Raphals (1992) considers that the 'aggregate lake would be one of the most mercury-polluted water bodies in the world and would remain a significant problem for 80–100 years.'

The problems caused by chlor-alkali plants, the use of mercury compounds in pulp mills and the use of mercurials as seed-dressings are now well known and should be things of the past. In contrast, problems that are likely to remain far into the future are the liberation of mercury into lakes when they become more acidic and the leaching out of mercury from the soil when areas are flooded for reservoirs and hydroelectric projects.

The best approach to safeguard human health would be surveys of mercury levels in fish. Such surveys would enable bans to be imposed in the case of commercial fisheries and safety warnings to be issued in the case of sport fishing. However, the cost of such a monitoring program would be high, since it involves the collection and analysis of fish from many lakes.

Besides effects on humans and other fish-eaters, the fish populations themselves may respond indirectly to effects on other parts of the food chain. Such indirect responses can complicate the interpretation of field observations. Zooplankton are particularly susceptible both to low pH and metals (Spry and Wiener, 1991). In the experimental acidification studies in Ontario, the rate of growth of lake trout first increased and then markedly decreased as pH was reduced (Mills *et al.*, 1987).

Based on these and other studies, it was considered that sulfate deposition should be limited to 9–14 kg $ha^{-1}yr^{-1}$ to protect the most sensitive aquatic ecosystems (Schindler, 1988). These values are far below the 20–50 kg $ha^{-1}yr^{-1}$ deposition measured over much of eastern North America and western Europe. Thus the standards suggested by Schindler could only be attained with very substantial reductions in the anthropogenic emissions of sulfur dioxide (SO_2).

A recent report by the Joint Nature Conservation Committee in the UK (Farmer and Bareham, 1993) concludes that, if Britain sticks to the target of a 60% reduction of the 1980 acidic emissions, 818 Sites of Special Scientific Interest (SSSIs) totaling 321 000 ha would remain vulnerable to acid rain. If the reduction were as much as 80%, the number of vulnerable sites would drop to 336. To put this problem into perspective, the number of SSSIs in England and Wales is 4520, and they total just over a million hectares.

Terrestrial environments

That pollution caused by coal burning can affect plants has long been known. The term 'acid rain' was coined by Angus Smith in 1852 in a paper delivered to the Literary and Philosophical Society of Manchester.

The earliest scientific papers on the subject seem to be those by Wislicenus in Germany in 1907 and Crowther and Ruston in the UK in 1911. Wislicenus' studies were carried out in the industrialized valleys of eastern Germany. The extensive forest devastation that was recorded there was largely attributed to SO_2. Both acute and chronic damage was noted. Coniferous trees proved to be more sensitive to chronic exposure, whereas deciduous trees reacted more to acute exposure.

Crowther and Ruston (1911) commented that the rain falling through the polluted atmosphere around Leeds became 'notably rich in suspended matters, chlorides, sulphates [often also other sulphur compounds, such as SO_2], nitrogenous compounds [notably ammonia] and free acid.' They found that these pollutants had direct adverse effects on leaves, but considered that the most serious effect was on nitrogen fixation in the soil.

Studies on terrestrial systems are a good deal more complicated than those on aquatic systems. The basic chemistry of the acidification of bodies of water is fairly well understood. So is the dependence of the rate of acidification on the rate of input and on the buffering capacity of bodies of water. The effect of pH on the survival and reproduction of aquatic organisms is easier to study than the corresponding effects on terrestrial organisms, especially long-lived organisms such as trees.

It is also clear that multiple pollutants and a multitude of other factors are involved. The pollutants include ozone and the heavy metals in addition to SO_x and NO_x. Effects may be direct (e.g. on the foliage of the tree) or indirect (e.g. an alteration of the nutrients in the soil). Air pollutants may stress forests beyond their ability to cope with natural stresses such as disease, insect attack, drought and other climatic factors. It can be very difficult to separate effects of pollutants from effects of natural factors. The most detailed studies have been made in Europe, especially in Germany.

Forests

Figure 9.2, taken from the 1988 *State of the World* Report issued by the Worldwatch Institute, shows the extent of damage to forests in Europe. The proportion of forest considered damaged is remarkably high. In several

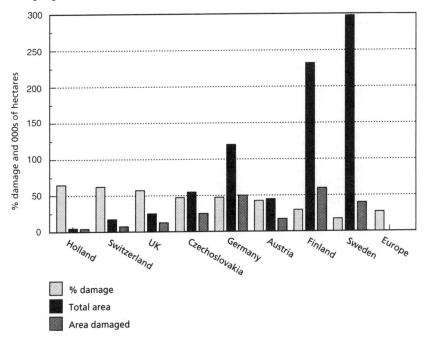

Figure 9.2 *Forest damage in Europe, 1986 (data from Brown, L.R.,* State of the World, *1988, Norton, New York).*

countries – Holland, Switzerland and the UK – it is at or just above 50%, and for Europe as a whole the figure is 22%.

Two basic points should be made in relation to these figures. First, they refer to damage from all sources. It is impossible to relate the damage to pollution, let alone any specific pollutant. Secondly, there is considerable disagreement about what measures of damage to use. Most surveys are based on foliage loss and degree of foliage discoloration. Nilsson and Duinker (1987) have pointed out that foliage discoloration and loss is caused by a variety of factors, including extreme temperature and drought, and is not well correlated with forest decline. These workers have produced maps of the damaged volume of forests in Europe expressed as a fraction of growing stock (Figure 9.3).

The loss of forests in Germany is now an important political issue and the word 'Waldsterben' – forest death – is now a household word. The history of the damage in Germany has been given by Blank (1985). Damage in the early 1970s was largely confined to the silver fir. This species is not important in German forests and, in any case, localized diebacks have been recorded over the past two centuries. However, by the mid-1970s silver fir decline occurred at many locations in Germany and in several other countries. From the late 1970s, damage was noted in several other species, notably the Norway and Scots pine and the common beech. Since these three species account for three-quarters of the total forest area of Germany, there was considerable alarm. In severely affected areas of Baden-Wurttemberg, the damage to spruce increased from 6% in 1981 to 94% in 1983.

The serious situation in Poland has been described by Mazurski (1990). He concluded that 'the structure of Polish industry remains unchanged and therefore ecological conditions continue to deteriorate, including the conditions necessary for healthy forests. In many areas forests are threatened with total extinction. This is especially noticeable in the western Sudetes Mountains. Therefore, resolute and rapid action is imperative to protect Poland's natural environment and that of neighboring countries.'

Potential causes of forest declines

1. Direct effect of acid. Direct effects of sulfuric and nitric acids are not considered a major factor. Levels needed experimentally to cause damage, even to sensitive plants, are much higher than those recorded. Further, there has been no recent increase in the acidity of rain in the affected areas.
2. SO_x and NO_x. Although these gases can affect plants, including trees, they have not been considered as a major cause in western Germany. Levels are usually well below those considered critical (25 µg m^{-3}

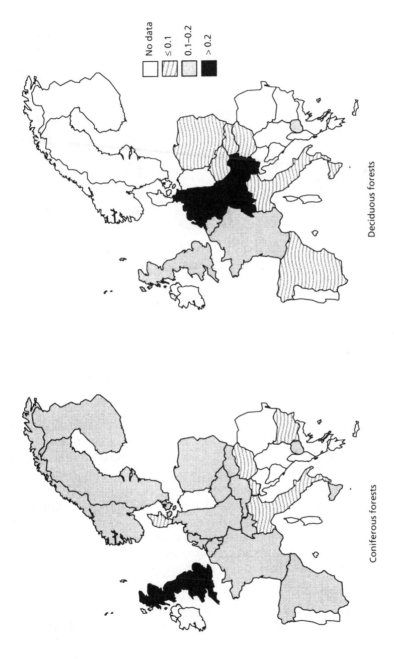

No data

≤ 0.1

0.1–0.2

> 0.2

Coniferous forests

Deciduous forests

Figure 9.3 *Volume of forest in Europe damaged as a fraction of growing stock (from Nilsson, S. and Duinker, P., Environment, 29(9), 4–31, 1987, with permission).*

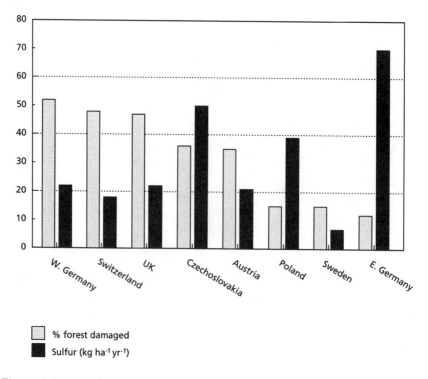

  ☐ % forest damaged
  ■ Sulfur (kg ha⁻¹ yr⁻¹)

Figure 9.4 *Forest damage and sulfur deposition in Europe, 1986–88 (data from Brown, L.R.,* State of the World, *1988, Norton, New York, and UNEP,* Environmental Data Report, *Blackwell, Oxford, 1992).*

overall mean, or daily levels of 50 μg m⁻³). In certain areas of eastern Germany and Czechoslovakia, however, levels greater than these critical ones have been observed. Certainly in Europe there is no good correlation between forest damage and sulfur deposition (Figure 9.4).

3. Ozone. A concentration of ozone of 100–200 μg m⁻³ has been found to damage trees when exposed for 6–8 hours a day for several days. This concentration is only 2–3 times background levels. In many rural areas of Europe average daily concentrations are regularly in this range and peak levels often exceed background levels six- to ten-fold. In Germany, scientists studying the dieback of conifers consider that ozone is the single most important factor (Postel, 1984).

A wide variety of other chemicals, such as peroxides and formaldehyde, have been identified in the atmosphere and may contribute to damage to plants (Gaffney *et al.*, 1987). Certainly it is clear that no single hypothesis can account for the varying patterns of forest decline that have been observed.

The problem of *'Waldsterben'* has recently been reviewed by Kandler (1992). He concluded that 'the suggestion that a large-scale decline of European forests (*Waldsterben*) has taken place since 1978/1980 is not corroborated by 10 years of intensive research.' He traced the effects of SO$_2$ damage, which started in the second half of the nineteenth century, and considered that, with a few exceptions, complete recolonization of devastated areas has been observed since 1960 and that there is now no correlation between the spatial and temporal distribution of air pollutants and damage to trees. In a companion paper (Mueller-Dombois, 1992), it is noted that forest declines of a similar type are also occurring in several Pacific forests unaffected by industrial pollution. Here dieback of the canopy is considered a natural phenomenon of forest dynamics.

It seems likely that a wide range of factors, both natural and unnatural, are at work on forest dynamics. From a chemical point of view, ozone seems the chemical most likely to do damage. This is supported by recent studies in the south-eastern United States (Richardson, Sasak and Fendick, 1992). Here no direct effects of acid precipitation were seen, but ozone greatly affected photosynthesis. These workers considered that even normal ozone levels in this region caused some reduction of photosynthesis and strong inhibition occurred at 2–3 times normal ambient levels.

Effects of acidic deposition on human health were considered by a Working Group of the World Health Organization, the findings of which were published in 1986 (WHO, 1986). The data available suggested a minimal risk to healthy individuals inhaling acidic aerosols at ambient concentration. However, some risk to sensitive groups (i.e. asthmatics) could not be ruled out. As far as indirect effects are concerned, the Group identified several potential problems caused by the ability of acidified water to leach out increasing quantities of metals (as discussed earlier in this chapter).

Climate change

The early history of the 'greenhouse' effect has been detailed by Gribbin (1990). The first description of the importance of the atmosphere in controlling the earth's temperature was made by the French mathematician Baron Jean Bapiste Fourier in 1827. He likened the atmosphere of the earth to the glass cover of a box. The role of water vapor and trace gases such as carbon dioxide in keeping the earth at a pleasant temperature was described by the British scientist John Tyndall in 1863. He appears to have been the first to coin the phrase 'greenhouse effect'. That industrial pollution might, over a period, increase the temperature of the earth was first put forward by the famous Swedish chemist Svante Arrhenius. While he is best known for his work on electrolysis, in 1886 he wrote a book, *Worlds in the Making,* that put forward the concept that the available energy of the universe is self-renewing. The book includes the role of various molecules in maintaining the earth's temperature. He expressed the hope that the greenhouse effect would improve the cold climate of his native Sweden.

By far the most important of the greenhouse gases is carbon dioxide. The steady, and so far inexorable, rise of the level of carbon dioxide in the atmosphere is one of the very few undisputed aspects of the greenhouse effect. The relative importance of fossil fuel combustion and land-use changes for emissions of CO_2 has already been mentioned in Chapter 3 (Figure 3.2). Large-scale felling of trees, as occurred in North America in the nineteenth century and Europe even earlier, is now happening in the tropical rainforest. This both increases the carbon dioxide emissions when the wood is burnt and decreases the amount of carbon dioxide taken up by photosynthesis. However, the burning of fossil fuel remains the main cause of the increase of atmospheric carbon dioxide.

The main point of debate has been, and still is, how large an effect this build-up of carbon dioxide will have on the global temperature. This issue cannot be debated here. However, most models agree that the temperature increase caused by a doubling of the carbon dioxide level (which if present rates continued would be by 2030) would be between 1.5 and 2.5 °C. Here we wish to consider the effect that this is likely to have on forestry and agriculture.

Forestry

Interesting evidence for global warming has come from a stand of pine trees on the island of Tasmania, off the southern coast of Australia. The Huon pine lives for up to 700 years in this unpolluted area of the world. Studies have shown that the width of the tree rings can be related to the temperature of the summer (Anderson, 1991). Both the cold spell of the early 1900s (when the pack ice of the Antarctic expanded and icebergs drifted much further north than usual) and the 'little ice age' of the seventeenth century can be clearly seen. Using this record of tree ring width, the temperature increase in the past 20 years is greater than anything seen in the previous 700 years. Effects of global warming would not be confined to a rise in sea-level. There would also be major effects on vegetation. The northern forests of Scandinavia, Russia, China and North America are dominated by evergreen conifers. Further south, except on high ground, the woods are dominated by deciduous forests of oak, beech and other broadleaved species. In the tropics we have the evergreen broadleaved rainforests. The boundaries of the different forest types have moved in the past as global temperatures have changed. During the last ice age, which involved a global temperature change of approximately 4 °C, this change occurred over 1000 years. One important aspect of the problem is the speed with which it is likely that these global warming effects will now occur. Palaeoecological studies on the boreal treeline in central Canada suggest that, during a period of climate warming some 4000 years ago, the initial transformation from tundra to forest occurred in about 150 years (MacDonald *et al.*, 1993). This suggests that changes in the past have occurred at similar speeds to those predicted from the present manmade global warming. Studies in the United States have suggested that the northern movement by as much as 300 or 400 miles of the southern boundary of the sugar maple could be associated with a doubling of the CO_2 concentration. The fossil record and present-day distributions suggest that some species have been able to shift successfully in response to climate change. In contrast, others have not, either because their rate of migration was too slow or because of the presence of barriers. A number of plant genera that had a circumpolar distribution in the Tertiary period became extinct in Europe while surviving in North America. These included magnolia, hemlocks, sweet gums and tulip trees; it is presumed that east–west barriers such as the Alps, Pyrenees and the Mediterranean were responsible (Peters and Darling, 1985). Of particular concern are nature reserves, which are frequently isolated areas of original habitat surrounded by large areas of human-dominated lands.

Altitudinal shifts of vegetation can also be expected. It has been calculated that a 3 °C increase in temperature would cause a 500 m altitudinal shift in species distribution. Species originating near to the peaks would not have any higher habitat to which they could shift as warming occurred and would be expected to become extinct. Extinctions of alpine plants occurred in Central and South America when vegetation zones moved upwards by 1000 m after the last glacial maximum (Peters and Darling, 1985).

The future possible effects of global warming, combined with the present-day effects of acid rain and depletion of the ozone layer, mean that it is difficult to predict accurately the effects on the ecosystems of the earth.

Agriculture

The possible impacts of global warming on agriculture can be considered under several headings: effects of temperature rise, rainfall changes, increase of carbon dioxide levels and the numbers and composition of pest species. These are considered in turn.

Temperature rise

The results from the various models devised to estimate the changes expected from the rise of carbon dioxide and other greenhouse gases vary considerably. Nevertheless, there is broad agreement that rises in temperature will occur. It has been calculated that increased temperatures would be advantageous for the rice crop of Thailand (Bachelet *et al.*, 1992). They would allow a northward expansion of rice-growing areas and lengthening of the growing season in those areas currently limited by low temperatures. In contrast, potatoes are favored by low temperatures and the areas suitable for this crop could be expected to decrease. The grain belt of North America, which provides much of the exportable food supply of the world, would be expected to shift northwards.

Rainfall changes

Water is the major limiting factor for agriculture. Regrettably, as far as global warming is concerned, the predictions on rainfall changes are more variable than they are for temperature. For example, when four predictive models were run to estimate the impact of doubling of CO_2 on the Thailand rice crop, there was general agreement among them for temperature but little agreement for rainfall. Increased temperature will certainly increase evapotranspiration which is likely to negate any increase in rainfall.

Effects of carbon dioxide levels

Plants can be divided into two groups (C_3 and C_4) based on the mechanism that they employ to fix carbon dioxide. Both groups have cells that control carbon dioxide, oxygen and water exchange across their surfaces. As the carbon dioxide level rises, both groups lose less water, but the C_4 plants have a pump that concentrates carbon dioxide near to the site of active photosynthesis. As carbon dioxide levels rise, this pump reduces oxygen binding. The C_3 plants do not have this mechanism and therefore are favored in high carbon dioxide levels. Nearly all trees and many major crops, including potatoes, rice and wheat, are C_3 plants.

Elevated carbon dioxide levels have long been used in greenhouses to accelerate plant growth, but it is doubtful if this advantageous effect would be widespread. One reason is that, although plants show increased photosynthesis initially, the rates soon drop to the same values as under normal CO_2 levels. Another problem is that increased rates of photosynthesis are only seen if other needs, such as nutrients and water, are not limiting. It is therefore not surprising that studies on the grasslands of the Arctic tundra have not shown increases of productivity with increasing CO_2 levels. There is also the problem that the community structure of plants could change if C_3 plants outcompete C_4 plants.

Effects on pests

There are many estimates of global losses of agricultural crops from insects, disease and weeds. Among these are figures of 12%, 12% and 11%, respectively (Pimentel, 1991). Increased temperature is expected to favor insects and weeds. In the former, increasing temperature would be expected to increase fecundity and number of generations by extension of the breeding season. Also, pests from the tropical areas could extend their range northwards. Weeds, on balance, are better adapted than crops to arid conditions, and could be expected to increase their competitive advantage with crops. Fungal plant diseases, on the other hand, should decrease in importance as dry conditions reduce their ability to attack crops.

In conclusion, it can be said that most models predict that global warming would increase problems for agricultural production. Modeling of climate changes suggests a 10–15% grain yield decline across a broad belt of more tropical countries in Africa, Latin America, India and South-East Asia. Some important crops that would be affected are millet, rice, sorghum, soybean and wheat.

While farming in northern marginal areas – sheep in Iceland, wine in England – could be expected to improve, the main concerns are effects on the mid-latitude grainbelt. Losses of 10–30% of unirrigated crops have been predicted for the grain-exporting regions of North America and Australia, unless offset by changes in agricultural practice. These changes could include increased irrigation, better control of pests and soil erosion, and the use of different varieties of plants. The 1988 Toronto Conference on The Changing Atmosphere concluded 'While averaged global food supplies may not be seriously threatened, unless appropriate action is taken to anticipate climate change and adapt to it, serious regional and year-to-year food shortage may result, with particular impact on the vulnerable.' Handling the problems caused by global warming will call for an integrated approach, not merely to pest management, but also to sound ecological agriculture.

Methane and other greenhouse gases

Carbon dioxide is not the only greenhouse gas, although it is the most important in that it contributes half the total greenhouse effect. Another is methane, which is produced naturally by bacteria in the absence of oxygen. This can happen in swamps (hence its old name 'marsh gas'), rice paddies and the guts of animals, especially of ruminants such as cattle. Methane is also produced when wood and vegetation are burnt and from leaks from natural gas and coal mines. The natural and anthropogenic sources of methane are shown in Figure 9.5. Methane is a more effective greenhouse gas than carbon dioxide, in that it is some 30 times more effective at trapping heat. However, it is much less stable, and is broken down by the ultraviolet light from the sun. It is estimated that methane currently contributes only about a fifth of the greenhouse effect. A

(a)

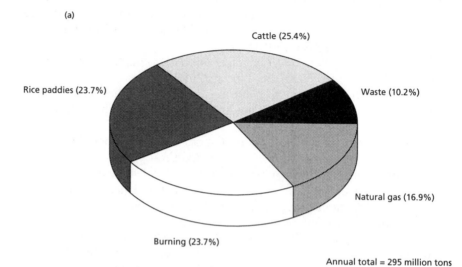

Cattle (25.4%)

Rice paddies (23.7%)

Waste (10.2%)

Natural gas (16.9%)

Burning (23.7%)

Annual total = 295 million tons

(b)

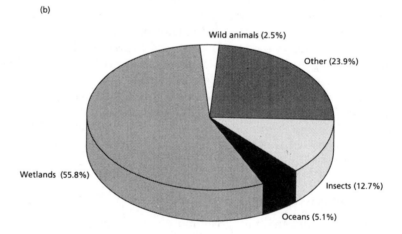

Wild animals (2.5%)

Other (23.9%)

Wetlands (55.8%)

Insects (12.7%)

Oceans (5.1%)

Annual total = 197 million tons

Figure 9.5 *(a) Anthropogenic and (b) natural sources of methane (data from OECD, The State of the Environment, OECD, Paris, 1991).*

particular concern is that, should the tundra of the far north begin to thaw, large quantities of methane could be released from the underlying layers of dead plants.

Nitrous oxide is an even stronger greenhouse gas, although it occurs at much lower concentrations. It is part of the natural nitrogen cycle, but almost half comes from man-made sources (combustion, use of fertilizers and from animal and human wastes).

The 'doomsday' scenario for a runaway greenhouse effect goes like this: as the temperature increases, large amounts of methane are released from the tundra. Simultaneously the increased ultraviolet radiation, caused by depletion of the ozone layer, damages the phytoplankton of the oceans. This removes an important 'sink' for carbon dioxide, further increasing global warming and thus the release of more methane. In the worst of all cases, the positive feedback mechanisms might combine to trigger an unstoppable warming. Greenpeace sent questionnaires to 400 relevant scientists. Of the 113 who replied, almost half thought a runaway greenhouse effect was possible and 13% thought it probable.

However, some atmospheric contaminants actually decrease the greenhouse effect. Sulfate particles thrown into the atmosphere by the burning of fossil fuel cool the earth by reflecting more of the sun's energy back into space. Similarly, noticeable cooling effects have followed volcanic eruptions such as Agung, El Chicon and, most recently, Pinatubo in the Philippines. The Pinatubo eruption ejected some 20 million tons of SO_2 into the stratosphere, making a layer that reduced the amount of solar heating that reached the earth's surface. It is estimated that the effects of the eruption will last 2–3 years.

Ozone/CFCs

The chlorofluorocarbons (CFCs) are, molecule for molecule, the strongest of the greenhouse gases. However, the main concern over these compounds is the depletion of the ozone layer. Ozone depletion is becoming increasingly well documented and here we will refer only to a few recent studies. Satellite mapping of total ozone concentrations started in 1979, and has revealed a steady decline globally. The data, showing a 1% per year loss over the period 1979–86, are shown in Figure 9.6 (Bowman 1988). Figure 9.7 shows the global pattern of decline of ozone based on satellite data up to March 1991 (Stolarski et al., 1992). The declines are greatest towards the polar regions, but decreases are found in all regions except near the equator. More recent studies (Hofmann et al., 1992) have shown a marked reduction in the ozone layers in the Antarctic at both 11–13 and 25–30 km altitude regions. This greater depletion at the lower altitude was considered due to aerosol particles from the eruption of Mount Pinatubo in June 1991.

Once the Antarctic hole was discovered, studies were undertaken to see if a similar situation was developing in the Arctic. The European Arctic Stratospheric Experiment was started in 1989 to study the problem. This is the largest ever European environmental science project and, despite its

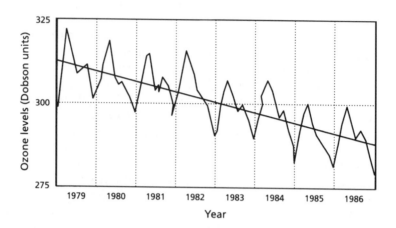

Figure 9.6 *Temporal trends in ozone levels over the Antarctic, 1979-86 (from Bowman, K.P., Science, NY, **239**, 48–50, 1988, with permission).*

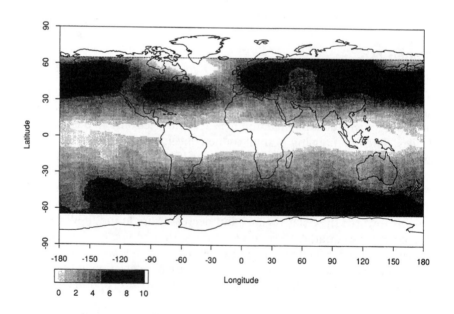

Figure 9.7 *Contours of total ozone mapping spectrometer data. Average trends in per cent per decade for December to March, 1978–91 (from Stolarski, R., Bojkov, R., Bishop, L. et al., Science, NY, **256**, 342–9, 1992, with permission).*

name, includes scientists from Canada, Japan, New Zealand and the United States. The data showed that while loss of ozone was occurring, this loss was much smaller than in the Arctic (10% compared to 60%). In the mid-latitudes, 30–60 ° north or south, however, the depletion was similar, at around 8% in late winter/early spring. The worrying feature here is that, whereas the mid-latitudes of the northern hemisphere contain large human populations, the equivalent band in the southern hemisphere is largely ocean. Although this reduces the likelihood of effects on human health, there are environmental concerns. There is no consensus on the likely effect of increased UV radiation on the southern oceans. The target of greatest concern is phytoplankton; experimental studies have shown that even modest increases of UV radiation can decrease productivity. The actual degree to which productivity is affected is difficult to measure. The possibility of serious damage is increased because the greatest depletion of the ozone layer (up to 60%) occurs in the southern spring. This is when the plankton is undergoing a great burst of productivity. The best estimates of the reduction of productivity of plankton are about 10–20%. This is small compared to the natural annual fluctuations that the system has been capable of surviving. The question of whether a 10–20% reduction in primary productivity is critical at the low point of natural fluctuations has not been resolved. It seems unlikely that higher organisms, such as fish, whales and seabirds would be affected directly, but obviously they would be affected if the Antarctic food web were disturbed.

Recent mesocosm studies (Bothwell, Sherbot and Pollock, 1994) on the differential response of algae and insect larvae to UV radiation suggest that there may indeed be such ecosystem effects. Initially the growth of algae is inhibited, but after a few weeks the effect on the insect population is so marked that the algae can increase due to the lower predatory pressure. The authors conclude that 'if differential sensitivity to UVB between algae and herbivores is a generalisable phenomenon, short-term measurements of UVB inhibition of algal photosynthesis, which have been the focus of much research during recent years, may not be addressing the most critical element of aquatic ecosystem sensitivity to elevations of UVB irradiance.'

The international aspects of the CFC problem are shown by two phenomena. Although 95% of the CFCs are released in the northern hemisphere, the redistribution is rapid enough that the concentrations in the southern hemisphere lag behind those in the north by only 10%. Also, by far the largest ozone depletion occurs over Antarctica.

Control of CFCs is a case where international action has been taken. In 1985 the United Nations Environment Programme came up with the Vienna Convention for the Protection of the Ozone Layer. This convention only dealt with international cooperation on observations and research and did not contain specific controls. Controls were, however, introduced in the Montreal Protocol adopted in 1987. The participants agreed to a freeze on production of CFCs at 1986 levels by 1990. They further agreed to a reduction of 20% by the beginning of 1994 and a reduction of an additional 30% by January 1999. Developing countries, however, received extensions of the deadlines and the Soviet Union also got an exemption. Overall it has been estimated that the reductions by the turn of the century would amount to 35% rather than 50%.

It is ironic that the biggest story, the 'ozone hole' over Antarctica, broke just 2 weeks before the Montreal meeting. This was too late for the information, which was anyhow still preliminary, to be evaluated by the scientists of the individual governments, for decisions to be made and for delegates to be instructed. However, the information did lead to amendments to the Montreal Protocol (London 1990). These increased the speed of reducing production of CFCs and called for zero production and consumption by the year 2000.

There is now abundant evidence that the size of the Antarctic ozone hole is increasing. The levels of ozone measured in the Antarctic spring have decreased sharply over the past 15 years and the duration of these decreases has increased. Evidence is now coming in about problems – not yet as severe – in the Arctic. That CFCs can survive in the stratosphere for 100 years (it varies from compound to compound, but a century is a reasonable average figure) means that, even with a complete ban on CFCs, the ozone layer would not be completely restored until the twenty-second century.

Are remedial measures possible? Several suggestions have been made. Two are pumping ozone into the stratosphere, and adding hydrocarbons which would react with the chlorine. The difficulties, and not just the technical difficulties, are enormous. Decisions would have to be based on the best available calculations. How serious will things really be if we do nothing? Will the solution really work? One would not know until afterwards. Who should make the decision? Somebody, some group, has to make the decision. All too obviously it is going to be a difficult decision, but we have to remember that no decision is also a decision.

References

Anderson, I. (1991) Global warmings ring true. New Scientist, September 21.
Bachelet, D., Brown, D., Böhm, M. and Russell, P. (1992) Climate change in Thailand and its potential impact on rice yield. Climatic Change, 21, 347–66.
Baker, L.A., Herlihy, A.T., Kaufmann, P.R. and Eilers, J.M. (1991) Acidic lakes and streams in the United States: The role of acidic deposition. Science, NY, 252, 1151–4.
Blank, I.W. (1985) A new type of forest decline in Germany. Nature, Lond., 314, 311–14.
Bodaly, R.A., Rosenburg, D.M., Gaboury, M.N. et al. (1984) Ecological effects of hydroelectric development in northern Manitoba, Canada: the Churchill–Nelson River diversion. In Effects of Pollutants at the Ecosystem Level, SCOPE, 22, 273–309.

Bothwell, M.L., Sherbot, D.M.J. and Pollock, C.M. (1994) Ecosystem response to solar ultraviolet-B radiation: influence of tropic-level interactions. *Science, NY*, **265**, 97–100.

Bowman, K.P. (1988) Global trends in total ozone. *Science, NY*, **239**, 48–50.

Crowther, C. and Ruston, A.G. (1911) The nature, distribution and effect upon vegetation of atmospheric impurities in and near an industrial town. *J. Agric. Sci.*, **4**, 25–55.

Farmer, A. and Bareham, S. (1993) *The Environmental Implications of UK Sulphur Emission Policy Options for England and Wales*, Report No. 176, JNCC, Peterborough.

Gaffney, J.S., Streit, G.E., Spall, W.D. and Hall, J.H. (1987) Beyond acid rain. *Environ. Sci. Technol.*, **21**, 519–24.

Gribbin, J. (1990) *Hothouse Earth: the Greenhouse Effect and Gaia*, Bantam, New York.

Haines, T.A. and Baker, J.P. (1986) Evidence of fish population responses to acidification in the eastern United States. *Water, Air, Soil Pollut.*, **31**, 605–29.

Hakanson, L., Nilsson, A. and Andersson, T. (1988) Mercury in fish from Swedish lakes. *Environ. Pollut.*, **49**, 145–62.

Henriksen A., Lien, L., Rosseland, B.O. *et al.* (1989). Lake acidification in Norway: Present and predicted fish status. *Ambio*, **18**, 314–21.

Hofmann, D.J., Oltmans, S.J., Harris, J.M. *et al.* (1992) Observation and possible causes of new ozone depletion in Antartica in 1991. *Nature, Lond.*, **359**, 283–7.

Kandler, O. (1992) Historical declines and diebacks of central European Forests and present conditions. *Environ. Toxicol. Chem.*, **11**, 1077–93.

MacDonald, G.M, Pienitz, R., Smol, J.P. *et al.* (1993) Rapid response of treeline vegetation and lakes to past climate warming. *Nature, Lond.*, **361**, 243–6.

Mazurski, K.R. (1990) Industrial pollution: the threat to Polish forests. *Ambio*, **19**, 70–4.

Mills, K.H., Chalanchuk, S.M., Mohr, L.C. and Davies, I.J. (1987) Responses of fish populations in Lake 223 to 8 years of experimental acidification. *Can. J. Fish. Aquat. Sci.*, **44**(Suppl. 1), 114–25.

Mueller–Dombois, D. (1992) A global perspective on forest decline. *Environ. Toxicol. Chem.*, **11**, 1069–76.

Nilsson, S. and Duinker, P. (1987) The extent of forest decline in Europe. *Environment*, **29**(9), 4–31.

Peters, P.L. and Darling, J.D.S. (1985) The greenhouse effect and nature reserves. *BioScience*, **35**, 707–17.

Pimentel, D. (1991) Global warming, population growth, and natural resources for food production. *Society and Natural Resources*, **4**, 347.

Postel, S. (1984) Air pollution, acid rain, and the future of forests. *Worldwatch Paper No. 58*.

Raphals, P. (1992) The hidden cost of Canada's cheap power. *New Scientist*, February 15, pp. 50–4.

Richardson, C.J., Sasek, T.W. and Fendick, E.A. (1992) Implications of physiological responses to chronic air pollution for forest decline in the southeastern United States. *Environ. Toxicol. Chem.*, **11**, 1105–14.

Rodhe, H., Galloway, J. and Dianwu, Z. (1992) Acidification in southwest Asia – prospects for the coming decade. *Ambio*, **21**, 148–50.

Scheuhammer, A.M. (1991) Effects of acidification on the availability of toxic metals and calcium to wild birds. *Environ. Pollut.*, **71**, 329–75.

Schindler, D.W. (1988) Effects of acid rain on freshwater ecosystems. *Science, NY*, **239**, 149–57.

Schindler, D.W., Mills, K.H., Malley, D.F. *et al.* (1985) Long-term ecosystem stress: the effects of years of experimental acidification on a small lake. *Science, NY*, **228**, 1395–401.

Spry, D.J. and Wiener, J.G. (1991) Metal bioavailability and toxicity to fish in low-alkaline lakes: a critical review. *Environ. Pollut.*, **71**, 243–304.
Stolarski, R., Bojkov, R., Bishop, L. *et al.* (1992) Measured trends in stratospheric ozone. *Science, NY*, **256**, 342–9.
WHO (1986) Health impact of acidic deposition. *Sci. Total Environ.*, **52**, 157–87.
WHO/UNEP (1992) *Urban Air Pollution in Megacities of the World*, Blackwell, Oxford.
Wislicenus, H. (ed.) (1907) *Sammlung von Abhandlungen über Abgase und Rauchschaden*, Paul Parey, Berlin.

CHAPTER 10

The international approach to integrated pest management and life-cycle analysis of chemicals

There is an increasing need for an international approach to environmental problems to tackle matters such as locust control, the spread of resistance among pest species and such environmental problems as acid rain, climate change and the 'hole' in the ozone layer. The international organizations created by the Conference on the Human Environment in 1972 and the deliberations of the recent United Nations Conference on Environment and Development are described.

Neither pests nor chemicals liberated into the environment respect international boundaries. The discovery, now over 30 years ago, of DDE in the fauna of Antarctica showed the global distribution of chemical pollutants. Certainly acid rain, climate change and the 'hole' in the ozone layer can only be tackled internationally.

The importance of international bodies in the regulation of chemicals has been increasing for decades.Organizations such as the World Health Organization (WHO) and the Food and Agriculture Organization (FAO) were set up under the auspices of the United Nations soon after the end of the Second World War. Both have played a major role in the regulation of pesticides. The Codex Alimentarius set up by FAO in 1962 to produce food safety standards has already been discussed in Chapter 3. A committee to examine pesticide residues in food was set up jointly by FAO and WHO and meets annually. So far they have evaluated more than 200 pesticides. In the 1970s the FAO broadened its work on chemical safety by looking towards harmonization of national requirements for registration of pesticides. This led to an International Code of Conduct on the Distribution and Use of Pesticides that was approved at an FAO conference in 1985.

From the environmental point of view the UN-sponsored Conference on the Human Environment in Stockholm in 1972, which led to the formation of UNEP, was an important turning point. Within UNEP, several important programs relating to our knowledge of environmental chemicals have been set up. The International Register of Potentially Toxic Chemicals (IRPTC) was started in 1976 and continues to collect and distribute information. Currently the IRPTC holds the toxicological profiles on some 800 compounds on computer. Additionally there are files on waste management, on chemicals currently being tested and on national regulations covering more than 8000 substances. The International Referral System for Sources of Environmental Information (INFOTERRA) was established in 1977 as a network of national references points for environmental questions. Its directory lists over 6500 institutions world-wide and it has access to some 600 databases.

The Global Environmental Monitoring System (GEMS) was created in 1975 to monitor the global environment and undertake periodic assessments of its health. The GEMS networks monitor changes in atmospheric composition, air and water pollution, deforestation and many issues related to diversity. Major programs include HEALS (Human Exposure Assessment Locations), which monitors the total human exposure to selected pollutants, and GRID (Global Resource Information Database), which provides environmental data in a readily accessible form. GEMS supports two research centers, the World Conservation Monitoring Centre at Cambridge and the Monitoring and Assessment Research Centre in London.

Other international agencies have also played an important part. The OECD Environment Committee was formed in 1970. By the mid-1970s, the OECD had set up six expert working groups on the minimum amount of data that should be available for every new chemical, the so-called MPD (minimum pre-market data). While this was not passed by the Council as a directive, it was used in most of the 24 member countries (such as the USEPA scoring system). It also formed the basis for the European Chemicals Control Act (79/831/EEC) adopted by the European Community in 1979. European legislation also calls for additional tests to be made when the tonnage first reaches certain levels.

Legislation calling for the manufacturer to produce a minimum data set before being allowed to market the material has proved much easier than tackling the problem of accumulating the basic data on existing chemicals. The lack of data on existing chemicals (p. 13) was highlighted by a US National Research Council Report in 1984.

Steps were undertaken by the OECD to obtain data on existing chemicals. A first step was to develop a registry of which chemicals were under investigation in member countries. Initially this enabled national bodies to get in touch with others involved with the same chemical, but was also a stepping-stone to the prevention of duplication of effort. This program, EXICHEM, now has some 13 500 entries on 5000 chemicals. Subsequently the OECD produced a list of chemicals that are produced in volumes of over 10 000 tons yr^{-1} in any member country (p. 13). These chemicals account for over 90% of the total global chemical production.

Air pollution has long been recognized as an international problem. Sweden reported to the Stockholm Conference on transboundary air pollution in Scandinavia. In the same year, the OECD began a cooperative program to measure long-range transport of air pollutants (LRTAP). This problem was subsequently taken up by the EEC, which led to the EEC Convention on LRTAP in 1979. Agreements on monitoring and research were more readily reached than agreements on the actual reductions of pollutants such as SO_x and NO_x. A UN protocol on the reduction of sulfur emissions by 30% was adopted in 1985. However, it has not been ratified by several major industrial nations, including the UK and the USA.

The oceans are another area where international cooperation is essential. An Intergovernmental Working Group on Marine Pollution was established at the Stockholm Conference. Scientific advice has since been provided by GESAMP (Group of Experts on the Scientific Aspects of Marine Pollution). The Convention on the Prevention of Marine Pollution by Dumping was agreed in London in 1972. A comprehensive new framework for the marine environment, the UN Convention on the Law of the Sea was agreed in 1982; it had been signed by 160 countries by the end of 1990. The difficulty with this agreement, as with so many other international agreements, is that agreements are freely entered, but the will to enforce them has often been lacking.

The UN Regional Seas Program has been more successful than some broader agreements since it requires several countries to work together to achieve common goals. Since the Convention for the Protection of the Mediterranean Sea against Pollution was signed in Barcelona in 1976, considerable progress has been made on the installation of pollution control around the Mediterranean. Much, however, remains to be done.

The most important event on the international scene of recent years has been the United Nations Conference on Environment and Development (UNCED) held in Rio de Janeiro in June 1992. It was attended by representatives of 115 countries and produced a document known as Agenda 21 to point the way forward into the twenty-first century.

From the point of view of this book the most important chapters of Agenda 21 are that on Changing Consumption Patterns (Chapter 4), Promoting Sustainable Agriculture and Rural Development (Chapter 14), the Environmentally Sound Management of Toxic Chemicals (Chapter 19), and the Environmentally Sound Management of Hazardous Wastes (Chapter 20).

The issue of changing consumption patterns is very broad and this chapter of Agenda 21 is more general than the other three chapters that we will be discussing. It is also more political in nature. Under 'Basis for action' it states, 'measures to be undertaken at the international level for the protection and enhancement of the environment must take fully into account the current imbalances in the global patterns of consumption and production.' Some of these imbalances, notably in terms of emissions of pollutants and usage of energy, have already been considered earlier in this book (Chapter 2).

Chapter 4 has described the role of international organizations, especially FAO and WHO, in pesticide regulation. Especially important has been the discouragement of the use of the organochlorine insecticides. However, the point has been made more than once in the previous chapters that it is perhaps a mistake both to 'tar' all organochlorines 'with the same brush' and to write them off for use in the developing world, ignoring the positive features several of them possess, such as the low acute mammalian toxicity and the ease with which application as a powder is possible in the absence of spraying equipment and water. FAO has, however, also attempted to promote IPM as a move towards sustainability of agriculture in developing countries. As well as aid for IPM projects, conferences and training programs have been mounted. Many aid agencies in developed countries, such as ODA in the UK and USAID in the USA, have also financed and advised IPM projects. This has often been done in collaboration with international agencies such as FAO or the World Bank.

International research stations (such as IITA, ICRISAT and IRRI) in the CGIAR network of research stations were referred to frequently in Chapter 7 in relation to the development of IPM in Asia, Africa and Latin America. After many years of focusing on plant breeding as their approach to crop protection, these research stations have now accepted IPM in their remit. Other international centers, but not linked to specified crops in the same way as CGIAR institutes, include ICIPE and IIBC (again see Chapter 7). Both are making major contributions to the development of IPM. The latter institute is specifically for biological control, but is increasingly including other elements of IPM in its thinking to improve the efficacy of natural control agents.

Consortia of European higher education institutions, with encouragement from the EEC, are developing training courses for IPM workers from developing countries. The courses will eventually be held in those countries, with the support of staff and training manuals from Europe.

The basic objectives of Chapter 4 in Agenda 21 are set out as:

1. All countries should strive to promote sustainable consumption patterns.
2. Developed countries should take the lead in achieving sustainable consumption patterns.
3. Developing countries should seek to achieve sustainable consumption patterns in their development process, guaranteeing the provisions of basic needs for the poor, while avoiding those unsustainable patterns, particularly in industrialized countries, generally recognized as unduly hazardous to the environment, inefficient and wasteful, in their development processes.

The proposed activities to reach these goals were given as:

1. encouraging greater efficiency in the use of energy and resources;
2. minimizing the generation of waste;
3. assisting individuals to make environmentally sound decisions;
4. exercising leadership through government purchase;
5. moving towards environmentally sound pricing;
6. reinforcing values that support sustainable consumption.

It is, as several activities listed in the UNED document indicate, going to take a fundamental change in attitude, a movement away from the 'throw-away society'. A Westerner walking through a market in a developing country is struck by the painstaking repair of old, apparently worthless articles. This maintenance of old equipment is environmentally friendly, but it is based on a plentiful supply of cheap labor, or in more blunt terms, poverty. The developed world has taken on board the concept of 'green goods'. For example, products are now labeled 'ozone friendly', and recycling is 'in', even though there are serious doubts about the effectiveness of some of these programs. However, there is little evidence that the developed world is prepared to make any real sacrifices. For example, the goal of 0.7% of GNP (gross national product) being devoted to foreign aid programs is receding rather than being realized. If we are going to raise the standards of the developing world significantly towards those of the developed world, it is (as we discussed in Chapter 3) going to take a drastic change in the use of energy and materials to produce goods and services.

Chapter 14 of Agenda 21 covers the very wide topic of 'Promoting sustainable agriculture and rural development'. This has the general theme that agriculture in developing countries must satisfy the demands of a growing population, largely by increasing production on land already under cultivation to avoid further inroads on land only marginally suitable for cropping. Much of the chapter is concerned with farming systems, genetic resources, etc., and emphasis is placed on low-input sustainable agricultural systems. This has implications on pesticides and IPM. The latter forms one of the 12 programs of the chapter. The whole chapter sets deadlines that appear to us to be totally unrealistic. For example, the entire agricultural policy review and establishment of the subsequent program have a 1995 deadline. By 2005, developing countries are expected to manage policy, program and planning activities. A key objective of the IPM program is 'to improve and implement plant protection and animal health services, including mechanisms to control the distribution and use of pesticides, also to implement the International Code of Conduct on the Distribution and Use of Pesticides,' not later than the year 2000! Another is 'to establish operational and interactive networks among farmers, researchers and extension services to promote and develop integrated pest management,' not later than 1998.

The program has as its bases:

1. the fact that losses caused by pests are estimated at between 25 and 50%;
2. that overuse of chemical control of agricultural pests has adversely affected farm budgets, human health and the environment; and
3. that integrated pest management is the best option for the future.

This could all have been written 30 years ago. However, a welcome and positive codicil is that the program also firmly includes continued use of pesticides and their appropriate management and regulation. Also welcome is the emphasis on farmer education into safe handling, application and disposal of chemicals.

As in other chapters in the document, considerable responsibility is placed on governments in the developing world to move forward rapidly in developing national research, coordination and development of human resources, though with the support of international organizations. In relation to pest management, UN agencies should specifically work with regional organizations to establish a system for collecting, analyzing and distributing data on pesticide use, to establish IPM networks and to develop proper IPM. There seems here little recognition of what has been achieved so far (Chapter 7 of this book). We feel it is all too vague and patronizing. It gives the impression that Rio, in relation to IPM, is a revolution rather than an intent to accelerate an evolution.

The cost of the program is estimated at US$1.9 billion annually between 1993 and 2000. National governments appear to be expected to meet all of this apart from US$0.3 billion annually. The latter would come in grant from the international community. Again, as in other chapters of Agenda 21, the developed world is prejudicing the success of its Rio initiative by its limited generosity.

The 'Green Revolution' was a phase of agricultural development in which high yields achieved by high inputs and new crop varieties (dependent on such inputs) were the goal. Particularly in the developing world, this revolution (no differently from political revolutions!) has proved unsustainable and so, after an initial peak, yields of many food crops are declining. This is due to many effects, including soil problems. However, in the context of this book, three important factors are the high suscepti-bility of the new high-yielding varieties, the side-effects of the overuse of pesticides, and the results of a decrease in biodiversity in agroecosytems.

There is no doubt in our minds, and in this we agree with the compilers of Agenda 21, that IPM is the future for sustainable crop protection. We also believe IPM is the way to a sustainable future for continued use of agrochemicals and therefore for the agrochemical industry. However, it has to be accepted that IPM will not give the levels of pest control achievable by sole reliance on pesticides in the early years of their use. Moreover, the pest-resistant crop varieties needed for IPM are unlikely to show the yield potential of the high-input-dependent varieties of the Green Revolution (p. 145). These two elements suggest that:

1. Although IPM will again raise yields where they have been depressed by overuse of pesticides, there has to be some yield sacrifice in comparison with the maximum yield achievements of the past.
2. The use of IPM may affect world redistribution of food production, since locally unimportant blemishes caused by pests may lead to significant losses from secondary infections during transport or cold storage.

There is therefore a dilemma in proposing IPM as the way forward in food production for the long term if one accepts the emphasis in Agenda 21 on increasing production per unit area as opposed to increasing the area cultivated in the developing world. Of course, in the short term, IPM will enable yield increases to be achieved on the large proportion of cultivated land where reliance on insecticides has been uneconomic and therefore losses from pests have been high. Eventually, however, it will probably be up to agronomists and cropping systems scientists, rather than crop protection specialists, to integrate IPM into their contributions for increasing production from the land already under cultivation.

The introduction to Chapter 19 of the UNCED document states that 'the two major problems, particularly in developing countries, are (a) lack of sufficient scientific information for the assessment of risks entailed by the use of a great number of chemicals, and (b) lack of resources for assessment of chemicals for which data are at hand.' Although these are major problems, it seems to us that they rank secondary to the lack of resources for remedial action once the assessments have been made and the problems identified.

The document clearly points the way forward with its suggestion for increased coordination between international bodies. It also reviews the problems of the number of chemicals involved, and the OECD initiatives that have already been considered in Chapter 2 of this book. The importance of international cooperation does not need to be stressed, but there is a need to centralize risk assessment as far as possible.

The report (Section 19.12) states: 'Risk assessment is resource-intensive. It could be made cost-effective by strengthening international cooperation and better coordination, thereby making the best use of available resources and avoiding unnecessary duplication of effort. However, each nation should have a critical mass of technical staff with experience in toxicity testing and exposure analysis, which are two important components of risk assessment.'

While this statement is fine, the following paragraphs refer to the fact that governments should 'give high priority to hazard assessment of chemicals' and should 'generate data necessary for assessing, building, *inter alia*, on international programs.' We would like to put forward the view that hazard assessment, especially the development of standards, should be done internationally. This would either be directly by international agencies or by 'farming out' the assessment to individual countries, on a chemical by chemical basis. This should not be considered paternalism by the developed world. It should be the norm within both the developed and the developing world. In the recent UNEP/WHO (1992) report on air pollution in the 'megacities' (those with populations of over 10 million) there is an appendix of the air pollution standards of various countries. Table 10.1 gives the values for 1 hour exposure to ozone for countries in Asia compared to the WHO standard. It will be seen that there are no significant variations and thus reliance on the international standard, without using resources to set national standards, should be acceptable. In this way scarce resources can be used at a national level to look at use patterns that could affect human exposure and potential environmental problems that are unique to the individual country.

Table 10.1 Standards for ozone in Asia (from UNEP/WHO (1992) *Urban Air Pollution in Megacities of the World*, Blackwell, Oxford, with permission)

Country	Air quality standard $\mu g\ m^{-3}h^{-1}$
China	120–200
Indonesia	160
Japan	120
Korea	200
Philippines	120
Thailand	200
WHO	150–200

Similarly there seems no need to 'establish, in conjunction with IRPTC, national registers and databases, including safety information, for chemicals' (Section 19.61). National availability of pre-existing databases, such as IRPTC, seems to be the need. Rather, national resources should be focused on such aspects of the problem as outlined in the next paragraph of the document, viz., 'generate field monitoring data for toxic chemicals of high environmental importance.'

In brief, hazard is a function of toxicity and exposure. It is suggested that, in general, toxicity is considered internationally and exposure is measured nationally.

The Basel Convention on the Control of Transboundary Movements of Hazardous Wastes and their Disposal was adopted in 1989. Although it was adopted by 116 countries, only a few have ratified it; 95% of the world's waste is generated by countries that have not ratified the convention. These include the USA, Japan and all the EEC countries except France. The convention will have little effect if the main industrial countries do not sign. At the moment it is 'business as usual' for traders in toxic waste (Anon., 1992).

The overwhelming problem facing us today, how to control the human population, is, fortunately for the authors, outside the scope of this book. Given that the human population will continue to increase, the problems are how to produce enough food and how to allow for an increase in economic growth without destroying the life systems of the planet.

Massive efforts are needed to improve efficiency of industrial processes so as to sustain output while decreasing both the use of energy and raw materials. Schmidt-Bleek (1992) puts it as follows, 'many times less virgin resources must be extracted from the earth per unit service delivered, including water and air, and many times less waste, emissions, and effluents must be generated and placed in contact with the environment per unit of service delivered. Waste disposal itself is inherently material and energy intensive.' In other words we have to move away, firmly and rapidly, from our short-life, over-packed, energy-wasteful approach.

There is a need to combine economic growth with sustainable development. Fifty years ago the Nobel prizewinner Sir John Hicks defined income as 'the maximum value which you can consume in a given time and still expect to be as well off as at the beginning.' Nevertheless the GNP, the most widely used current indicator by such organizations as the World Bank does *not* take this into account. Environmental degradation can well boost GNP in the short term, and most politicians think only in the short term.

To be sustainable, an economy must maintain its total capital stock. If its population increases, then to maintain constant income *per capita*, either the total capital stock must be increased, or production efficiency per unit of capital must increase. Although the United Nations Statistical Division (UNSD) started to work on an international standard system of accounts in 1968, progress was slow. In 1991 the UNSD reworked the accounting system after consultation with the World Bank and UNEP. The results have been published in a handbook entitled *Concepts and Methods of Environmental Statistics: Statistics of the Natural Environment*. It is clearly difficult to make such computations with great accuracy. The approach for non-renewable resources starts by estimating the value of the resource, the percentage used and the environmental cost of that use. Thus, if one compares coal to oil, the value of oil would be somewhat higher than coal (more energy per ton). The percentage of the resource used would be lower (there are larger reserves of coal than oil) and the cost of environmental damage would be higher with coal (although both produce CO_2, coal also produces SO_x and NO_x). If part of the revenues from coal or oil sales were invested in renewable forms of energy production, this would contribute to the modified GNP. A recent survey (Pearce and Atkinson, 1992; cited in Duthie, 1993) estimated environmental damage ranging from 1% of GNP in the Netherlands to 17% of GNP in Indonesia and Nigeria.

Despite the difficulties in putting environmental costs into national and international accounting, it is important that it be done as well as possible. A greater awareness of environmental costs will not only help the environment, but may also decrease the likelihood of human conflict. At the conference on Environmental Change and Acute Conflict held in Toronto (Homer-Dixon, Boutwell and Rathjens, 1993), it was concluded that 'scarcities of renewable resources are already contributing to violent conflicts in many parts of the developing world' and that 'these conflicts may foreshadow a surge of similar violence in coming decades.'

References

Anon. (1992) *New Scientist*, **December 12**, 9.
Duthie, D. (1993) How to grow a green economy. *New Scientist*, **January 30**, 39–43.

Homer-Dixon, T.F., Boutwell, J.H. and Rathjens, G.W. (1993) Environmental change and violent conflict. *Scientific American*, **February**, 16–23.

Schmidt-Bleek, F. (1992) Will Germany remain a good place for industry? The ecological side of the coin. *Fresenius Environ. Bull.*, **1**, 417–22.

UNEP/WHO (1992) *Urban Pollution in Magacities of the World*, Blackwell, Oxford.

CHAPTER 11

A look into the future

The major issues of pesticides, IPM and industrial pollution raised in this book are briefly re-examined in relation to likely future developments. Some 'crystal-gazing' 20 years ahead is attempted.

Silent Spring concludes with a chapter entitled 'The other road'. It is a strong advocacy of biological control. The various approaches for biological control put forward by Rachel Carson have already been reviewed in Chapter 3 and described in more detail in Chapter 5. As pointed out in Chapter 3, the term 'integrated pest management' had not been coined at the time that *Silent Spring* was written. In 'The other road' it is implied, but not specifically stated, that control of pests can be achieved without pesticides. It is true that advocates of IPM want to move away from 'the chemical barrage'. However, most consider that pesticides remain, and are likely to remain, an important tool.

Silent Spring is largely concerned with the effects of organochlorine pesticides on man and the environment. Its index shows many entries for all of the well-known organochlorine pesticides. However, there are very few for others such as the carbamates, organophosphates and pyrethroids. Since *Silent Spring* was written, there has been a major shift from the persistent, bioaccumulating organochlorines to much less persistent pesticides without bioaccumulating properties. This shift is now virtually complete in the developed world. There have also been some improvements in selectivity and delivery systems. Examples are pesticides that block chitin synthesis and the ULV (ultra-low volume) techniques that have decreased the amount of pesticide needed per unit area. Nevertheless, as Table 2.2 shows, large amounts of pesticides are still used. Organochlorine pesticides, largely phased out in developed countries, are rapidly being replaced by more modern insecticides in developing countries also. However, as pointed out in Chapters 4 and 7, fairly large quantities are still used in India and China, for both crop pest and vector control. In both these countries, much of the organochlorine is produced locally. We have found it difficult to obtain up-to-date figures for organochlorine usage in developing countries. The figures we have (Table 11.1) stem from as far back as 1988 in some instances. The organochlorines for which we have figures are limited to aldrin, chlordane, dieldrin, dienochlor, endosulfan and lindane for crop application; for public health they are aldrin, DDT, lindane/γ-HCH, heptachlor and chlordane. Table 11.1 confirms our impression, based on anecdote, that only a few 100 tons are still used in Africa, though more may be used for vector control. The stocks were mainly bought in the past, though small amounts of fresh HCH are imported for treating seeds or as a soil treatment for termites. In Asia, usage in India, and we suspect China, remains considerable. However, only small amounts are used over much of the rest of the region. Use in Latin America is also probably not a serious problem.

Rachel Carson contrasted two roads, the insecticide road and 'The other road'. The future, however, belongs to a third road – the 'middle road' of IPM. After long gestation since the foundations of IPM surfaced in the late 1950s and early 1960s (Chapter 3), IPM has advanced rapidly in the past

Table 11.1 Metric tons of organochlorine insecticide used in the developing world (data courtesy of Landell Mills Market Research Limited)

Country	Year	Field Application	Year	Public Health
Asia				
India	1993	9039	–	–
Indonesia	1991	210	1991	11 800
Malaysia	1991	93	1991	237
Pakistan	1991	188	–	–
Philippines	1991	394	1991	195
Singapore	–	–	1991	49
South Korea	1992	161	1991	0
Taiwan	1993	18	–	–
Thailand	1993	252	1991	677
Middle East and Africa				
Iran	1992	111	–	–
Ivory Coast	1992	87	–	–
Kenya	1992	15	–	–
Morocco	1992	0	–	–
Nigeria	1989	93	–	–
Sudan	1992	125	–	–
Zimbabwe	1992	42	–	–
South America				
Argentina	1993	0	1988	67
Brazil	1993	1229	1988	1 432
Chile	1990	23	1988	Very little
Colombia	1993	190	1988	171
Costa Rica	1993	12	–	–
Ecuador	1990	13	–	–
Guatemala	1993	81	–	–
Honduras	1993	17	–	–
Nicaragua	1990	11	–	–
Paraguay	1990	23	–	–
Peru	1993	95	–	–
Venezuela	1990	76	1988	23

–, Not surveyed.

decade. This has happened both in developed and developing countries. Even the initial stage of basing pesticide use on economic thresholds has reduced the frequency of applications and conserved indigenous biological control (Chapter 6). It has proved possible to introduce such measures on many crops and in many countries without the need for extensive preliminary research. We envisage that future research will increasingly progress four lines of development, described in more detail in Chapter 5:

1. It will become increasingly simple for farmers to operate economic thresholds. There will be more involvement of national advisory services in providing forecasting services to replace crop inspection. Also, initially mainly in developed countries, farmers will increasingly look to IPM consultants to manage their crop protection.

2. There will be a gradual spread of IPM protocols for different crops, evolving towards fully fledged menu systems. For some crops, artificial intelligence programming of computers will provide 'farmer-friendly' decision-making tools.
3. Such protocols and menus will, if incentives to overcome market limitations can be introduced, make increasing use of biopesticides (pathogens and plant-derived toxins) for a wide range of crop protection problems, also of semiochemicals for insect control. The next 20 years will also see the appearance of new pesticides with novel modes of action. These compounds will also show greater selectivity and high safety for human health and the environment.
4. There will be further improvements in pesticide application technology, making controlled droplet application a more practical and farmer-operable system.

In the present book, we have attempted a wider canvas than just pesticides. For that reason, also because many problems discussed were not identified in 1962, comparisons with the views put forward in *Silent Spring* cannot be made.

With several major environmental concerns, there is going to be a need for continued measurement and analysis. A prime example is climate change. Here just about the only thing on which there is agreement is the rise in the levels of atmospheric carbon dioxide. The consequences are far from clear, and the necessity for taking strong, and expensive, action is not persuasive. There is a difficulty – if those who claim that severe, even disastrous, effects will occur are right, it will be too late by the time this has been proved. If it is considered necessary to stop the rise of atmospheric carbon dioxide, it will take very strong measures indeed. Current discussions hinge on holding emissions to the 1990 levels, but this would still cause a steady rise in the carbon dioxide levels.

There is no doubt about the decrease of tropospheric ozone, but again there is doubt about the seriousness of the increase of UV radiation. Here, however, the solution is easier and less expensive. It is possible for mankind to stop using chlorofluorocarbons. The restrictions on these compounds (Chapter 8) have been a success story of international cooperation.

With acid rain there is no doubt that damage has been done and this has led to some agreements for the reduction of SO_x and NO_x emissions. How much should the reduction be, since it gets progressively more expensive, remains the big question. As already cited in Chapter 8, a recent report of the UK Joint Nature Conservation Committee gives estimates of how much damage to SSSIs would be prevented by various degrees of emission control. All these problems are a combination of science and politics. It is necessary to have as firm a scientific base as possible, which includes an honest assessment of the uncertainties. Once this is obtained, it is a decision by society whether the costs outweigh the benefits.

The information available on chemicals is, thanks to international cooperation, improving at a good pace. It is important to avoid duplication, but this can only be done if the information is freely available. It is vital that problems involved with 'freedom of information' in respect of industrial data are solved. Internationally compiled, comprehensive and readily available data on the toxicology of chemicals are needed. These would enable national authorities to concentrate on the exposure side of the hazard assessment, before making the final judgment on the acceptability or otherwise of the risk.

Life-cycle analysis, or the 'cradle to grave' approach for the handling of chemicals, is gaining support. As discussed in Chapter 3, there is a need to expand this concept. Schmidt-Bleek (1992) has recently examined the ecological side of the changes in industrial practices that he considers necessary in the future. He states that 'continued economic development requires drastic adjustments of the energy and material intensity for products and services during the coming decades.' The reductions that Schmidt-Bleek considers necessary are startling. They are a 50% reduction in the material streams of production within 25 years and 90% within 50 years. Some areas that he considers should be tackled are listed in Table 11.2. The changes proposed should help to decrease CO_2 emissions. However, if stabilization of levels of atmospheric CO_2 is considered necessary, then other measures – decreased use of fossil fuels in energy production, increases in the area of forests – will have to be implemented.

What will be the focus of the third major UN Conference in 2012, if the sequence is continued? The Stockholm Conference of 1972, which set up the UNEP and led to many international programs and conventions, can be counted as a success. It is far too early to make any judgment of the impact of the Rio Conference in 1992, which had 'sustainable development' as its theme. The north–south divide was both a major factor and a friction point at Rio.

It is diverting to make some predictions (and safe enough since, if the authors are still alive 20 years from now, their active careers will be over!):

1. There will still be a developed–developing world divide, but there will be a decrease in the size and the depth of that divide. The *per capita* usages of resources of the developed world will have decreased to a level below that currently used in western Europe. This applies especially to the current prolific usage in the United States. Although some changes in human life styles will be entailed, there will still be a comfortable standard of living.

2. Several countries now classed as 'developing' will have taken their place among the developed countries. Thus there will be a divide, but

Table 11.2 Changes needed in material intensity of industrial processes (after Schmidt-Bleek, F. (1992) *Fresenius Environ. Bull.*, **1**, 417–22)

1. A drastic, life-cycle dematerialization of industrial processes
2. Substantial improvement of retention of natural resources in circulation
3. Packing and transport intensities of goods and services must be sharply curtailed
4. Many new types of products, vehicles, materials, etc. must be available within 10–30 years
5. Economic incentive policies by governments are essential
6. International cooperation is vital for eco-restructuring

the developed group will be larger. One could argue two ways. Either this will make the 'haves' stronger in resisting the demands of the 'have-nots', or the resources of the developed world will be better able to help the developing world.

3. In crop protection, chemical herbicides will still be used extensively. However, use of insecticides and fungicides will have fallen considerably (and asymptotically) to about 10–20% of the present level.

4. Data on climate change will be much stronger. Although there will still be debate, scientific opinion on the severity of the effects will have moved towards the low end of the present scale.

5. There will be evidence that the bans of CFCs are having an effect and that the concentration of ozone in the troposphere, although lower than 30 years before, has become essentially stable.

6. The earlier predictions of the 'gloom and doom' lobby about imminent starvation, such as Paul Ehrlich's *The Population Bomb* published in 1968, have been incorrect. However, increasing human population and increasing *per capita* use of resources in the developing world will, despite pressures from certain groups, cause the word 'population' to be in the title of the 2012 conference.

Certainly, we live in interesting times.

References

Ehrlich, P. (1968) *The Population Bomb*, Ballantine, New York.
Schmidt-Bleek, F. (1992) Will Germany remain a good place for industry? The ecological side of the coin. *Fresenius Environ. Bull.*, **1**, 417–22.

Index